TECHNICIAN
PROCESSES AND MATERIALS

TECHNICIAN WORKSHOP PROCESSES AND MATERIALS

R. T. Pritchard

Lecturer in Mechanical Engineering,
Garretts Green Technical College, Birmingham

HODDER AND STOUGHTON
LONDON SYDNEY AUCKLAND TORONTO

British Library Cataloguing in Publication Data
Pritchard, Reginald Thomas
  Technician workshop processes and materials.
  1. Machine-shop practice
  I. Title
  621.7'5   TJ1125

ISBN 0-340-22100-3

First published 1979

Copyright © R. T. Pritchard
All Rights Reserved. No part of this publication may be reproduced or transmitted in any form or by any means, electronic or mechanical including photocopy, recording, or any information storage, or retrieval system, without permission in writing from the publisher.

Printed in Great Britain for Hodder and Stoughton Educational, a division of Hodder and Stoughton Ltd., Mill Road, Dunton Green, Sevenoaks, Kent by Richard Clay (The Chaucer Press) Ltd., Bungay, Suffolk

PREFACE

This volume provides full coverage of the TEC Standard Unit for Workshop Processes and Materials *I* for the first year of the Mechanical Engineering Technicians Course. It is intended as a class book for students, and an attempt has been made to keep the written text to a minimum and to make the fullest use of clear and informative drawings. Drawings and diagrams are the language of manufacturing engineering; it is important that the student technician makes continual and constant contact with them. The assimilation exercises at the end of each unit will prove a useful yardstick or measure of unit comprehension.

At all times the book proceeds from basic principles of engineering manufacture, and any student who diligently reads and studies the unit areas of this volume increases his potential as an Engineer.

Sutton Coldfield   R.T. Pritchard

CONTENTS

Preface	v
SI Units	ix
Introduction	1
1 Hazards in the Workshop	3

Identifiable and non-identifiable dangers, Health and Safety Act 1974, dangers of moving parts, grinding, turning, milling, electrical power, isolating switches, stop buttons, eye protection, head protection, feet protection, skin protection.

2 Hand Processes   17

Surface plates, scribers, scribing blocks, dividers, odd-leg calipers, centre punches, hacksaws, files, chisels, taps and dies, powered hand tools, electric drills, air drills, safety precautions, principles of marking out, datum faces, centres, principles of measurement, steel rule, calipers, protractors, line standards, Vernier principle, micrometer principle, comparators, end standards, types of comparators, drilling machines, twist drills, reamers, sheet metalwork.

3 Machine Tools   112

Centre lathe, workholding, turning examples, screw cutting, gear trains, taper turning, safety precautions, capstan lathe, automatic lathe, shaping machine, shaping techniques, cutting tool theory, rake and clearance angles, tool grinding, types of tools, cutting fluids, coolants, lubricants, cutting oils.

4 Fastening and Joining   211

Threaded fasteners, locking devices, riveting, soft soldering, sweating, fluxes, soft solders, hard soldering, basic welding, adhesives, safety precautions.

5 Engineering Materials   247

Ferrous metals, non-ferrous metals, non-ferrous alloys, physical properties, density, electrical conductivity, thermal conductivity, mechanical properties, ultimate tensile strength, elastic limit, ductility, hardness testing, toughness testing, use of grey cast iron, use of wrought iron, use of carbon steels, heat treatment of carbon steels, use of non-ferrous metals, bearing materials, use of plastics, corrosion, galvanising, tinning, electro-plating, metal spraying, rust proofing.

6 Working in Plastics 307
Methods of joining, adhesive bonding, solvent welding, heated-tool welding, high-frequency welding, bending and casting, drilling, turning, encapsulating.

Index 319

SI Units

In this volume, SI metric units have been used throughout. To assist the student, the more common SI units are listed in the following table.

Physical Quantity	SI Unit	Unit Symbol
mass	kilogramme	kg
length	metre	m
area	square metre	m^2
volume	cubic metre	m^3
time	second	s
force	newton	N
temperature	degree Celsius	°C

The following multiples and submultiples are used for all physical quantities.

Prefix	Multiplying Factor	Symbol
tera	$1\,000\,000\,000\,000 = 10^{12}$	T
giga	$1\,000\,000\,000 = 10^{9}$	G
mega	$1\,000\,000 = 10^{6}$	M
kilo	$1000 = 10^{3}$	k
hecto	$100 = 10^{2}$	h
deca	$10 = 10^{1}$	da
deci	$0{\cdot}1 = 10^{-1}$	d
centi	$0{\cdot}01 = 10^{-2}$	c
milli	$0{\cdot}001 = 10^{-3}$	m
micro	$0{\cdot}000\,001 = 10^{-6}$	μ
nano	$0{\cdot}000\,000\,001 = 10^{-9}$	n
pico	$0{\cdot}000\,000\,000\,001 = 10^{-12}$	p

For example: 2000 metres = 2 kilometres = 2 km
0·001 metre = 1 millimetre = 1 mm
0·000 001 metre = 1 micrometre = 1 μm.

Note: All dimensions on mechanical engineering drawings are usually expressed in millimetres (mm).

INTRODUCTION

It is the business of engineers to make things. Very few people outside the engineering profession appreciate the complexity of manufacture, the wide range of workshop processes and materials, and the high degree of engineering skill required in the manufacture of commonplace components.

Before embarking on a course of study, it is as well to have clearly defined, not only the necessity for the course, but also the reasons for the choice and content of the subject-matter. The following notes will serve to illustrate the essential knowledge required by those who aspire to gain technician status in the engineering industry, and also as an introduction to the subject-matter which forms the unit areas of this book.

Hazards in the workshop

There is no truer saying than that a good worker is a safe worker. Perhaps one of the first lessons to be learned by the potential engineering technician is that the safety and welfare of his fellow workers takes precedence at all times over the needs and demands of manufacture and construction. Chapter 1 is designed to show the potential causes of injury and accident likely to be encountered when engaged in engineering manufacture.

Hand processes

If we remember that all machine tool operations are merely mechanised or automated methods of carrying out the tasks or jobs so laboriously performed by hand methods, then it is clear that the engineering technician who is not only familiar, but also reasonably skilled in the use of hand tools, is better able to appreciate the very great advantages that the use of machine tools offer.

Chapter 2 is designed to show the importance of correct selection and use of the hand tools needed to mark out and shape metal, and also the choice and use of simple measuring instruments needed to check the accuracy of the hand work produced.

Machine tools

Machine tools are the producers of industry, making possible the high standard of living enjoyed by those nations that make the fullest use of them. Although concerned only with the centre lathe and shaping machine in Chapter 3, the aim has been to acquaint the technician with the basic principles underlying the production of the geometrical surfaces possessed by all engineering or manufactured components, together with the design and use of single-point cutting tools used in these machine tools.

Fastening and joining

It would take some time to count the number of components which when joined together or assembled, make a complete motor car. The aim of Chapter 4 is to show not only the use of conventional joining devices, but also the ever increasing applications of new techniques.

Materials

The correct choice and use of materials is an important decision for all engineers. Metals need to be mined and then smelted, and the constant increases in metal costs mean that the greatest economy needs to be exercised at all times, in order to keep the cost of the finished product at a competitive level. Chapter 5 reviews the properties and uses of the more common engineering metals, and the newer plastics materials, together with the methods employed to protect metals from atmospheric corrosion.

Working in plastics

The use of plastics materials increases year by year. Plastics have a great advantage over the more conventional metals, because it is possible to produce a plastic material to meet specific conditions of stress and usage. Chapter 6 gives an indication of the techniques to adopt when machining plastics, and shows also the basic principles underlying plastics joining, welding and encapsulating.

1
HAZARDS IN THE WORKSHOP

Objectives—The principles and applications of:
1. Identifiable and non-identifiable dangers
2. The dangers associated with moving parts
3. The dangers associated with the use of electrical power
4. Environmental dangers

1.1 Identifiable and non-identifiable dangers

There are few people who, at some time or another, have not lifted the bonnet of a motor car while the engine is running, and Fig. 1.1 shows in a simple manner what may be seen. The crankshaft pulley, rotating in a clockwise direction drives both the fan and generator pulleys by means of a Vee belt.

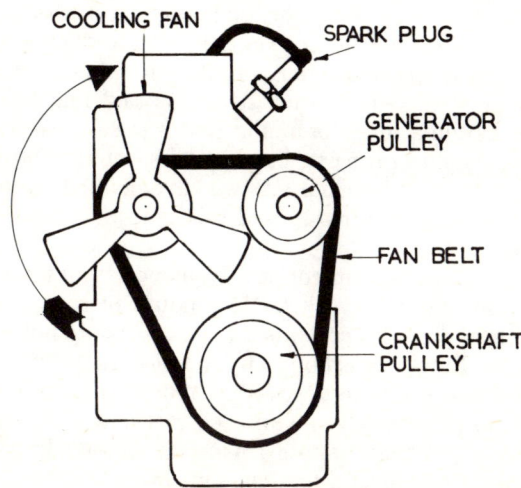

Fig. 1.1 Identifiable and non-identifiable hazards

Clearly there are dangers here. Although modern production techniques favour the use of plastic fans, serious injury will result if the hand comes in contact with the revolving blades. This cooling fan may be considered as an identifiable danger, that is to say, the average person is aware that the rotating blades possess energy and are thus capable of inflicting injury. Motor car manufacturers are aware of the danger of a fast-rotating fan; it is standard practice to make them a bright red or yellow colour, so that they are more easily seen against the darker background of the engine.

It is evident that moving parts are readily identified as sources of danger. For example, the revolving crankshaft and generator pulleys, together with the fast moving Vee belt are all capable of causing injury because they transmit rotational power from the engine. Yet there are other dangers not so easily identifiable. The cylinder block or exhaust manifold may be hot enough to cause a burn; touching a spark plug may result in a series of unpleasant shocks; removing the radiator filler cap may lead to serious scalding if for some reason or another the engine is overheated.

It should be clear at this point, that dangers or hazards are not always obvious, and we will do well to classify them as follows:
(i) identifiable hazards,
(ii) non-identifiable hazards.

It is because of the non-identifiable hazards that so many accidents take place, and this means that each and every one of us has a duty to adopt the correct attitude to safety precautions at all times.

It is the business of engineers to make things, and it is the duty of engineers to ensure that all work is carried out in a safe and responsible manner. It is very necessary for the young apprentice, at the start of his working life to adopt the correct approach to safety precautions, for no more effort is required to learn to do a job the safe way, than it is to learn to do it the wrong way. It can be said that safe working results from a certain attitude of mind. That is to say, respect not only for one's own person or property, but equal respect also for the person and property of fellow workmates. As we have pointed out, workshop dangers are not easily identifiable, and this calls for a development of general awareness.

This awareness results from constant vigilance at all times, producing an observant and safe worker. No amount of rules, regulations, posters or lectures prevent accidents if the workers are not safety conscious. While the Employer is bound by Act of Parliament to maintain a safe working place, Employees have a moral duty to work safely; that is to say, to perform all jobs in a correct and safe manner. The young apprentice technician who learns to do all jobs the correct and safe way has the makings of a first-class engineer.

Hazards in the Workshop 5

1.1.1 Health and Safety Act 1974

This Act covers all people at work except domestic servants in private employment. This is the first time that the general public are included in the legislation, and strict control will be maintained over any industrial activity that may put at risk, the health or safety of the general public.

Employers
The duty of employers may be summarised briefly as follows:
To provide (i) safe plant,
 (ii) safe premises,
 (iii) safe systems of work,
 (iv) adequate training, instruction and supervision.

Employees
Employees will have a duty to take all reasonable care to avoid injury to themselves or to others by their work activities, and are to co-operate with employers and others in meeting statutory requirements.

Health and Safety at Work booklets
These useful booklets are designed to give up-to-date facts and advice about best practices in health and safety concerning industrial and other employment. The following list gives a few selected titles from the Series:

No. 24 Electrical limit switches and their applications.
No. 29 Carbon monoxide poisoning: causes and prevention.
No. 31 Safety in electrical testing.
No. 35 Basic rules for safety and health at work.
No. 43 Safety in mechanical handling.
No. 47 Safety in the stacking of materials.

1.2 Dangers of moving parts

1.2.1 Grinding machines
Figure 1.2 shows a piece of equipment likely to be found in any workshop. It is a pedestal or off-hand grinding machine designed for hand grinding cutting tools such as cold chisels, twist drills and punches. The grinding wheels which revolve at high speeds are good examples of clearly identifiable dangers. Driven by a self-contained electric motor, these wheels are capable of inflicting very severe injuries to hands or fingers coming in contact with them. Grinding wheels that are able to remove hardened steel with a shower of sparks will make short work of human fingers.

Such contact can be made accidently, perhaps tripping and falling

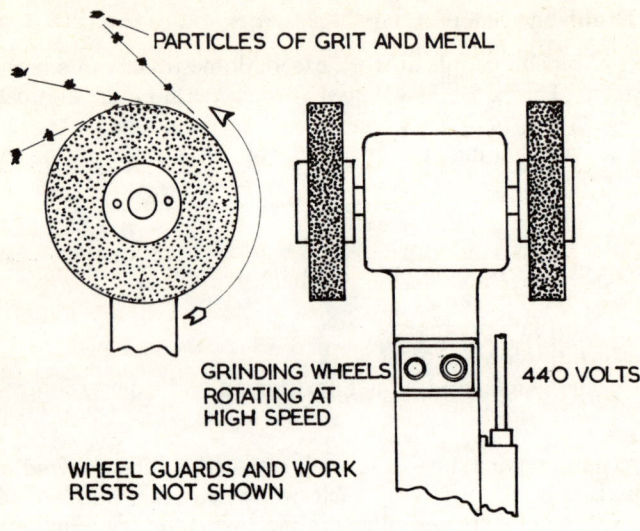

Fig. 1.2 Hazards when grinding

against a pedestal grinder and putting out a hand to protect against a fall. Clearly it is imperative that the grinding wheels are properly and adequately guarded, and this is the rule that must be applied to all moving parts. A non-identifiable danger present when grinding a hand tool may be recognised by reference to Fig. 1.2. Particles of both the grinding wheel and metal being ground leave the rotating wheel on tangential paths, moving at high speeds and quite capable of causing eye injury. An adequate guard, properly fitted, ensures that no flying particles will be present in the vicinity of the operator's face, who will further reduce the risk of eye injury by wearing safety glasses.

It must be remembered also that great care is needed when fitting new grinding wheels. Always test a wheel by suspending it and tapping lightly with a metal bar. A sound wheel will have a distinct ring, and can be fitted with complete safety, while a faulty or cracked wheel will have a dead sound. Do not return such a wheel to the Stores, but break it up immediately. This will prevent someone else, less informed than yourself, from fitting it to a grinding machine with possible serious consequences. A wheel must never be forced on the spindle of a grinding machine. A nice free fit is needed here, easily obtained by scraping the lead or nylon bush at the wheel centre as required. We see now that the fitting of new grinding wheels to even the simplest type of pedestal grinder is no job for the thoughtless or careless worker, and the following list indicates the safety rules to be observed when operating or setting grinding machines.

Hazards in the Workshop

1. Do not change a grinding wheel unless you have received proper training and hold written confirmation from your employer that you are competent to carry out such duties.
2. Ensure that the guard and tool rest are reasonably close to the grinding wheel.
3. Do not use undue force when grinding and never use the side of a straight grinding wheel.
4. Wear goggles or safety glasses at all times.
5. Allow a newly fitted wheel to run for several minutes, ensuring that the space around the machine is kept clear.
6. Never allow a wheel to be rotated in excess of the maximum speed which is stamped on the wheel label.
7. In any case of doubt, stop and isolate the machine. Inform someone in authority without delay.

1.2.2 Lathes

There are many types of lathes, for example centre, capstan, turret and automatic, but in general a lathe may be described as a machine tool, power-operated and capable of producing cylindrical surfaces. This is achieved by a simple principle called generating, requiring rotation of the workpiece about its centre line, and movement of the cutting tool parallel to the centre line of the revolving workpiece. This basic principle is simply illustrated at Fig. 1.3, and it is clear that the following identifiable dangers exist:

(i) Entanglement of operator's clothing around revolving workpiece.

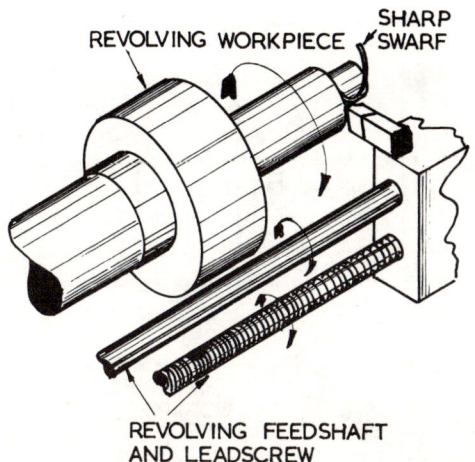

Fig. 1.3 Hazards when turning

(ii) Continuous lengths of sharp edged swarf leaving workpiece.
(iii) Small particles of swarf entering operator's eye.
(iv) Careless use of files and emery cloth on rotating workpiece.
(v) Careless leaving of tools and equipment on the headstock, saddle or lathe bed during a turning operation.

There is little doubt that projecting work represents the most dangerous aspect of lathe turning, and adequate guarding of capstan, turret and automatic lathes is a legal obligation on the employer. At the same time it is the duty of the employee to adopt safe working practices, for many moving parts associated with lathework cannot be properly guarded. Typical examples include revolving feed shafts, leadscrews, chucks and faceplates. Loose clothing, dangling ties and long hair have no place in the vicinity of centre lathe turning, and great care is needed in the removal of swarf or the fitting of a heavy duty four-jaw chuck.

It is vital that the lathe is isolated when tool or chuck changing, and the space around the lathe must be kept free from all obstructions, not to mention oil or water seepage.

1.2.3 Drilling machines

The drilling machine is one of the simplest machine tools in use. Once the drill is secured in the chuck or fitted into the taper, vertical feed only is required for the rotating drill to generate an internal cylindrical surface, namely a hole. Drilling machines are available in many varieties, including bench, pillar and radial machines. In all cases however, the basic drilling operation is the same, as shown in Fig. 1.4,

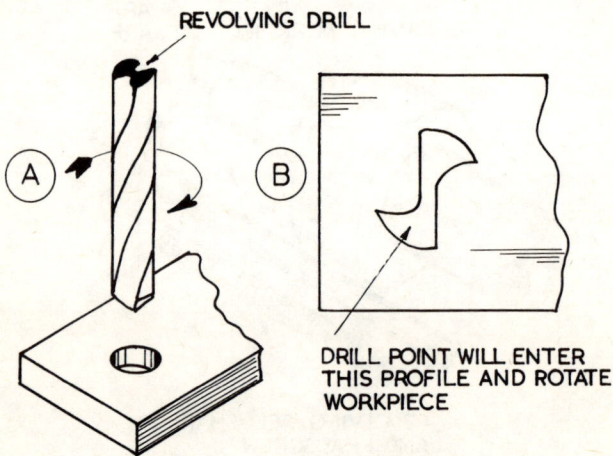

Fig. 1.4 Hazards when drilling

Hazards in the Workshop

and it can be stated at once that more accidents take place when drilling, than when operating any other machine tool. Observance of the following safety precautions will assist in reducing the risk of injury when operating drilling machines of all types:

1. Always clamp the work. Never hold by hand, not even the smallest job.
2. Never wear gloves during a drilling operation, they increase the risk of injury.
3. Avoid all loose sleeves, long ties, scarves when drilling. Ensure that all long hair is securely tucked in.
4. Do not remove swarf with bare hands; use a brush or stick.
5. Do not remove any guards that may be fixed to the drilling machine.
6. Do not attempt to save time by stopping the spindle by hand.
7. Take special care when drilling thin metal sheets. On breaking through, the profile shown at Fig. 1.4B is produced, and a twist drill is able to screw itself through this hole, resulting in rotation of the workpiece when reaching the end of the flutes. A thin steel sheet, rotating at speed is quite capable of severing several fingers.

1.2.4 Milling machines

Milling machines are available in many types, including horizontal, vertical and universal. Their operating principle however is identical, namely the production of a plane surface by feeding the workpiece against a revolving cutter. This principle is simply illustrated at Fig.

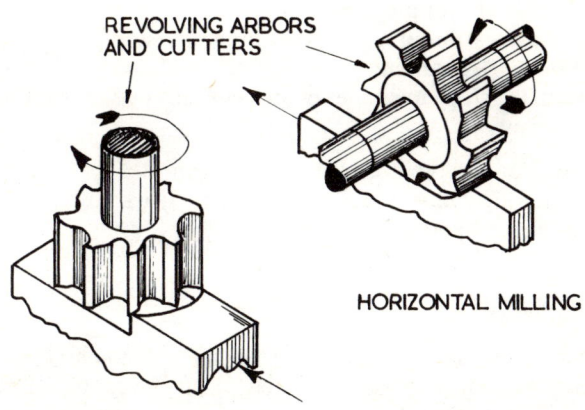

Fig. 1.5 Hazards when milling

1.5, and it will be noted that it is a difficult matter to provide efficient guarding. The general rule is to ensure that as much of the cutter as possible is covered by the guard, and once again it is the duty of the milling operator to adopt safe working practices to suit the type of milling operation at hand. The precautions that follow will help to reduce the risk of accidents when operating milling machines.

1. Avoid wearing loose clothing, gloves or scarves.
2. Take particular care with long hair and bandages.
3. Keep the work table free from all tools and instruments.
4. Stop the machine to remove swarf, and use a hook or stick.
5. Isolate the machine for any tool changing.

1.3 Electrical power

A machine tool may be described as a power driven device designed to produce a given geometrical surface. The old-fashioned method consisted of driving several machine tools from overhead shafts, and the fast moving leather belts represented a very real hazard to the careless operator.

The modern technique of using self-contained electric motors has undoubtedly made the workshop a safer place, but hazards still exist, for electricity is a good example of an unidentifiable danger. That is to say, should a fault fevelop in an electrical device, it is impossible to know whether or not the device is capable of transmitting an electrical shock. This means that great care is needed, not only in the use of electrical power but also in its control. With regard to the safety of the machine tool operator, the following systems of control are essential:

(i) Control over stopping and starting.
(ii) Positive isolation of machine from mains supply for purposes of tool or work changing.
(iii) Ability to stop machine or to cut off power supply in an emergency.

A simple diagram of these essentials is shown at Fig. 1.6A, and it is worth while taking a closer look at the means by which both the safety and welfare of the operator are ensured.

1.3.1 Isolating switches

The purpose of an isolating switch is to isolate the machine completely from the mains supply, in other words cut off the source of power or energy to the machine. While the stop button effectively stops the machine by cutting off the power supply to the motor, the power supply is still present, and there is the possibility that if the machine is

Hazards in the Workshop

accidently started, serious injury will result should the operator be engaged on changing the cutting tool.

The rules for the correct use of the isolating switch should never be forgotten, and they are as follows. Isolate the machine when:

1. Making any adjustments to the tool.
2. Making any adjustments to the workholding device.
3. Making any adjustments to the guarding device.
4. Changing the tool.
5. Cleaning down the machine.
6. Removing any covers or plates for purposes of gear changing or oil replenishment.

Isolating switches are available in many types, but all have the same basic purpose of breaking the power supply to the machine in a positive manner; the usual method is by hand movement of a small lever or switch. It is standard practice in all workshops to isolate, at the end of the shift, any machine tool on which you have been working.

1.3.2 Stop and start buttons

With the isolating switch in the *on* position, pressing the start button will supply power to the electric motor, and only when the operator is satisfied that it is safe to do so, will he start the machine. It is vital that this button is not pushed by accidental means, and in the interests of safety all start buttons are green in colour and well recessed in their

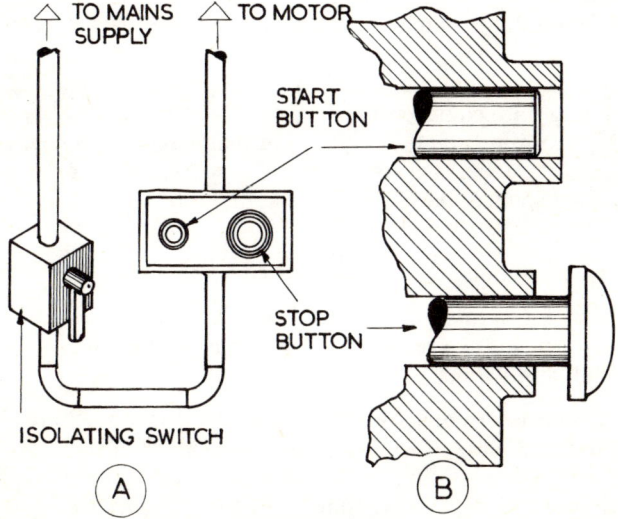

Fig. 1.6 Control of electrical power

holders, as shown at Fig. 1.6B. This makes it practically impossible to push them in accidentally.

On the other hand, it may be noted by reference to Fig. 1.6B that the stop button has a large cap or head, thus ensuring that it is easily found and pressed when the motor is to be stopped. This button of course, will be red in colour. On small machine tools, both start and stop buttons may be housed in the same unit, so that they are readily at hand to the work position of the operator. Larger machine tools, such as horizontal borers, require widely spaced working positions, and it is necessary to provide additional stop buttons or a movable control panel, enabling the operator to have complete control of the machining at any working position.

1.3.3 Emergency stop buttons

The purpose of emergency stop buttons is to allow immediate cut-off of the power supply to all the machines in a workshop. In the event of an operator becoming entangled in moving machinery, and his plight being noticed at the other end of the workshop, valuable time is saved by stopping the power supply at the nearest emergency stop button. For this reason all emergency stop buttons are placed at convenient positions around the shop, and it is the duty of workers to know their locations and to ensure that there is easy access to them.

1.4 Environmental dangers

Environmental dangers may be described as those hazards connected with the sort of working conditions surrounding one's occupation. For example on a building site, there is always the possibility of heavy objects falling from considerable heights. The wearing of safety helmets is therefore compulsory. In engineering manufacture there are many environmental dangers; in other words, a worker may meet with an accident totally unconnected with his actual job.

1.4.1 Eye protection

The danger of grinding dust or swarf entering eyes is well known, but there are other sources leading to eye damage which are not so well appreciated, and we may classify these as follows:

(i) Damage to eyeball by splinters of steel, wood or any foreign body hard enough to pierce the outer layers of the eyeball.
(ii) Damage by exposure to the high intensity light given off when arc or oxy-acetylene welding.
(iii) Damage by contact with acids or corrosive vapours.

Hazards in the Workshop 13

In the first case, complete protection of eyes is obtained by ensuring that safety glasses, goggles or transparent visors are always worn. This rule applies not to the actual operatives only, but also to all visitors who may have occasion to visit or pass through the area where eye protection is needed. There are many excellent eye or face visors available made from toughened plastic, which take only a moment to slip on one's head to afford complete protection. This precaution cannot be over emphasised where arc or oxy-acetylene welding is carried out together with flame-cutting. These processes emit an intense glare at the heat or flame source, and cause permanent damage to the eyesight if they are watched with the naked eye; it is vital that the correct tinted protective goggles or visors be worn.

Eyes can also be permanently damaged by splashes from acids or from contact with corrosive vapours or gases. If at any time one suspects that something in the atmosphere is causing eye irritation, immediate advice must be sought from the first aid room and someone in authority informed.

1.4.2 Head protection

Safety headgear is not required in an ordinary workshop. It is possible however, that an engineering apprentice may be required to pass through a part of the factory where components are carried on overhead conveyor systems, in which case head protection may be necessary. There are many types of safety helmets available; light in weight and comfortable to wear they provide positive protection against the possibility of a head injury. It must be remembered at all times that damage to the human skull can have the most serious consequences.

1.4.3 Feet protection

Not enough attention is given to the need for protective footwear in industrial manufacture. It is a mistake to think that heavy, ungainly and uncomfortable industrial boots must be worn in order to prevent injury. Such injuries arise from three main causes:

(i) Heavy piece of equipment falling on foot.
(ii) Sharp object, such as a nail piercing sole of shoe.
(iii) Sharp object, such as corner of steel sheet cutting ankle.

A wide range of industrial safety footwear is now available including both shoes and boots. The safety shoes are attractive in appearance and comfortable to wear. The soles can be heat and oil resistant and also tough enough to prevent piercing by nails. This is achieved by having a moulded-in steel intersole, and added protection is given by a moulded-in steel toe-cap. The material for the sole is also an excellent electrical insulator, as well as having first-class non-slip properties. In appearance, they compare quite favourably with an ordinary walking-out shoe, except that they provide the fullest protection against almost any kind of possible foot injury.

Safety boots are more likely to be worn by those workers engaged on constructional sites, or engaged in heavy engineering trades such as ship-building or foundry work.

1.4.4 Skin protection

Many machine tools such as millers and drillers use a cutting fluid to carry away the heat and swarf produced by the action of the cutting tool. These cutting fluids can cause skin irritation leading to a skin complaint known as *dermatitis* which is both unsightly and unpleasant. The most important safety precaution to adopt is personal cleanliness; that is to say, thorough washing of hands and arms with soap and water, before, during and at the end of the working day.

In particular, wash before the lunch or tea breaks, and before visiting the toilets. Especial care needs to be taken if any cuts or abrasions are present on fingers or hands, and should oil or fluid penetrate the clothing it is advisable to change the oil-soaked clothing which is in contact with the skin. It is always good practice to use a recommended barrier or protective cream, which should be applied before commencing work, and the use of solvents or abrasive powders to clean hands is not advised.

It is certain that no matter in which particular department the engineering apprentice is engaged, some hazardous conditions will exist. We have seen in the previous sections that some dangers are readily identifiable, while other dangers are not immediately obvious. There is no substitute for practical experience in these matters, but the apprentice technician engineer is well advised to develop a general awareness of his immediate surroundings. In other words, he must be alive, not only to his normal environment, but also to any changes that may take place in it. For example, a machine tool may be in the process of renovation or repair, and oil has leaked from the gearbox, forming a pool on the workshop floor. The presence of this oily (and slippery) pool represents a source of danger and it is essential that the matter be immediately reported or steps taken to remove the oil and cover the floor with dry sand or sawdust.

Similarly, metal sheets may have been carelessly stacked with one or more sheets projecting out of alignment, presenting sharp edges capable of cutting the unwary passer-by. Once again the matter should be reported or steps taken to line-up the steel sheets and perhaps cover them with cloth or sacking.

Perhaps a start or stop button appears to stick on occasions; there may be a strange smell or odour in the workshop; part of a machine

Hazards in the Workshop 15

tool may seem to be warmer than usual. In short, anything that seems different to normal should be investigated. If the cause cannot be determined the matter should be reported immediately.

Finally, Fig. 1.7 indicates in a simple manner most of the protective

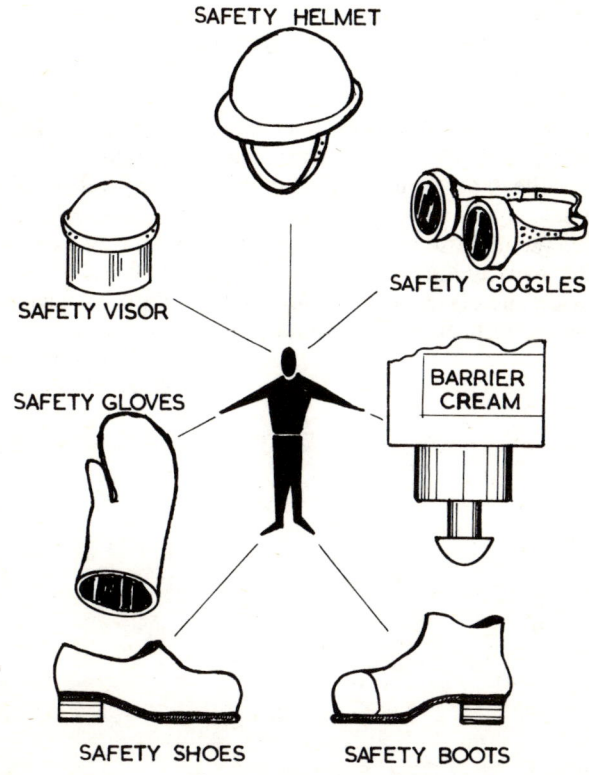

Fig. 1.7 Protective devices

devices which are available, and the engineering technician apprentice who makes it a fixed rule to wear the appropriate protective device at all times has the makings of a first-class engineer.

Chapter 1 Assimilation Exercises

1 Explain why it is important that an engineering apprentice adopts a safety-conscious attitude as soon as he takes up industrial employment.
2 Give an example showing how carelessness by one worker could result in injury to a workmate.

3 Outline two examples of non-identifiable dangers when operating a bench drilling machine.
4 Explain the need for safety precautions when regrinding a cold chisel at a pedestal grinding machine.
5 State why emergency push buttons are required in an engineering workshop.
6 With neat diagrams show the design differences between start and stop buttons for a machine tool.
7 Roof repairs are in progress above the workbenches in a toolroom. State any likely causes of injury to the toolmakers, and outline the safety precautions that should be taken.
8 Outline the main sources of danger when operating a milling machine.
9 Outline the main sources of danger when operating a centre lathe.
10 Explain why less accidents are likely in a well-set-out workshop with wide and clearly marked gangways and proper provision for the storage of tools and equipment.

2
HAND PROCESSES

Objectives—The principles and application of:
1. Use of hand tools
2. Methods of marking out
3. Technique of measurement
4. Using the sensitive drilling machine
5. Performing basic sheet metal operations

2.1 The use of hand tools

The engineering apprentice technician will do well to remember that there is an important and subtle difference between producing components on a machine tool such as a centre lathe, and producing components on the bench.

If we consider that a centre lathe costs around £3000, and is specifically designed to produce cylindrical work, then there is no particular merit in just turning down a bar of metal, and the importance of the operator diminishes accordingly. If, however, the operator possesses the skill and experience to make the maximum use of a machine tool, producing well-finished accurate work in minimum time, then he reassumes his importance.

The selection of hand tools involves bench work, and it must be appreciated at once that greater skill is required if accurate work is to be produced at the bench using hand tools. Added to this, there is the matter of manual effort to be considered, for a machine tool has an electric motor which provides the considerable energy required to remove metal while the bench hand must rely on his own muscular ability. This brings us to our first rule concerning the use of hand tools at the bench, and this is, obtaining the maximum metal removal with the minimum of effort.

This is not to suggest that the object is the avoidance of effort, but to emphasise that the bench hand and machine operator have the same goal, namely the production of accurate well-finished work in minimum time, and more important, with minimum effort. Always remember that a tired worker is an inefficient worker.

2.1.1 Marking-out tools

The main purpose of hand tools is to remove metal and Fig. 2.1 shows how the different types can be classified into either marking-out or metal removing tools. In all cases metal is removed, for the purpose of a scriber is to provide a permanent visual line, indicating the profile of

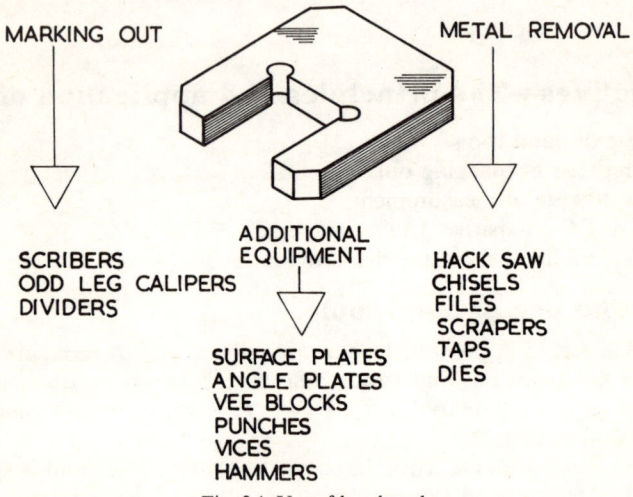

Fig. 2.1 Use of hand tools

the job, and this is possible only if the scriber makes a slight groove in the metal. We see also, in Fig. 2.1, that additional equipment is needed for both marking out and metal removal. For example, vices are needed to ensure that the work is properly and safely held, while hammers are needed to provide the hand blows required when using chisels or punches. The following notes give an indication of the best use of standard marking-out tools.

Surface plates

A surface plate is a precision piece of equipment providing a flat surface or reference plane from which a number of dimensions may be marked off or scribed. Figure 2.2A shows a surface plate used for marking-off a number of holes in a simple aluminium chassis from which a small radio receiver is to be constructed. A pictorial view of the finished job is shown at Fig. 2.2B, where it may be seen that the hole centres are in line with the edges of the chassis. The advantages

Hand Processes

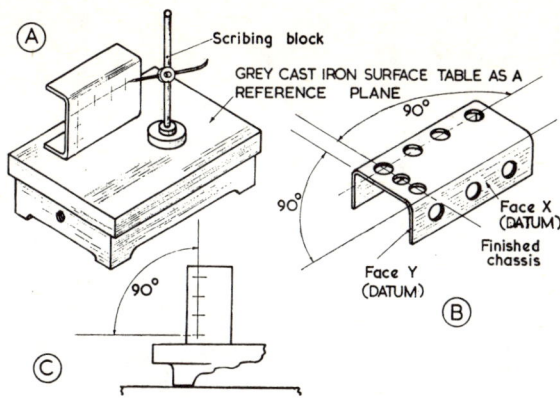

Fig. 2.2 Use of the surface plate

gained by using a surface plate when marking out engineering components should be clear on close study of Fig. 2.2. Note that at C the chassis is set on its side; provided that this side is 90° to the face marked x, all lines scribed at the set-up shown at C will be 90° to the lines scribed at the set-up shown at A. Remember that both x and y are datum faces, that is to say they are faces or reference planes from which measurements are taken, and they rest on the datum face provided by the flat surface of the surface plate.

Surface plates are available in a variety of sizes. A toolmaker uses a small surface plate called a toolmaker's flat, no more than 100 mm diameter, when engaged on high precision work. The surface of this plate is lapped to a high degree of accuracy giving it a mirror finish. At the other end of the scale, a large casting would be marked out using a surface table having an area of up to 20 square metres.

Care of surface plates

A surface plate is a precision piece of equipment and must be treated as such. It must be used only for marking out and measuring purposes, and should be protected by a well-fitting wooden cover when not in use. It must not be used for hammering, riveting or chiselling. It is permissible to smear the surface lightly with engineers blue for purposes of flatness testing, but care should be taken to clean the surface thoroughly when the blue is no longer required.

Scribing blocks

Two typical scribing blocks are illustrated at Fig. 2.3. At A we see a relatively simple type, while at B is shown a better type having provision for fine adjustment. The correct technique when using a scribing block is to keep the scriber horizontal, allowing only the minimum

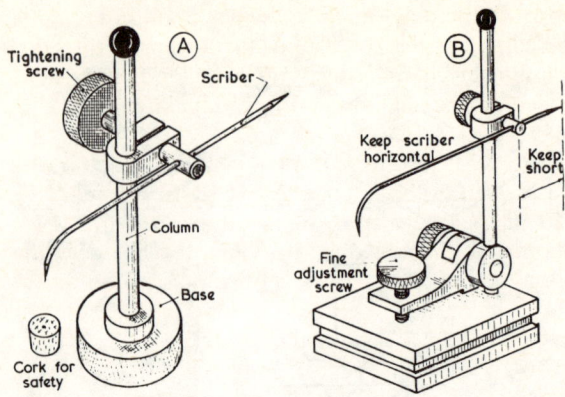

Fig. 2.3 Scribing blocks

length to protrude as shown at Fig. 2.3B. It is of the greatest importance that the line is scribed once only, care being taken to ensure that firm constant pressure is applied during the scribing stroke. Ensure also that the other point of the scriber is covered with a small cork. This useful safety precaution gives excellent protection during the marking out operation, as scriber points are exceedingly sharp and thus capable of causing serious injury.

Hand scribers
As far as possible, all marking out should be carried out using a surface plate as a datum, but occasions may arise when a line must be scribed using a hand scriber. Figure 2.4 shows the correct method of scribing a line on a metal bar; note the use of a try-square to ensure that the line will be 90° to the side of the bar. Once again, scribe the line once only, keeping the scriber about 60° in the direction of scrib-

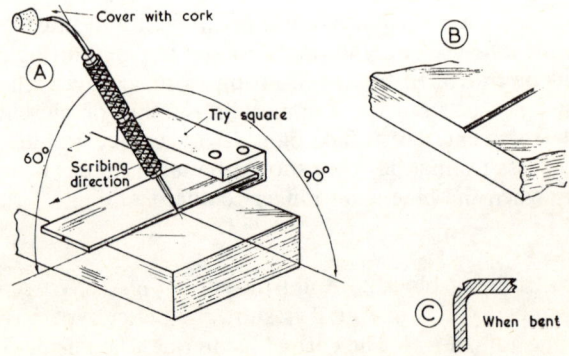

Fig. 2.4 Hand scribing

Hand Processes

ing, as shown in the diagram. The scriber point, if correctly hardened and tempered is capable of producing a clean visual line in most metals, including bright mild steel, and it is seldom necessary to resort to the use of a colouring medium in order to make the line show up more clearly.

If a line is to be scribed to indicate a bending position, it is unwise to use a hand scriber, and reference to Fig. 2.4C will make this clear. A well-scribed line actually removes metal, as shown at Fig. 2.4B, and bending the metal tends to open or widen the line, leading to certain cracking along the bend as shown at Fig. 2.4C. The softer the metal the deeper will be the scribed line, and it is good practice to scribe all bend lines using a hand pencil such as 2H or 3H.

Care of scribers

The points of scribers must be protected by corks at all times, and the points must be kept sharp. It is considered bad practice to re-sharpen a scriber point using a pedestal grinder; this is necessary only if the point has broken off and needs re-grinding. A blunt scriber point is best sharpened using a fine emery cloth or a smooth abrasive stone. Use the sharp straight point for scribing lines on mild steel, when fairly heavy pressure is needed to produce a clean well-defined line. To scribe lines on softer metals such as copper and aluminium, use the curved point of the scriber with less pressure applied.

Engineer's dividers

We may consider dividers as a pair of joined scribers which can be set at predetermined distances, making possible the scribing of circles.

The best types have a threaded screw, allowing for fine adjustment,

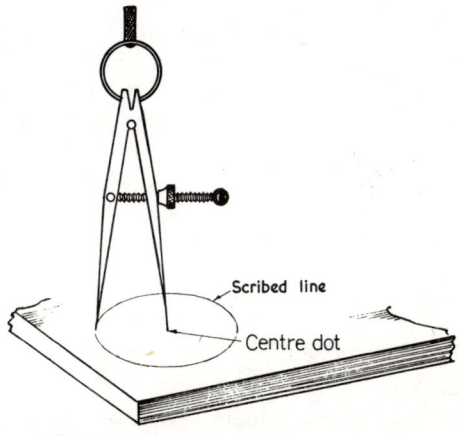

Fig. 2.5 Engineer's dividers

such a type is shown in Fig. 2.5. When scribing a circle, one point is located in a small dot made by a 60° centre punch, and with the dividers set to the required radius the circle is scribed once only. The apprentice technician is advised to master this vital marking out technique of scribing a line once, and once only. It is bad practice to scribe a line several times; to need to do so is a sure indication that a blunt tool is being used. Engineer's dividers require the same care as that described for scribers.

Odd-leg calipers
Sometimes referred to as hermaphrodite firm-joint calipers, this marking-out tool is a combination of scriber point and locating edge. Their main application is when a line is to be scribed parallel to an adjacent face as shown in Fig. 2.6. Some care is required when using this marking-out tool, and reference to the diagram shows that it is

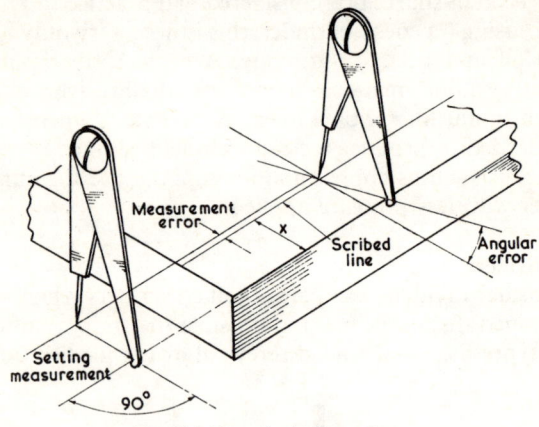

Fig. 2.6 Use of odd-leg calipers

essential to keep the scribing point at 90° to the datum face. Otherwise the distance shown as x will differ from the setting between the scriber point and the location edge. When parallel lines are required to be marked out it is advisable to use a scribing block and surface plate. Remember that odd-leg calipers require the same care and protection as that given to scribers and dividers, so keep the scriber point protected with a small cork.

2.1.2 Additional equipment for marking out

Centre punches
Centre punches serve two main purposes. The first is to centre-dot the intersection of two scribed lines in order to mark out a circle, using

Hand Processes

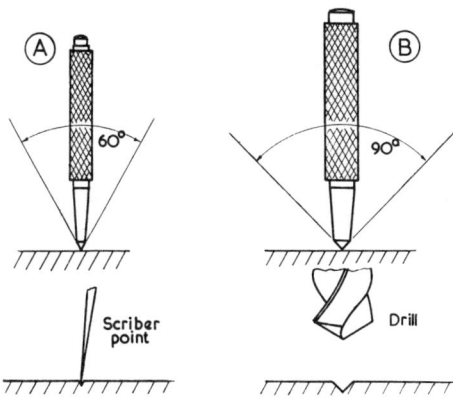

Fig. 2.7 Use of centre punches

dividers with the centre dot as the circle centre. A centre punch suitable for this purpose is illustrated at Fig. 2.7A; note that the working point is ground to an included angle of 60°, providing a snug fit for the location of the divider point. The centre punch shown at Fig. 2.7B is of more robust construction with the angle of the point increased to 90°. It is used for the second purpose, which is to provide a start for a twist drill as shown in the diagram.

It is good practice to first use the 60° centre punch, and when satisfied with the accuracy of the dotted intersection, enlarge the dot carefully using the 90° centre punch. This will ensure that the drill commences drilling at the precise intersection of the scribed lines.

Vee blocks
Vee blocks (sometimes called simply V blocks) are used for holding or locating cylindrical work, and Fig. 2.8 shows a typical set-up. Note that the vee blocks must be a matched pair, and care must be taken to ensure that they are always kept as a pair and not interchanged with others.

2.1.3 Metal-removing hand tools

The hacksaw
The use of a hacksaw is the most rapid way of removing metal while working at the bench. The great advantage is, that for a relatively thin cut involving a small amount of metal removal, a large amount of waste material is removed in one piece. This is in accordance with our stated principle of maximum metal removal with minimum effort. A hacksaw can be used for a wide range of work, including sawing pipes, bars and other sections to prescribed lengths. If, however, maximum

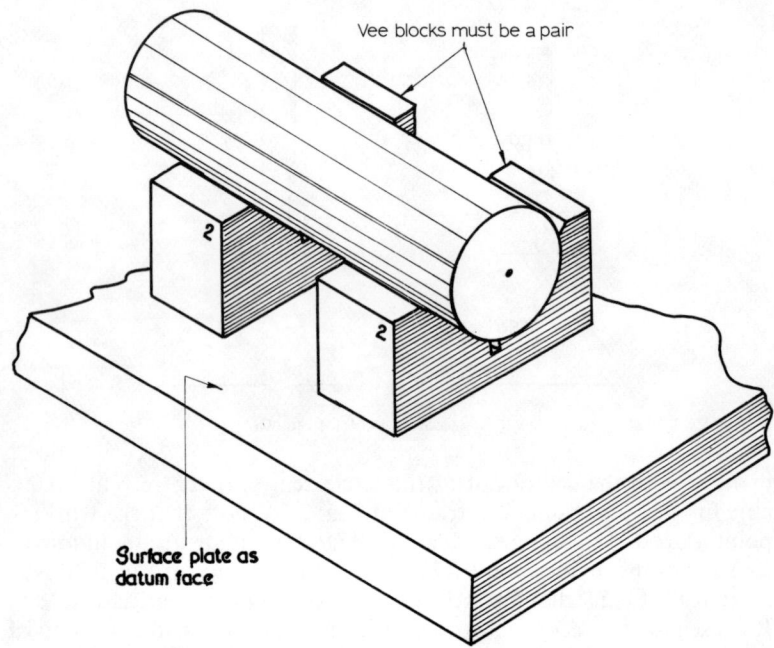

Fig. 2.8 Use of Vee blocks

efficiency is to be obtained from the hacksawing process, it is essential to ensure that the correct blade is chosen for the job in hand, and that the blade is properly mounted in the hacksaw frame.

Types of blades

Hacksaw blades are available in three main types:
  (i) all-hard,
  (ii) flexible,
  (iii) 'spring-back'.

'All-hard' blades are hardened throughout their whole section and cannot be bent without the risk of breakage. This rigidity makes for very accurate sawing, and considerable skill is needed if blade breakage is to be avoided. These blades are best used for reasonably thick work requiring straight cuts; for example sawing through a mild steel bar of 25-mm square section. The flexible blade is hardened along the teeth only, and thus has a considerable degree of flexibility, making it very much suitable for the inexperienced operator. The 'spring-back' blade can be considered as a compromise between the all-hard and the flexibles. It has sufficient flexibility or spring to reduce the possibility of breakage, and it is almost certainly the most popular all-purpose hacksaw blade in general use.

Hand Processes

Tooth pitch
The three types of hand blades mentioned above are available with different tooth pitch, and it is important that the correct tooth pitch be chosen to suit the job in hand. Table 2.1 gives the recommended values for tooth pitch according to the metal requiring sawing. It may be noted that the softer the material to be cut, the smaller is the number of teeth per centimetre.

Table 2.1 Hacksaw blade pitch and metal thickness

Material	Tooth pitch (per cm)	Metal thickness (mm)
Wrought iron, mild steel	10 8	up to 6 6–25
Tool steels, alloy steels	14 10	up to 6 6–25
Grey cast iron	10 8	up to 12 12–75
Copper, brass, bronze	6 or 8	
Steel and copper tube, conduit and sheet	10 or 14	

All the above blades are referred to as single-edge hand blades, with two lengths available—250 and 300 mm. The blades are 12 mm wide with the following tooth pitches usually available:

> 250-mm blades: 8, 10, 14
> 300-mm blades: 6, 8, 10, 14

Setting the blade
A typical hacksaw is illustrated on Fig. 2.9A. Note the adjustment for 250- or 300-mm blades, also the tensioning device. The blade teeth must point away from the handle as shown, and just tension should be applied to give rigidity to the blade. Press downwards on the forward stroke only using the full length of the blade, making about 50 to 60 strokes per minute.

Simple hacksawing techniques
The following hints will assist in the correct use of a hacksaw.

(i) Keep the cutting line as close as possible to the vice jaws.
(ii) Use a new blade when sawing the soft non-ferrous metals such

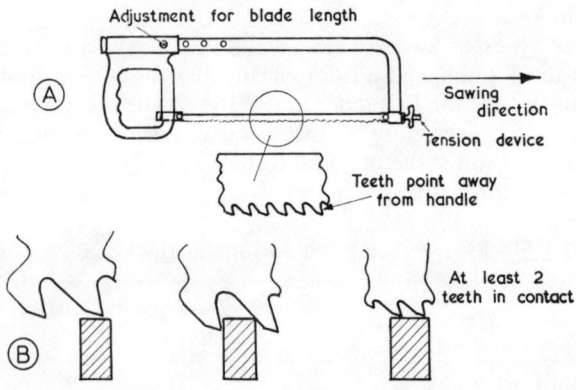

Fig. 2.9 Use of hacksaw blades

as copper or brass, and keep these used blades for later use on mild steel.
(iii) Apply light pressure at the start of a cut and also at the finish of a cut.
(iv) Make sure that a fine-pitch blade is used when sawing thin-walled tube. Figure 2.9B shows how a coarse-pitch blade leads to tooth breakage.

Files

Files are made from high carbon steel hardened and tempered. New files should be kept for soft non-ferrous metals such as copper, aluminium and brass, and then passed on for use on mild steel. It is very difficult and tiring to attempt to file copper or brass using a worn file.

It is best to regard a file as a surface-producing hand tool. All surfaces produced in engineering manufacture are geometrical (typical

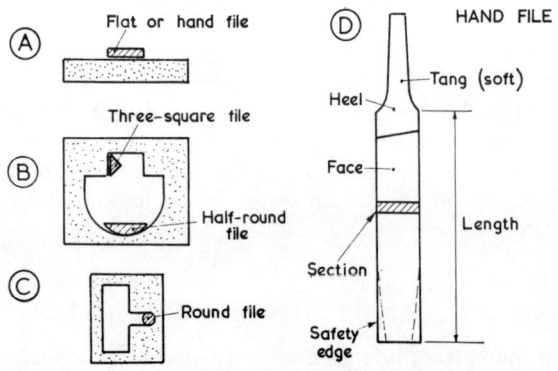

Fig. 2.10 Files and their uses

Hand Processes

examples include plane, cylindrical and conical), and this means that a wide range of file sections are needed. In Fig. 2.10 we see typical examples of file sections together with their applications.

Roughing and finishing
It is best to separate all hand filing into two distinct operations:

(i) roughing,
(ii) finishing.

The object of roughing is to remove the maximum amount of metal in minimum time, with less regard for finish and accuracy. This means that a file with coarse or large teeth is needed, having very good metal removing properties but unable to leave a good surface finish.

Such files have a double-cut tooth-form as shown at Fig. 2.11A, with teeth fairly widely spaced. When roughing always choose the biggest file possible, that is to say the file on which the length shown in Fig. 2.10D is the maximum available. Hand files are available from 100 to 500 mm in length; the longer the file, the more weight can be applied to it, and the more metal removed with each stroke. The skilled method is to lean on the file during the forward stroke, ease off on the return stroke and file at the rate of about 50–60 strokes per minute.

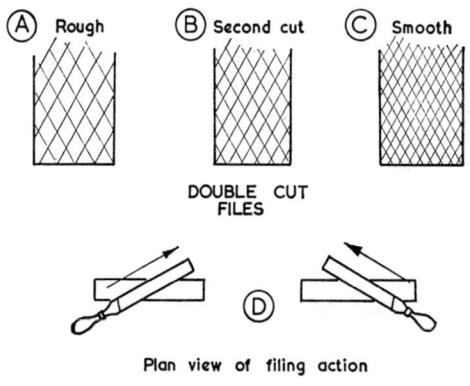

Fig. 2.11 Rough filing

Keep the file in a diagonal position along the length of the job as shown at Fig. 2.11D, changing direction from time to time. The object is to keep the maximum surface area of file in contact with the metal; this assists in producing a flat surface.

Finishing
The object of finishing is to obtain the desired accuracy together with a good surface finish. A small amount of metal only must be left for

finishing, and arm pressure is now used, permitting closer control over the filing action. A smooth file must be used called a single-cut file, having the teeth cut in one direction only as shown in Fig. 2.12A. This

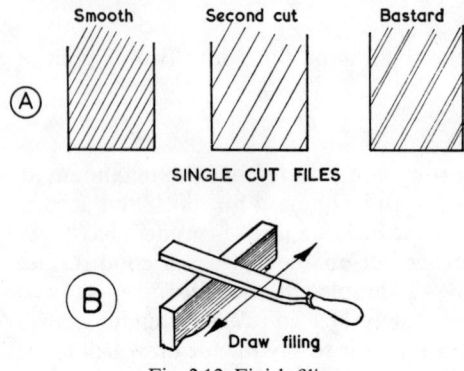

Fig. 2.12 Finish filing

type of file gives an improved finish, and best results are obtained when the surface is finished off using the technique called draw filing. This is shown in Fig. 2.12B; the file is moved across the surface of the job as shown, taking care to keep it at 90° to the vertical. A very pleasing finish is quickly obtained, and it is not necessary to polish the surface using emery cloth.

Safety precautions when filing
Although a hand file is not normally considered a dangerous tool, there are certain precautions that must be taken. Never use a file without a properly fitted handle, and this is especially important if the file is to be used on a rotating bar in a centre lathe. Replace any handle that is split, or lacks the metal ferrule. Keep the files in a proper file rack, and remember that it is a good plan to reserve new files for the softer non-ferrous metals such as copper or brass, and later use them for mild steel or cast iron. Never use a file for any purpose other than filing; on no account should a file be used to eject a taper-shank twist drill from the spindle of a pillar drilling machine.

Scrapers
Scrapers are used because of their ability to remove small amounts of metal. The face of a file is provided with cutting teeth, and it is extremely difficult, if not impossible to file a small amount of metal from the centre of a job as shown in Fig. 2.13A, using a hand file. The removal of small amounts of metal in awkward places is no problem, provided a flat hand scraper is available, and there are two types in common use. Figure 2.13B shows the push-type scraper, while at C we

Hand Processes

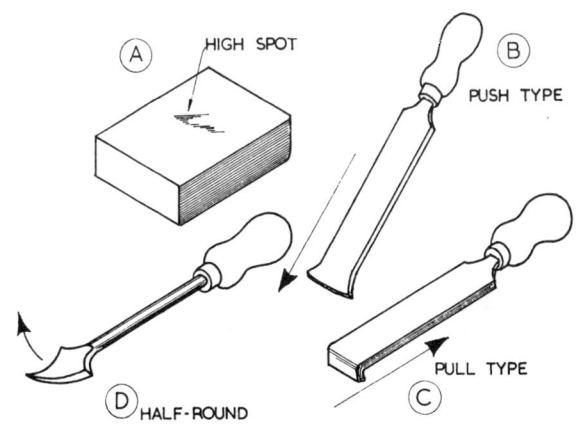

Fig. 2.13 Use of scrapers

see the pull-type scraper. The cutting edges must be extremely keen or sharp; an India stone is used for sharpening the edge, which must be protected when not in use. On no account must a scraper be sharpened using a pedestal grinding machine, and it must be remembered that a surface needs to be reasonably flat before any attempt is made to scrape it. Internal cylindrical surfaces such as bored holes or bearings for shafts are scraped with the half-round scraper shown in Fig. 2.13D. It must be appreciated at the outset, that hand scraping is a skilled and lengthy business, and the best scrapers have carbide tipped blades giving them an extremely keen edge, very necessary for all types of scraping.

Cold chisels
Chisels are used to chip away surplus or unwanted metal. If used on red-hot work they are called hot chisels, while the chisels used at the engineer's work bench are known as cold chisels.

A cold chisel makes use of a chipping technique; that is to say the surplus metal is removed in small chips as the chisel is struck with a hammer. This means that the chisel needs to be held and guided with one hand, and struck with a hammer held in the other.

Chiselling, then, is a difficult operation; the rate of metal removal is slow, and accurate work is very difficult to produce. A cold chisel however, can prove a most useful tool for certain operations, although a considerable amount of skill and care is required with each of the following types of cold chisels.

Flat cold chisel
This is the most widely used cold chisel. A typical flat cold chisel is

illustrated in Fig. 2.14A; remember that these chisels are available in a range of sizes. They are made from good quality cast steel (high carbon steel) and their section is usually hexagonal or octagonal; this helps to prevent them from rolling off the work bench. Figure 2.14B shows a

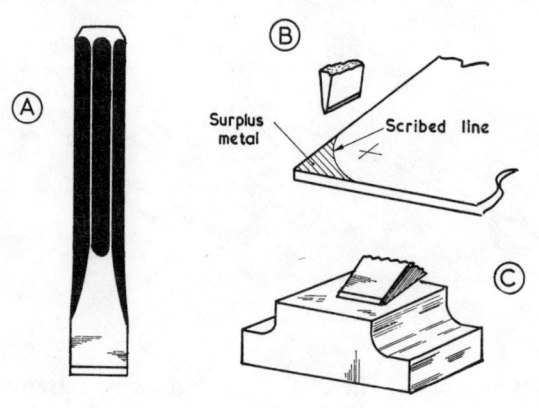

Fig. 2.14 Flat cold chisel

typical application for a flat cold chisel—the removal of surplus metal prior to filing a radius. In this example, no chipping takes place; the purpose of the flat chisel is to shear the metal and separate the unwanted portion, the relatively short cutting edge of the chisel allowing fairly close following of the scribed arc. A flat cold chisel may be used for producing a small flat surface as shown in Fig. 2.14C, or splitting rusted nuts and chipping off the heads of rivets.

Cross-cut chisel
Somewhat similar in appearance to a flat cold chisel, a cross-cut has a narrow cutting edge, as shown in Fig. 2.15A. This is a useful chisel for

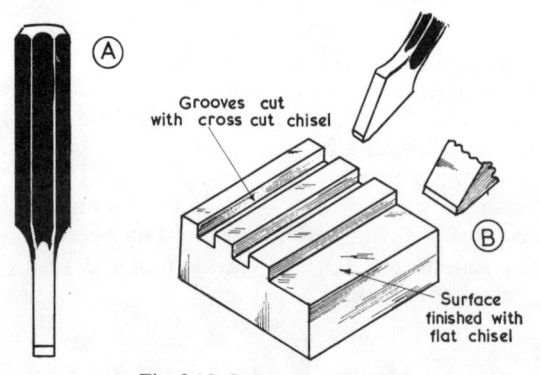

Fig. 2.15 Cross-cut cold chisel

Hand Processes

cutting off small radii, or breaking down a flat surface prior to the use of a flat chisel. This technique is shown in Fig. 2.15B.

Diamond-point chisel
The point of this chisel gives rise to its name. As can be seen from Fig. 2.16A, it is very suitable for cutting small grooves or for simple engraving such as the chiselling of letters or numbers on metal components. It may be used also to provide a start or guide for chiselling oil grooves using a round-nose chisel.

Round-nose chisel
This chisel is illustrated in Fig. 2.16B, where it can be seen that the cutting edge has a radiused face, making it very suitable for chiselling simple oil grooves in phosphor bronze, or white-metal bearings.

Safety precautions when chiselling

As previously stated, chiselling is essentially a chipping process, and the metal chips can cause severe injury to the operator's eye, or injure other workmates close by. To prevent injury to others a chipping

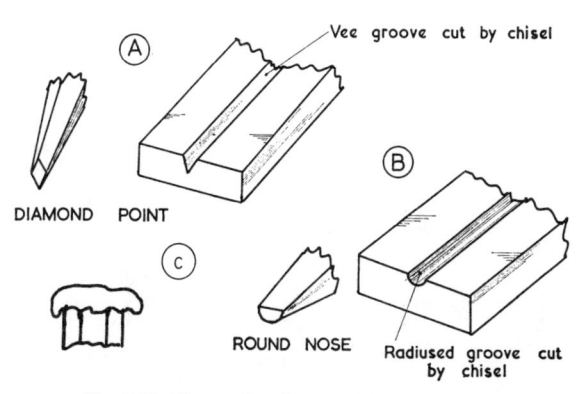

Fig. 2.16 Diamond-point and round-nose chisels

guard should be used, positioned so that it faces the chiselling operation. Such a guard is simply a flat sheet of metal equipped with legs, so that it stands upright and prevents the chipped metal flying across the bench or workplace.

It is essential also, to ensure that the cutting edge of a chisel is kept in a sharp condition. A blunt chisel can be dangerous, for excessive force is required from the hammer blows in order to chip the metal, and this can easily result in the chisel or hammer flying out of control across the bench. Avoid using chisels that have the condition known as mushroom head. This is shown in Fig. 2.16C. When such a chisel is struck with a hammer small fragments are likely to fly off and cause injury. The chisel must be re-ground or returned to the stores.

2.1.4 Screw cutting at the bench

The removal of metal in order to produce an internal or external thread is often carried out at the bench. The hand tools used to cut an external thread are called taps, and a set of three is required if an efficient thread is to be produced. The principle of roughing and finishing as *two* separate operations has already been described, and the same technique is used when cutting internal threads with taps. Since it is not possible to adjust a tap, the roughing out is performed by one taper tap, followed by a second; finally a plug or finishing tap is used to bring the thread to size and also give a good finish.

These taps are shown in Fig. 2.17A, and it is important to remember to start the taper tap so that the axis is 90° to surface of the work. This alignment is readily checked with a small try-square, as shown in Fig. 2.17B. Always have a plentiful supply of cutting fluid to hand, and do

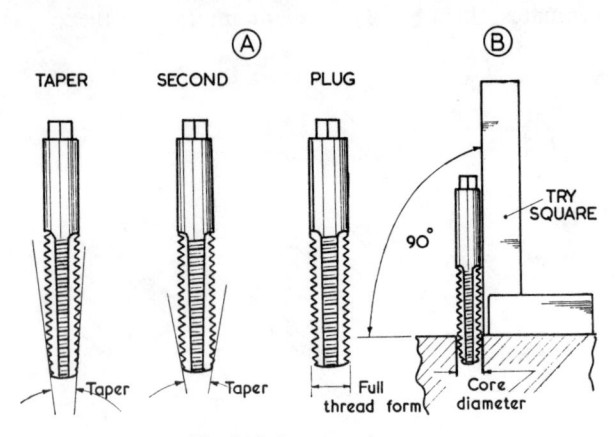

Fig. 2.17 Screw-cutting taps

not use too large a tap wrench; take special care when tapping blind holes, rotating the tap in an anti-clockwise direction after each cutting turn. This helps to free the swarf from the cutting edges of the tap and reduces the risk of tap breakage. Remember that a tapping drill must be used; the diameter of this drill is equal to the core diameter of the thread. Table 2.2 gives the tapping drills for the metric threads likely to be used in an engineering workshop.

Hand Processes 33

Table 2.2 Tapping drills for ISO Metric Coarse series threads

Thread diameter (mm)	Pitch (mm)	Tapping drill (mm)
6	1·0	4·7
8	1·25	6·8
10	1·5	8·1
12	1·75	9·8
14	2·0	11·5
16	2·0	13·5
20	2·5	16·75
22	2·5	18·75
24	3·0	20·25

Cutting external threads

The tool used to cut an external thread is called a die, and the device used to hold this die in order to give it a turning movement, or torque, is called a stock. Stocks and dies are widely used for hand cutting external screw threads at the bench, although sometimes a hardened die nut is used to clean up a damaged thread, and such a die nut is shown in Fig. 2.18A.

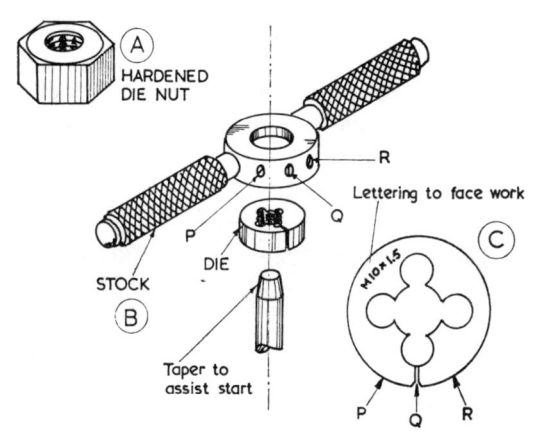

Fig. 2.18 Stocks and dies

In Fig. 2.18B we see a small stock together with a split or button die. Note the three tightening screws in the stock, shown as P, Q and R. Correct tightening of these screws provides slight adjustment of the split die, and this is achieved as follows.

The split in the die, faces the screw Q, and for the first cut the die is inserted with the chamfered edge facing the bar and screw Q tightened slightly. This tends to open the die, as shown in Fig. 2.18C, and the screws P and R are tightened sufficiently to hold the die firmly in place. A small taper at the end of the bar to be threaded assists in making a correct start with the die, and with a plentiful supply of cutting fluid, the die is run down the bar. Once again the die must be rotated backwards after each cutting stroke to ensure swarf breakage and so reduce the risk of damage to the thread or to the cutting teeth of the die. It is certain that a nut will not screw easily on the thread produced, so a finishing cut can now be taken after slackening off screw Q and tightening screws P and R. This has the effect of slightly closing the die. Running the die back down the thread is equivalent to a finishing cut, for a small amount of metal will now be removed. This process can be repeated until the required fit is obtained, and with practice and experience, well-finished accurate screw threads may be cut at the bench.

Large-diameter and fine-pitch threads

A different thread cutting technique is needed if large diameter threads are to be cut at the bench, for example on conduit pipe used in electrical installation work. It is of the greatest importance that the thread is cut in close alignment with the axis of the tube; for the tube wall is relatively thin, and a misaligned thread is likely to cut through it.

A special type of stock is used, fitted with a pilot bush as shown in Fig. 2.19A; note the use of loose dies. The pilot bush ensures that the thread is cut parallel to the pipe axis, and adjustment of the loose dies enables the cut to be started while the dies are in full contact with the pipe. The thread is brought to size by tightening the adjusting screw; a plan view of the stock is shown in Fig. 2.19B.

Loose dies are also used in the stock shown in Fig. 2.19C. This type

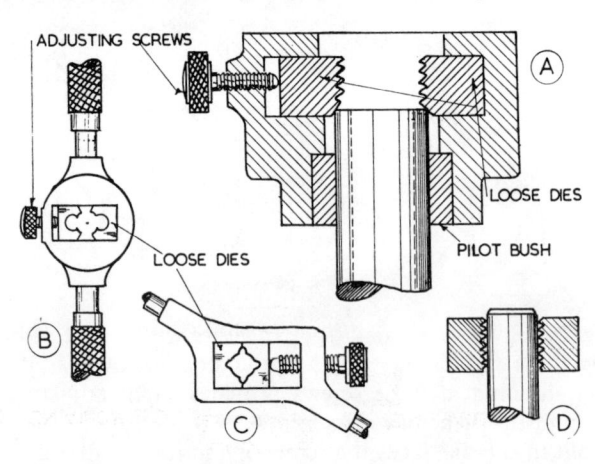

Fig. 2.19. Loose dies and pilot bushes

Hand Processes 35

of stock is best used for fine pitch threads, which can be started as shown at Fig. 2.19D. The tightening screw is used to increase the depth of thread, the die run down the bar until the required fit is obtained.

2.1.5 Use of powered hand tools

All the hand tools so far described require muscular effort in the form of hand pressure or hammer blows. The continuous application of effort on the part of the user soon leads to fatigue and loss of concentration, and these are precisely the conditions that not only lead to inaccurate work but also increase the risk of accidents. As far as possible, the skilled worker should be relieved of prolonged physical effort, and this means that a selection of powered hand tools needs to be available. The operator is now able to concentrate his efforts on guiding and controlling the powered device; in this way he is better able to produce accurate well-finished work over fairly long working periods. It follows then, that all powered hand tools are portable, that is to say they can be taken to where they may be required, and this means that they must be of reasonable weight and comfortably handled.

Two main sources of external power are used;

(i) electrical power,
(ii) compressed air.

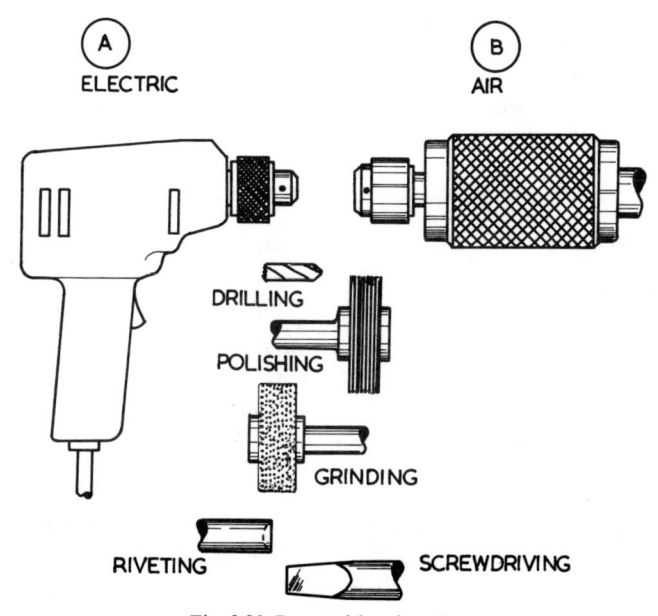

Fig. 2.20 Powered hand tools

Electrically powered hand tools are the most popular because electrical power is easily available and relatively simple to instal. Figure 2.20 shows the basic operations that can be carried out using powered hand tools.

Powered hand drilling

If a component cannot be taken to a drilling machine, then it is essential that a powered hand driller be available, for the drilling of holes using a hand or breast drill is a most ardous and time consuming task. In Fig. 2.20 we see that holes may be drilled using either electrical or air powered drills.

Electric drills

Generally speaking electric drills are not suitable for large diameter holes because the power required results in a heavy and unwieldy piece of equipment, difficult to both handle and control. Providing, however, that the diameter of the hole to be drilled does not exceed about 10 mm, it is not difficult to drill it using an electric drill. Make sure that the position of the drilled hole is deeply punched using the 90° point punch illustrated in Fig. 2.7B, for unlike a drilling machine which has a fixed spindle, the drill point of a drill held in an electric drill is free to wander. Make sure also, that the drill is kept in a truly vertical position, and it is unwise to attempt to drill holes of less than 5 mm diameter, because such drills are very easily broken due to the difficulty of keeping the electric drill in a rigid position. Great care needs to be taken also of the pressure or feed exerted on the drill when drilling the hole, especially if it is of small diameter. The movement and vibration of the revolving armature within the electric drill, and the general lack of rigidity in the drilling set-up, makes impossible any sense of feel or sensitivity, and undue pressure soon leads to drill breakage.

Air drills

Air drills are less commonly used than electric drills, and this is mainly due to the fact that the power required to rotate the drill spindle is provided by the high pressure of the air supplied to the air drill. It is both difficult and costly to compress and circulate compressed air, because extensive piping and tubing is required. If, however, compressed air is available, air drills are very suitable for holes of small diameter in relatively thin metal, as in sheet metal work.

As can be seen in Fig. 2.20B, the air drill is more compact and fits quite easily into the user's hand. The spindle speed is much higher than that obtained from electric drills, making the air drill very suitable for smaller diameter holes. Once again it is difficult to avoid drill breakage

Hand Processes

if very small diameters need to be drilled; this is due to the lack of sensitivity during the drilling operation, for the air drill tends to vibrate to some extent when in use, and it is not easy to keep one's hand steady.

Polishing

As may be seen in Fig. 2.20, it is possible to hold a small polishing mop in the chuck of an air or electric drill. In this way the polishing process is greatly reduced in terms of time and effort required. A polishing mop is made up of layers of cloth, and best results are obtained when using an air drill, because of the higher spindle speed available.

Hammering

Small diameter rivets are quite easily finished off using a suitable riveting head attachment as shown in Fig. 2.20. The riveting head or snap has a reciprocating axial movement and both electric and air drills are suitable for riveting, although the electric drill is preferred because of the extra power available.

Screwdriving

Powered screwdriving is quite possible using either air or electric drills provided the spindle or attachment is fitted with a slipping clutch. This device ensures that when the correct tightening pressure is reached, the screwdriver blade ceases to turn the screw, a loud clicking noise warning the oerator that the screw is tightened sufficiently. A similar principle is adopted to tighten nuts or studs on a production basis. Application of a hand powered tool rapidly tightens home the nuts on car wheels, and in many cases as many as five wheel nuts are tightened simultaneously, all to the correct tightening pressure or torque.

Grinding

Both electric and air powered tools can be used for grinding. Figure 2.21A shows the use of an air powered hand grinder, cleaning the inside radius of a hardened precision die. High spindle speeds and light pressures are needed for this work, and the air powered tool is ideal for this kind of portable grinding. In Fig. 2.21B we see a more popular application of hand grinding, namely the removal of surplus metal from a casting; a process known as *fettling*. Because of the additional force required, electrical power is used, and it is certain that this portable grinder will be balanced with a counterweight. In this way the operator is relieved of the fatigue of lifting and supporting the portable grinder during the fettling operation, and is able to concentrate his efforts on the actual metal-removing operation. This important, fatigue-saving principle is simply illustrated in Fig. 2.21B.

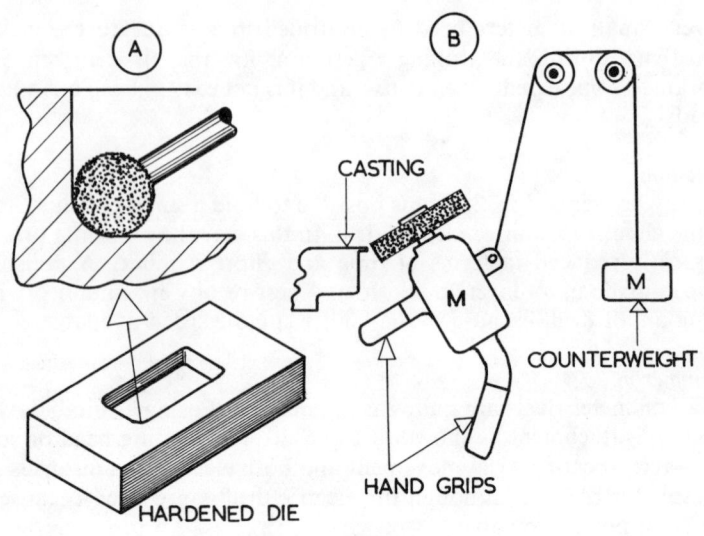

Fig. 2.21 Portable grinders

Safety precautions with powered tools
Both air and electrical powered hand tools are real sources of danger and injury to the operator, although less danger is present with air operated tools. Leaking connections or joints, while reducing the pressure and hence the efficiency of the system, present no real danger to the operator. Most accidents with air operated tools are caused by practical jokes and it needs to be remembered at all times that high-pressure air must be treated with respect.

Air operated tools are much less bulky than their electrical counterparts and when not in use, are easily hooked above a workbench out of harm's way. On the other hand, the use of electrical hand powered tools presents many safety problems. These tools are much heavier and bulkier, and as previously stated it is necessary to use counterpoise weights so that they may be handled and used with minimum effort by the operator.

Great care needs to be exercised in the positioning and checking of the power cables to the tool. The cable must of necessity be flexible, and cables that are allowed to run along the floor, are very easily and quickly damaged. As far as possible, always ensure that the cables are neatly secured with overhead clips, and pay regular attention to the connecting sockets and plugs.

Because water is a conductor of electricity wet or damp working conditions are to be avoided at all costs, since the risk of injury from electric shock is greatly increased. Most electrical power tools operate on a 240-volt circuit, never on a 440-volt circuit, and the safest system

Hand Processes 39

is that which operates on a low-volt basis, say 110 volts. The earthing arrangements on all portable electrical tools must be checked at regular intervals.

Advantages of powered tools
Although the use of powered hand tools increases the possibility of accident and injury to the operator, there are several important advantages to be gained. Firstly there is the matter of speed of production, or more efficient use of the operator's time. The more external power a worker has at his disposal, the faster he is able to perform a given job. Power may be described as the rate of doing work; for example a certain amount of power is required to say, tighten the four wheel nuts of a motor car. Clearly, a worker using a hand brace could tighten the four nuts perhaps in one minute, but he could not possibly keep this up over a period of hours.

It is well established that a tired worker is not only inefficient but also unsafe, and lack of powered equipment must result in poor workmanship, expensive products and high accident rates. It follows therefore, that the actual amount of power available to workers of any type, can be taken as a measure of the efficiency of the manufacturing system, providing always that the powered tooling is well designed, properly installed and adequately manufactured.

These are the responsibilities of the engineering technician, and with the increasing amount of applied technology, powered tooling is increasing year by year as are the duties and responsibilities of the engineering technician.

2.2 Marking-out principles

Marking-out is the name given to the process of making an outline of the component required. It is carried out in many manufacturing industries, for example:

Textile:	chalk on cloth
Wood working:	pencil on wood.
Engineering:	pointed tool on metal.

It is evident that the engineering industry has the greatest problem, because metals are hard, and it is not easy to make permanent lines on them. We have seen, earlier in this Unit, that the tool used to mark metal is called a scriber, and that there are many other tools used in marking-out. However, the correct choice of marking-out tools will come to nothing, if the basic principles underlying marking-out are not understood and followed. *Two* basic principles are followed when marking out, depending on the profile of the component. These are:

(i) working from datum faces,
(ii) working from centre-lines.

2.2.1 Working from datum faces

Figure 2.22 shows an engineering component. It is a lever from an overhead conveyor system, and an immediate replacement is required, for the conveyor system is out of action causing a total stoppage of production. This is a typical example of the need for accurate marking-out, because marking-out at the bench is a skilled operation, and is carried out only for the manufacture of one-off components. Another look at the component in Fig. 2.22 shows that it is made up of straight

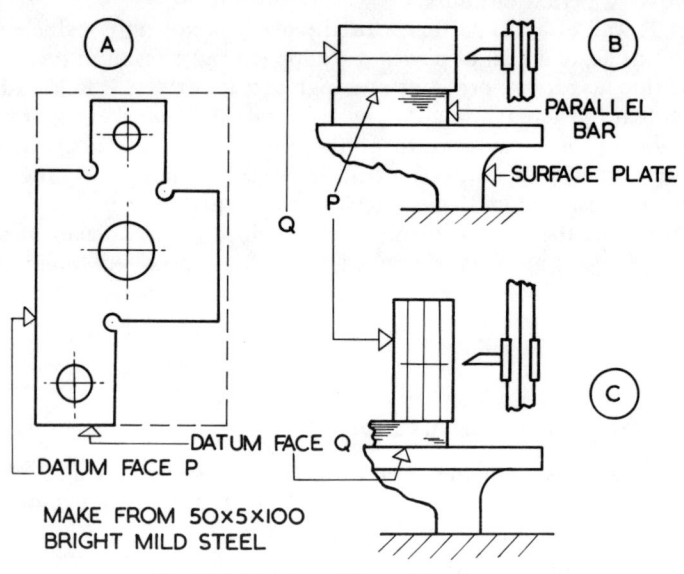

Fig. 2.22 Marking-off from datum faces

flat surfaces at 90° to each other; this is ideally suited for marking-out using the basic principle of working from datum surfaces. The following notes give the sequence of operations necessary for the correct, speedy and yet accurate marking-out of this component. The dotted lines in Fig. 2.22A indicate the initial size of the material.

Op. 1 Obtain from metal stores a length of 50 × 5 bright mild steel. Using a power hacksaw, cut off a length of 100 mm. Clean and remove all sharp edges.

Hand Processes

Op. 2 Lightly rub one surface with medium grade emery cloth, and heat slowly using a gas–air blowpipe until surface turns brown. Quench in water or oil. This brown oxide film makes an excellent aid to marking-out, with no effect on the metal structure.

Op. 3 File the longest edge flat. This is shown in the diagram as P, and represents the first datum face.

Op. 4 File the second face flat and 90° to first face; this is shown as Q and is the second datum face.

Op. 5 Locating from datum face P, and using parallel bar, surface table, scribing block or Vernier height gauge, mark off all the lines required; the set-up is shown in Fig. 2.22B. Support job with angle plate.

Op. 6 Turn job 90° and locating from datum face Q mark off all remaining lines.

Op. 7 Using a 60° included angle centre punch, centre-dot intersecting lines to indicate hole centres, and with engineer's dividers set at required radii, scribe the diameters of the holes; this helps to ensure that they are drilled in their correct positions. We will deal with the hole drilling technique later on this Unit. Note that a 90° included angle centre punch must be used to open up the centre-dots prior to drilling.

Op. 8 Drill all holes.

Op. 9 Place job in vice so that profile lines are vertical and as close to the vice jaws as possible. Using a hacksaw, saw downwards keeping as close to the scribed line as possible. Too great a distance from the scribed line, means that much more effort and time will be involved when filing down to the line.

Op. 10 With a good roughing file, file down as close to the line as possible; with practice it is possible to file half a well-scribed line away, the process assisted by using a small jeweller's eyeglass. Under no circumstances must a line be filed away completely. Finish off job by draw filing all around and removing sharp edges using a smooth file. The sharp edges of the drilled holes can be removed by hand, using a twist drill of larger diameter than the hole to be chamfered.

Using marking-out tools, scribing the line once only, and working from datum faces using a surface plate, the marking-out of engineering components is not difficult. All lines scribed will be parallel to the datum faces, and also at 90° to each other, when the principle of working to datum faces is adopted.

2.2.2 Working from centre lines

Figure 2.23A shows another engineering component which is required under similar conditions to the previous component. We see however, that there are practically no straight lines present; the profile is made up of curves or radii. Clearly there is little to be gained by filing two datum faces, for these will have to be removed when the component is filed to the marked-out profile.

In Fig. 2.23B, we see the best technique to adopt, with the plate clamped to an angle plate resting on a surface plate. The position of the centre line P must be about centre of the plate, which is levelled as shown in the diagram. When this line is scribed, the angle plate is turned 90°, with the set-up as shown in Fig. 2.23C.

The centre lines Q and R can now be scribed, and with the intersec-

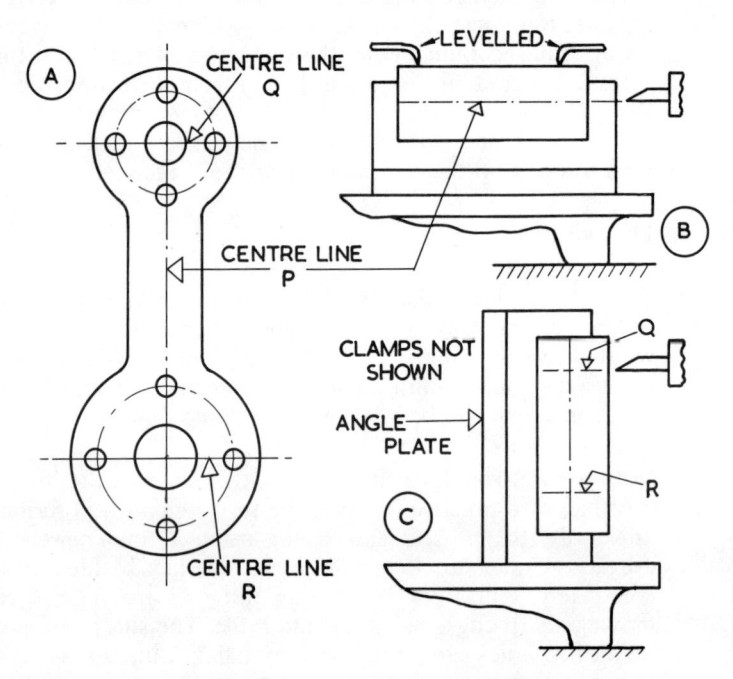

Fig. 2.23 Marking-off from centre-lines

Hand Processes

tions centre-dotted, circles are scribed using engineer's dividers. These circles represent the pitch circle diameter of the drilled holes; the intersections again centre-dotted to provide a position for each hole centre. The sequence of operations to bring the component to the required profile is identical to that described for the previous component. The job should be roughed, finished and draw-filed, with all sharp edges removed.

We will see later on in this Unit how accurate marking-out to plus and minus 0·02 mm is possible using a Vernier height gauge, which is the most suitable marking-out instrument whether working to datum faces or working to centre-lines. It is another well-established rule that as far as possible, all marking-out is carried out on a surface plate using a Vernier height gauge; the scribing of lines at the bench using steel rules is not recommended except for rough work.

2.2.3 The use of Vee blocks

Figure 2.24A shows a spindle which has been turned on a centre lathe, and is now to be marked out for the two slots shown as P and Q in the diagram. The best principle to adopt for this kind of job is working to centre-lines, the centre-line in this case being the axis of the spindle.

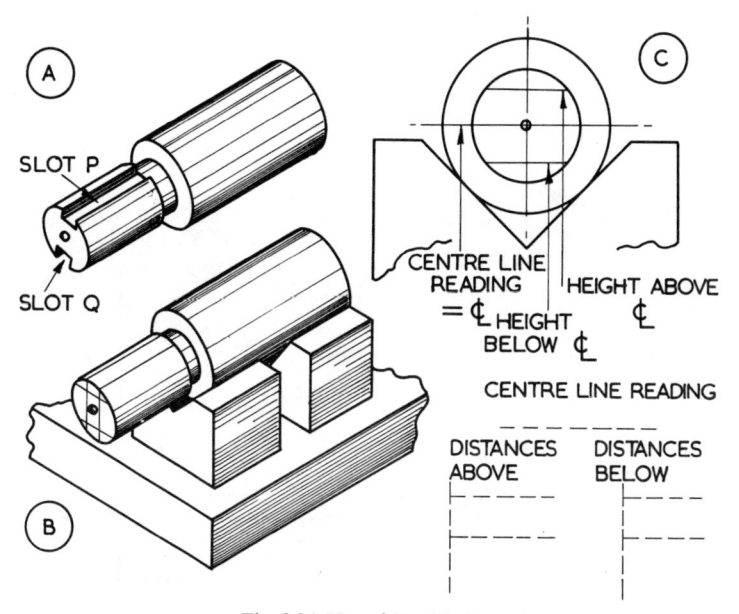

Fig. 2.24 Use of Vee blocks

This is a typical application of the use of Vee blocks for marking-out, and once again the set-up takes place on a surface plate. This is shown in Fig. 2.24B and it will be noted that the Vee blocks are a matched pair, always used together. Note the technique adopted to find the centre of the spindle using a Vernier height gauge, and when carrying out the calculations always keep a pencil and notebook convenient. It is bad practice to chalk on the actual surface plate, for it is a datum surface and must be treated as such. A neat writing-up of the calculations is an essential part of marking-out, especially if the component is of intricate shape requiring many scribed lines. The best method is to keep all additions to one side, and all subtractions to the other side, with the centre-line dimension in the middle. In the event of an error, it is not difficult to check back through the calculations and find where the error has taken place. This technique is simply illustrated in Fig. 2.24C.

2.3 Engineering measurement

Consider the pulley shown in Fig. 2.25. We have seen it previously in Chapter 1 when considering the dangers of moving parts; it is the generator pulley which transmits the turning movement or torque necessary for a motor car engine to generate the electrical energy needed to charge the battery. If this pulley is to do its work properly it is important that the following conditions be observed.

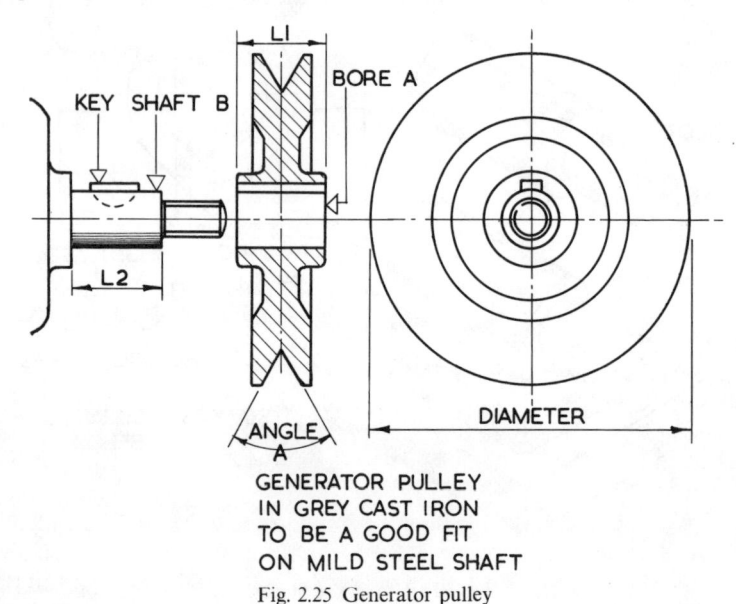

GENERATOR PULLEY
IN GREY CAST IRON
TO BE A GOOD FIT
ON MILD STEEL SHAFT

Fig. 2.25 Generator pulley

1. Bore A to be a good fit on shaft B.
2. Key in shaft to be a good fit in keyway in bore.
3. Length L1 to be slightly longer than length L2.
4. Angle A must be correct.
5. Pulley must run true when assembled to shaft.
6. All parts must assemble, for example any nut chosen at random must screw on any shaft also chosen at random.

Clearly these are all problems connected with precise measurement; in other words the sizes of the mating parts need to be machined within prescribed limits in order to bring about the sort of fit required. Unless close control is maintained over these sizes when the dimensions are machined, assembly becomes a difficult if not impossible matter. The art of ensuring that all dimensions are machined to within acceptable limits is known as *dimensional control*, and the principles involved are the same, irrespective of the number of components produced, or the rate of manufacture.

2.3.1 Elements of measurement

Reference to the pulley in Fig. 2.25 shows that *three* main measuring elements are present.

1. Linear dimensions (measurement of length).
2. Angular dimensions (measurement of angular inclination).
3. Non-linear functions (concentricity, flatness, roundness).

We need not, at the moment, concern ourselves unduly with the non-linear functions of roundness, concentricity and flatness, except perhaps to appreciate that when the pulley is assembled to the shaft, it must run true; that is to say there must be no wobble or eccentric motion. We need, however, to take a much closer look at the principles underlying the measurement of lengths and angles. These principles are important when the dimensions or sizes are *functional*.

Functional and non-functional dimensions
Reference back to Fig. 2.25 shows that the pulley must assemble to the shaft, yet not be too slack a fit. The actual type of fit needed is a matter for the engineering designer and once the fit is chosen, the relevant diameters of the pulley bore and shaft are obtained from tables of limits and fits. Clearly the diameters of the shafts and holes are *functional*: unless they are machined to the correct dimensions the required fit will not be obtained.

Similarly the angle of the taper opening in the pulley must be identical to the taper of the Vee belt; any variation will result in excessive belt wear leading to early failure of the Vee belt. In the same way, if the dimension L_2 on the shaft is longer than the dimension L_1 on the pulley, when the nut is tightened the pulley will still be slack having a slight axial movement; the pulley will not be held tightly by the nut.

On the other hand, the overall diameter and the width of the pulley are *non-functional*. It is not necessary to machine these dimensions to a high degree of accuracy; such a procedure would be wasteful of money and time. It must be remembered at all times, that the machining and measuring of engineering dimensions to close limits of accuracy calls for a high degree of skill, together with expensive machine tools and costly measuring equipment. It is pointless, wasteful and uneconomical, to machine to fine limits, dimensions which have no effect on the quality or performance of the engineering component or assembly.

2.3.2 Measurement of linear dimensions

We have seen that the accuracy to which a particular dimension is machined depends mainly on whether the dimension is functional or non-functional. In the case of functional dimensions such as the diameter of the shaft in Fig. 2.25, a high degree of accuracy is required. Unfortunately it is impossible to machine any dimension to an *exact* size, a certain amount of variation from the required size is permitted.

This is known as the *tolerance*, and it is a basic rule in engineering manufacture that functional dimensions are given small tolerances, while non-functional dimensions are given large tolerances. For example let us consider the measurement of the length of key used in the assembly shown in Fig. 2.25. The precise length of the key is a non-functional dimension, only a reasonable degree of accuracy is needed, and for measurements of this kind an engineer's steel rule is ideal.

Engineer's steel rule

An engineer's steel rule is a precision measuring instrument and must be treated as such, and kept in a nicely polished condition. Available in 150-, 300- and 600-mm lengths, Fig. 2.26A shows a typical 150-mm engineer's steel rule, together with a list of its desirable qualities.

When using a rule it is of the greatest importance that the principle of *common datums* is employed, identical in all respects to the use of a surface plate in both Figs. 2.22 and 2.23. This principle is shown in Fig. 2.26B, the set-up indicating the correct method of measuring the length of the key. Once again, a surface plate is used as a datum face; although it is possible that the term *reference plane* may be used instead of 'datum face'. Irrespective of what term is used for the flat surface, its purpose is to provide a common location or position from which the measurement can be made. Note that both the rule and key are at 90° to the working surface of the surface plate, and the use of an angle plate simplifies the set-up.

Hand Processes

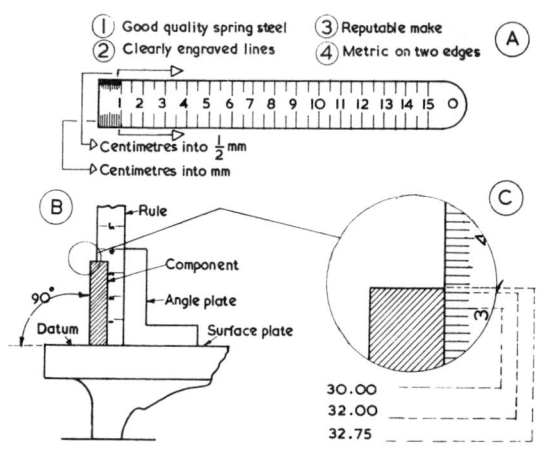

Fig. 2.26 Engineer's steel rule

With practice is is possible to determine linear dimensions to an accuracy of about plus and minus one tenth of a millimetre using a 150-mm engineer's steel rule; the correct technique is simply illustrated in Fig. 2.26C. Let us assume that close examination of the top of the key when compared with the engraved lines on the steel rule (a magnifying glass or watchmaker's eyeglass can be used to great advantage) shows the reading enlarged in Fig. 2.26C. Clearly the length of the key is the sum of the following:

$$3 \text{ cm} + 2 \text{ mm} + \tfrac{3}{4} \text{ mm}.$$

Note that we have estimated a *part of a millimetre*; it is the user's skill in this estimation that determines the accuracy possible from a steel rule.

Changing the above to millimetres and adding:

3 cm	30·00
2 mm	2·00
$\tfrac{3}{4} \times 1$ mm	0·75
	32·75

Hence the length of the key is measured as 32·75 mm. We shall see, later on in this Unit how the use of a Vernier device eliminates the need for guessing fractional parts.

The caliper principle
It is not always possible to make accurate comparison between the

work and the engraved lines on a steel rule. Fig. 2.27A shows that the pulley boss prevents the rule from being applied at the correct diameter, indicated by the broken line xx. It is essential that the measurement is made along the largest distance across the pulley, for this is the true diameter. Even if we were able to apply a rule directly at this point, there is no means by which we could be sure we were measuring along the largest distance or true diameter.

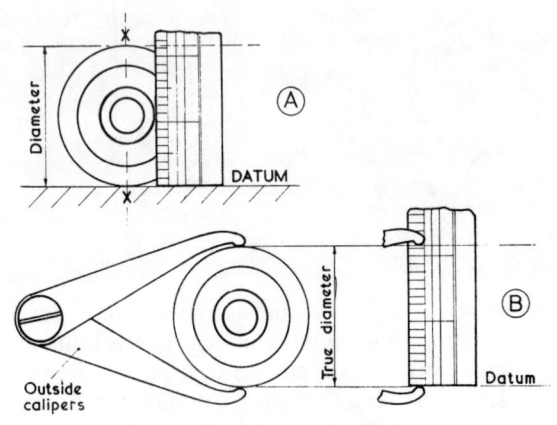

Fig. 2.27 Use of outside calipers

Application of the caliper principle overcomes this problem of measuring diameters, and the principle is simply shown in Fig. 2.27B. The calipers shown are known as *firm joint* calipers; the correct procedure is to adjust the calipers until they just nip or bite when passed over the diameter to be measured. This nip or bite, is more commonly referred to as *feel*, and considerable skill, gained through experience, is needed to cultivate a nice sense of touch, which is very necessary when using all kinds of measuring instruments.

Note the method of transferring the caliper reading to the rule. The bottom edge of the rule now acts as a datum face, and once again it may be necessary to estimate a part of a millimetre in order to make an accurate determination of the diameter.

A similar technique is adopted when using inside calipers; very useful for the determination of inside diameters as shown in Fig. 2.28A. Note the continuing use of a datum face on reference plane when transferring the caliper reading to the engineer's steel rule (Fig. 2.28B). The caliper principle is also useful when measuring dimensions still held on machine tools, for example when turning a diameter at a centre lathe, or milling a slot at a horizontal milling machine. Once the correct feel is obtained, the transference to a steel rule can be made as a separate and distinct operation. In this way, very accurate work is

Hand Processes

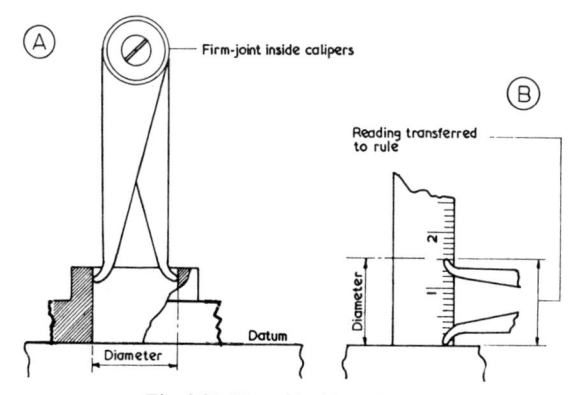

Fig. 2.28 Use of inside calipers

possible using simple measuring equipment, without the need to remove the component from the machine tool.

2.3.3 Measurement of angular dimensions

It is not generally appreciated that 90° is an angular measurement, and there are very few engineering components that do not have faces or edges at 90° to each other. We may define an angle as the amount of inclination of one face, plane or centre-line to another, and because the 90° relationship is so widely used, the engineering try-square is an essential and much used instrument. Whenever possible, it should be used from a datum face, and Fig. 2.29A shows a typical application. For more accurate work a master square is used, and the use of white

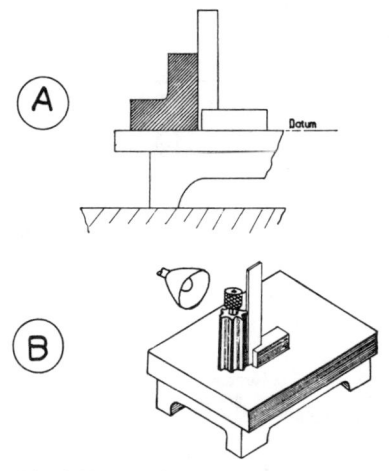

Fig. 2.29 Use of try and master squares

light greatly assists in the resultant accuracy obtained. As seen in the diagram at Fig. 2.29B, the master square must be used only on a good quality surface plate, whilst a try-square can be applied readily to various parts of an engineering component.

A try-square is a measuring instrument and must be treated as such, and should be checked against a master square at regular intervals. A white light placed behind the set-up, as shown in Fig. 2.29B, assists in the testing process, for white light is not able to pass through a gap of less than 0·002 mm. The appearance of blue light, between the try-square blade and master square, indicates that at no place does the try-square blade deviate more than 0·002 from the vertical or 90° position. Such a try-square can be considered to have a high degree of accuracy.

Engineer's plain protractor

When angles other than 90° need to be marked out or measured, a protractor must be used. Provided the angular tolerance is reasonable, say, plus and minus half a degree, an engineer's plain protractor will be found very suitable. Such a protractor is simply illustrated in Fig. 2.30A, and it may be noted that the design of this instrument has a limiting effect on its usefulness.

For example it could not be used to check the taper on the component shown in Fig. 2.30B because the blade has a fixed length from the pivot. This difficulty is simply overcome by making use of a *bevel gauge* as shown in the diagram. The angle required is first set on the bevel gauge by transference from an engineer's plain protractor and the bevel gauge applied to the work. The blades are adjustable for both angle and length, and are readily applied to a variety of measuring applications.

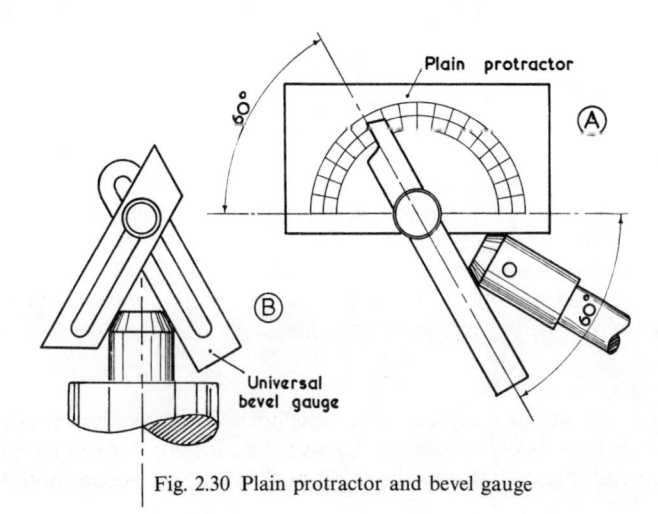

Fig. 2.30 Plain protractor and bevel gauge

Hand Processes

2.3.4 Line standards

All the measuring instruments so far described may be considered as *line standards*. This is because the actual determination of both linear and angular dimensions has been achieved by the comparison of the edge of a component against a graduated scale on the measuring instrument used. This scale is essentially a line standard, made up of engraved lines, as are found on engineers' steel rules or plain protractors. If now, we wish to measure the diameter of a shaft to within say, plus or minus 0·002 mm, then clearly this kind of accuracy is not possible using a line standard such as an engineer's rule. We need a more precise method of estimating a part or fraction of a millimetre, and this is possible using a Vernier device.

The Vernier principle

The Vernier principle owes its name to Pierre Vernier, who patented his ingenious device as far back as 1631. It is widely used for accurate measurements of many kinds, not necessarily connected with engineering manufacture. For example, navigators at sea need to determine the angular inclination of the noon sun to a high degree of accuracy, and a sextant fitted with a Vernier device is used. Also a barometer fitted with a Vernier device may be used to detect very small changes in the atmospheric pressure, and in the same way, precise measurements are possible by having Vernier devices fitted to standard measuring instruments such as engineers' steel rules and protractors.

Verniers applied to line standards

We have seen already that an engineer's rule is a line standard, that is to say the measuring scale is made up of graduations or dimensions machine-divided on the rule. All steel rules can be considered as part copies of the Imperial Standard Yard which is the legal unit of length in this country. First introduced in 1824, it was destroyed by fire in 1834, and a new bronze bar was introduced and legalised in 1855.

One Imperial Standard Yard was taken to be the distance between two lines scribed on two gold plugs inserted at each end of the bar when at 68°F or 20°C. However, a Weights and Measures Act, passed in July 1963 paved the way for conversion to the Metric System. This Act stipulates that the Imperial Standard Yard shall be 0·9144 of a metre, the conversion assisted by the use of the wave length of an atom-radiated light as a basic standard.

Vernier calipers

Figure 2.31 shows a typical Vernier caliper. It consists essentially of a steel rule modified to measure diameters by having a fixed caliper jaw at one end. Note that a range of diameters may be accommodated by

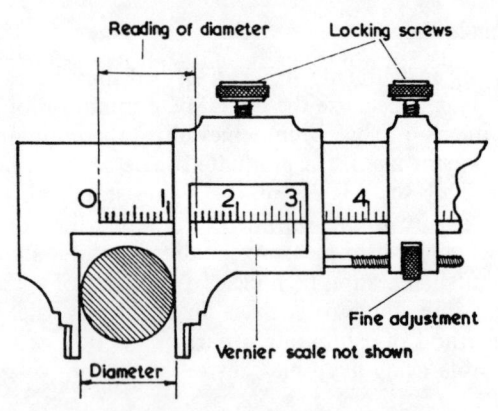

Fig. 2.31 Vernier calipers

sliding the movable caliper jaw along the rule or fixed member, which is calibrated in the normal manner. The position of the zero on the sliding jaw gives the diameter of the bar under test and this amount is known as the reading.

The 0·1 mm Vernier scale

This scale is engraved on the moving jaw and in Fig. 2.32 we see a simplified Vernier caliper having both a fixed and moving scale; note

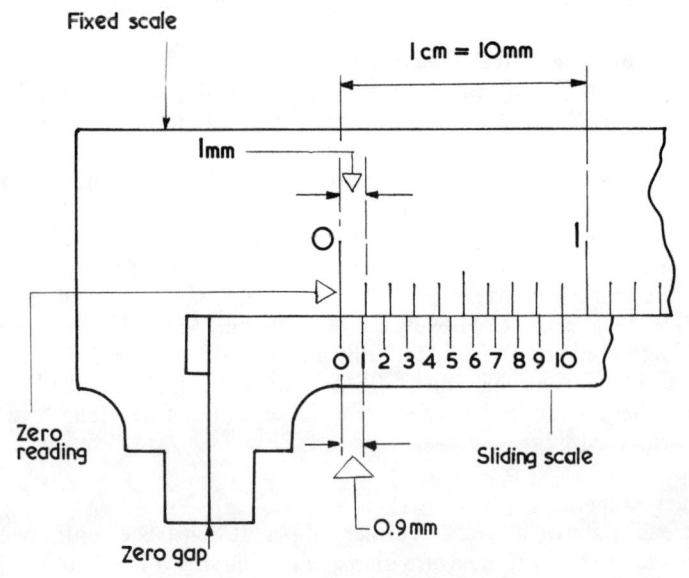

Fig. 2.32 Principle of 0·1 mm Vernier

Hand Processes

that the fixed scale is graduated in millimetres. There are ten divisions on the Vernier scale, the total length of which is 9 mm, therefore the width of one division on the sliding scale is:

$$\tfrac{1}{10} \text{ of } 9 \text{ mm} = 0.9 \text{ mm}$$

Hence the difference in width between one division on the fixed scale and one division on the Vernier scale is:

$$1.0 - 0.9 = 0.1 \text{ mm}$$

This means that if the first line on the Vernier scale, (marked 1) is made to coincide with the line marking the end of the first millimetre on the fixed scale, the sliding jaw has moved 0.1 mm from the zero position. The lines on the Vernier scale are numbered from 1 to 9, therefore when the jaws are closed or in the zero position, each number gives the amount that the sliding scale must be moved to bring the numbered line coincident with the adjacent line on the fixed member.

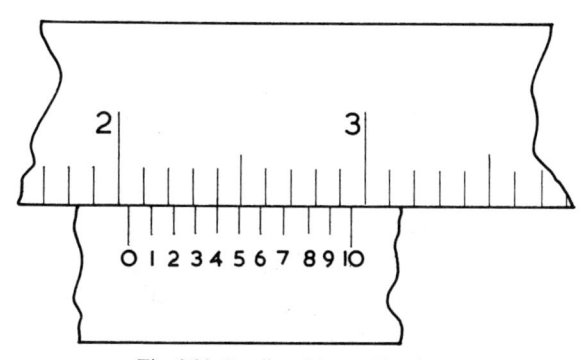

Fig. 2.33 Reading 0.1 mm Vernier

Perhaps a simple example will show the method adopted to read a 0.1 mm Metric Vernier caliper, as shown in Fig. 2.33. Remember always that the reading is the position of the zero on the Vernier scale with respect to the fixed scale. It is evident that the Vernier zero is between 2.0 and 2.1 cm, or between 20 and 21 mm. Further examination shows that the number 4 line on the Vernier scale is coincident with a line on the fixed scale, and this means that 0.4 mm must be added to the reading. Hence the reading is made up as follows:

Total number of mm = 20
Coincident line = 4
4 × 0.1 = 0.4
‾‾‾
20.4 = Reading
‾‾‾

The 0·02 mm Vernier scale
The 0·01 mm Vernier scale just described suffers from two defects:

(i) the accuracy is only 0·1 mm
(ii) the Vernier scale is small and difficult to read.

This means that a 0·1 mm Vernier has little practical use in engineering measurement or marking-out, but Verniers reading to 0·02 mm are of real value and widely used.

The 0·02 mm Vernier
The accuracy of a Vernier is the difference between the width of a division of the fixed scale, and the width of a division on the moving scale. In order to read to 0·02 mm this must be the difference between the fixed scale and Vernier scale divisions. There are two types of 0·02-mm Verniers in general use, the first commonly known as the small scale, and the other known as the large scale.

Small scale 0·02 mm Vernier
Shown in Fig. 2.34A it may be seen that the width of the graduations of the fixed scale are 0·5 mm. There are 25 divisions on the Vernier scale taking up a distance of 12 mm, making the width of one division on the Vernier scale

$$\tfrac{1}{25} \text{ of } 12 \text{ mm} = 0.48 \text{ mm}.$$

Thus the difference between a division on the fixed scale and a division on the Vernier scale is

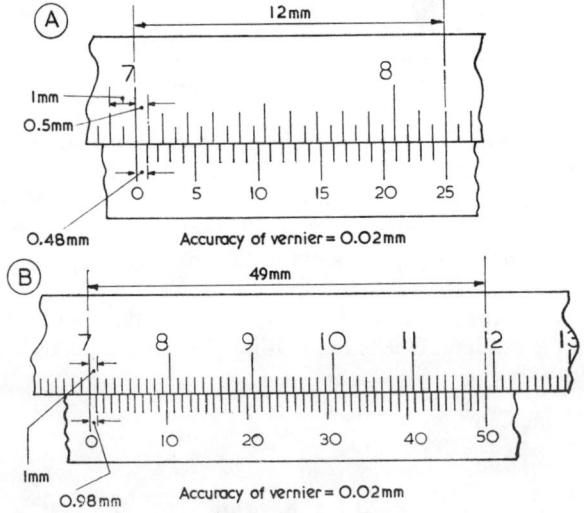

Fig. 2.34 Principle of 0·02 mm Vernier

Hand Processes 55

$$0.5 - 0.48 = 0.02 \text{ mm}.$$

This means, looking at Fig. 2.34A, that if the sliding part is moved to the right so that the fifteenth line on the Vernier scale becomes coincident with the adjacent line on the fixed scale, the reading will be as follows

$$\begin{aligned}7 \text{ cm} &= 70 \text{ mm} + 15 \times 0.02 \\ &= 70 \qquad + 0.30 \\ &= 70.30 \text{ mm}.\end{aligned}$$

It must be remembered at all times that as the accuracy of this Vernier is 0·02 mm, or two hundredths of a millimetre, the value of the coincident line on the Vernier scale must be *doubled*. For example, if the thirteenth line is coincident then thirteen times two is equal to twenty-six hundredths of a millimetre (0·26 mm) must be added.

Applications of the small scale 0·02 mm Vernier
Because the Vernier scale covers a fairly small distance, namely 12 mm, it is not easy to read, and a small jeweller's eyeglass should be used to enlarge the scale. This type of scale is restricted to the smaller type measuring instruments such as Vernier calipers and Vernier depth gauges. Figure 2.35 shows typical examples of small scale metric Verniers used to measure linear dimensions to an accuracy of 0·02 mm. With regard to the calipers, note the radii on the external faces of the measuring jaws, these radii extend the use of the calipers by enabling them to be used for the measurement of internal diameters as shown in the diagram. It is essential to add to the Vernier reading, the width of the measuring jaws; these are usually engraved or stamped where they can be clearly seen. The Vernier depth gauge shown in Fig. 2.35B is ideally suited for measuring the depth of slots to an accuracy of 0·02 mm.

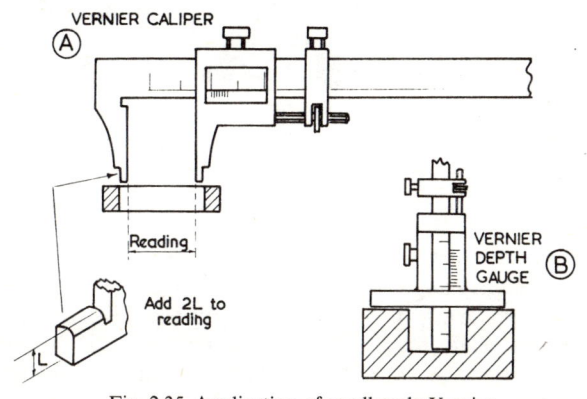

Fig. 2.35 Application of small-scale Vernier

Large scale 0·02 mm Vernier
Reference back to Fig. 2.34B shows the principle of the large scale 0·02 mm Vernier. It will be seen that the width of the main scale divisions are 1 mm, while 50 divisions on the Vernier scale take up a distance of 49 mm. Hence the width of a division on the Vernier scale is

$$\tfrac{1}{50} \text{ of } 49 \text{ mm} = 0.98 \text{ mm.}$$

The accuracy of this type of Vernier will again be the difference between the width of the divisions on the fixed and Vernier scales, so accuracy is

$$1.00 - 0.98 = 0.02 \text{ mm.}$$

This is the same accuracy as that obtained from the small scale Vernier, therefore the principle of operation is identical, and may be summarised as follows:

1. Write down total number of divisions.
2. Add half divisions (if any).
3. Find coincident line on Vernier scale.
4. Multiply coincident line number by two.
5. Express product as hundredths of a millimetre and add.

Referring back to Fig. 2.34B, let us assume that the twenty-second line on the Vernier scale is coincident with the next line on the fixed scale. We may now read the Vernier scale as follows, remembering always that we are determining the position of the zero line on the Vernier scale with respect to the fixed scale.

Divisions on main scale = 7 cm = 70·00 mm
No half divisions = 0·00
Coincident line = 22
22 × 2 = 0·44

Vernier Reading = 70·44 mm

This type of Vernier which has a large easily read scale, tends to increase the degree of accuracy possible, although the use of a simple optical device to enlarge the reading is still recommended.

Applications of the large scale 0·02 mm Vernier
The Vernier height gauge shown in Fig. 2.36A is a widely used marking out and measuring instrument, the large scale 0·02 mm Vernier scale making it easily read and accurate in use. Note the English or Imperial Scale opposite the Metric scale, and the removable scriber, which enables different attachments to be used. For example, the

Hand Processes

depths of slots or recesses are readily obtained by using the depth rod attachment shown in Fig. 2.36B, thus converting the height gauge into a depth gauge capable of working to an accuracy of 0·02 mm. This Vernier height gauge is a first-class example of a precision measuring

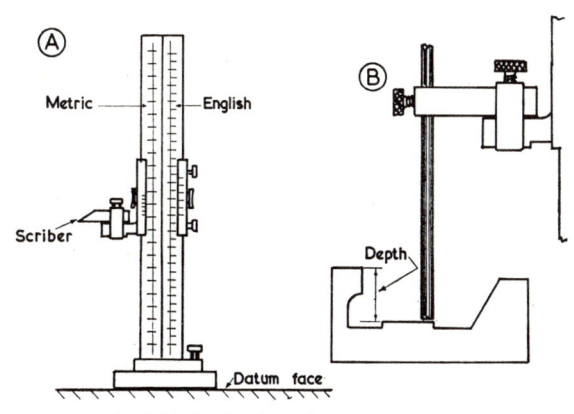

Fig. 2.36 Application of large-scale Vernier

instrument, and in the hands of a skilled and experienced technician, very accurate work is possible.

The Vernier protractor

A Vernier protractor is often called a bevel protractor, and Fig. 2.37 illustrates this instrument. The design of the universal bevel type considerably increases the scope of angular measurement, for the blade is adjustable and the protractor can be indexed through 360°, with the acute angle attachment simplifying the measurement of small angles.

Fig. 2.37 Universal bevel Vernier protractor

The accuracy of a Vernier protractor is five minutes of arc, and the engineering method of showing angular inclination is given below:

Twenty degrees – 20°
fifteen minutes – 15′
twelve seconds – 12″

There are sixty seconds in one minute, and sixty minutes in one degree, hence one second is $\frac{1}{360}°$, a very small angle indeed.

Principle of the Vernier protractor
Figure 2.38 shows a part view of a Vernier protractor, including the Vernier scale calibrated in degrees; this scale extends both sides of the

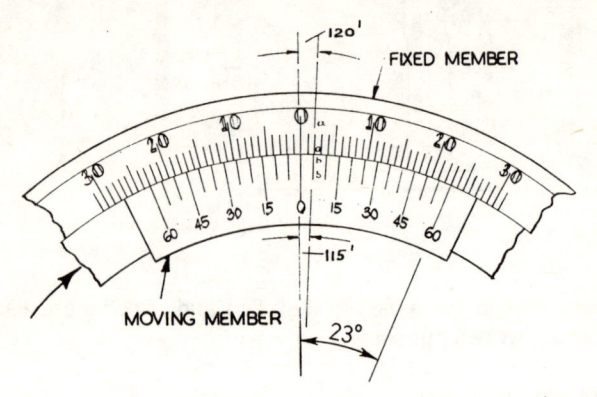

Fig. 2.38 Principle of the Vernier protractor

zero line on the fixed scale. The Vernier scale has 12 divisions each side of the zero line, and these take up a total of 46°, or 23° each side of the zero on the fixed scale. Thus the angle of one division on the Vernier scale is

$\frac{1}{12}$ of 23° = $\frac{1}{12} \times 23 \times 60$ minutes
= 115 minutes.

Reference to the diagram shows that one division on the Vernier scale is slightly less than two divisions on the fixed scale, in other words the accuracy of the Vernier is the angular difference between one Vernier scale division and two fixed scale divisions, or

2° − 115′ = (120 − 115)
= 5′.

This means that if the moving member is slightly turned so that the lines indicated as aa and bb are coincident, the moving member has moved through an angle of five minutes.

Reading the Vernier protractor

Reference to Fig. 2.39A will help to show the correct procedure to adopt when taking a reading with the Vernier protractor. The actual angle is indicated by the position of the zero line on the Vernier scale, with respect to the fixed scale. We see from the diagram that the zero

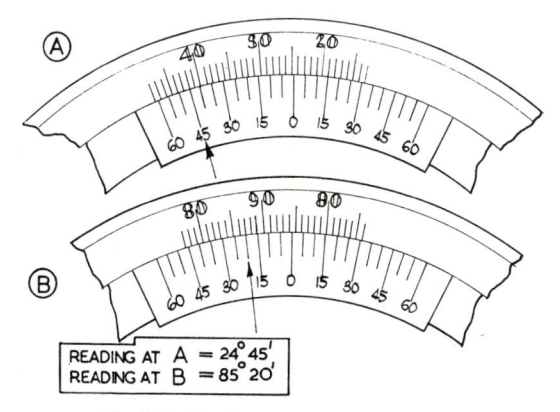

Fig. 2.39 Reading the Vernier protractor

line falls between 24° and 25°, and the purpose of the Vernier scale is to give us the amount of arc or angle in excess of 24°.

This small angle is found by close scrutiny of the Vernier scale, preferably with a jeweller's eyeglass, and it may be seen that the coincident line is 45; this must be added to 24°, hence the reading is 24° 45′. In Fig. 2.39B we see another example of a reading on a Vernier protractor, and we may enumerate this reading as follows:

1 Position of zero on main scale = 85°.
2 Coincident line = 20.
3 Reading = 85° 20′.

Uses of the universal bevel Vernier protractor

Figure 2.40 gives an indication of the great versatility of the universal bevel Vernier protractor when determining the angular inclinations of engineering components. At A we see the method adopted to check the inside bevelled face of a ground die, and at B the inside face of a Vee block. The advantage of the acute angle attachment is clearly seen at C where the small acute angle is under test. In all of the angular measuring examples shown, the accuracy will be within five minutes of arc.

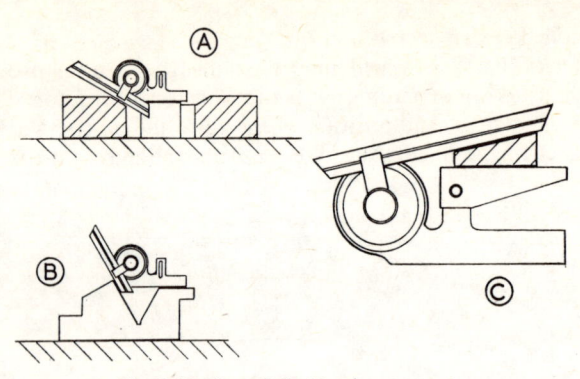

Fig. 2.40 Uses of the Vernier protractor

2.3.5 The micrometer principle

All the measuring instruments so far described are *line standards*. The standard is the scale which is engraved on the body of the instrument, and because engraved lines possess the linear quality of *width*, then precise measurement is only possible if the centre of the engraved line is used. Clearly this is not practicable, for the lines are extremely narrow, and considerable magnification would be needed using optical methods.

This problem of making consistent measurements to close limits of accuracy without the aid of optical magnification was solved when a device called 'Système Palmer' was patented in France in 1848. Essentially, the device obtained the magnification needed, not optically, but by subdivision of the rotation of an accurate screw thread or lead screw. This device was known as a micrometer caliper, and an early model is illustrated in Fig. 2.41A. Known at the time of its introduc-

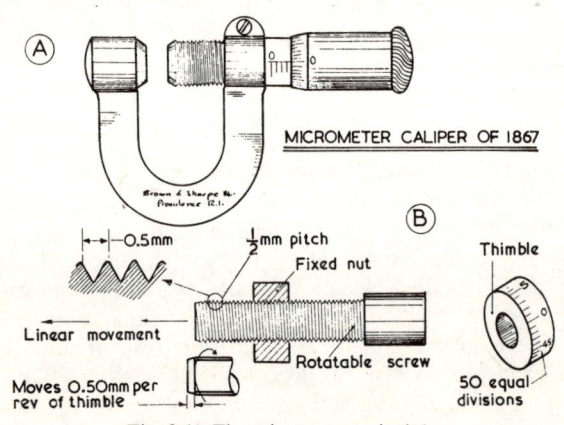

Fig. 2.41 The micrometer principle

Hand Processes

tion as a 'pocket sheet metal gauge', this little measuring instrument provided a positive and rapid method of checking the thickness of wire, sheet and strip, always to a consistent accuracy of within 0·02 mm. The principle of this measuring device is simply illustrated in Fig. 2.41B, where it may be seen that the lead screw has a pitch of 0·5 mm, so that one complete turn of this screw gives it an axial movement of 0·5 mm. If now, a thimble evenly graduated with 50 divisions around its circumference, is fixed to the end of the lead screw then turning the thimble through one division results in an axial movement of

$$\tfrac{1}{50} \text{ of } 0\cdot 5 \text{ mm} = \frac{0\cdot 5}{50}$$
$$= 0\cdot 01 \text{ mm}.$$

This is twice the accuracy obtained from all Vernier calipers and height gauges, for as we have seen the accuracy of the Metric Vernier is only 0·02 mm, with a greater degree of skill required to ensure that the correct coincident line is obtained. The micrometer calipers shown in Fig. 2.41 proved an instant success, and all modern micrometers stem from this simple principle. Fig. 2.42 shows a modern micrometer with the main parts named. The lead screw of this micrometer has an accurate ground precision thread of 0·5-mm pitch; the thimble has 50

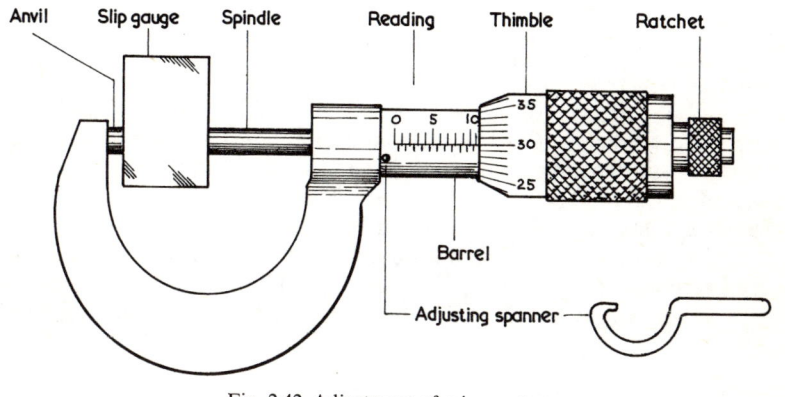

Fig. 2.42 Adjustment of micrometers

divisions and is rotated by a ratchet device which ensures that excessive tightening pressure is not applied to the lead screw. Note the clamping device which locks the lead screw at any required reading, also the small hole in the barrel. The purpose of this hole is to allow the micrometer to be set to zero, and the procedure is as follows:

1 Clean anvils and place a 10-mm slip gauge between the anvils.
2 Turn thimble using the ratchet until two ratchet clicks are heard.

3 Note the reading, and if the horizontal line on the barrel does not coincide with the zero line on the thimble, use the adjusting spanner to correct the error with slight rotation of the barrel.

It is important to appreciate at this stage that the micrometer principle makes use of a magnifying technique, that is to say a gap of 0·01 mm between the anvils is equivalent to a division width on the thimble of about 1 mm. Thus the actual distance is magnified approximately 100 times. The greater the diameter of the thimble, the greater the magnification possible, and Fig. 2.43 shows a *fiducial* bench micrometer. Note

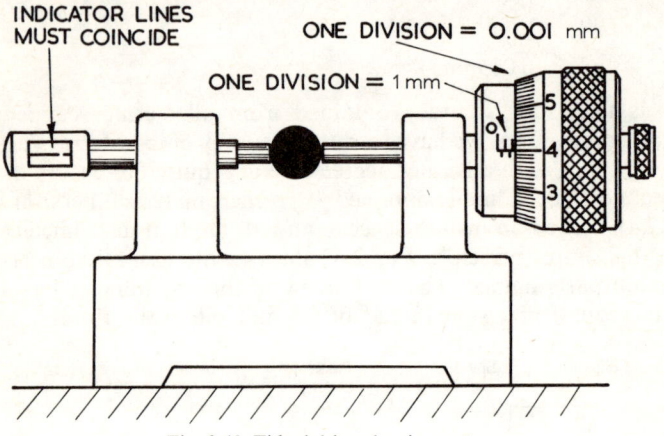

Fig. 2.43 Fiducial bench micrometer

the large diameter of the thimble, and although the pitch of the lead screw is still 0·5 mm, one small division on the thimble represents a gap of 0·001 mm between the anvils. The name fiducial is obtained from the method adopted of ensuring that a correct or faithful reading is always obtained; the line on the indicator shown in Fig. 2.43, must read zero when the reading is taken. This technique may be considered as a visual ratchet, ensuring that the correct tightening pressure is used at all times.

Applications of the micrometer principle
The ease and simplicity obtained when using the micrometer principle for precise measurement has led to a development of alternative micrometer instruments and these are simply illustrated in Fig. 2.44. Providing the diameter of a machined bore is between 50 and 250 mm, it is readily checked using an inside micrometer. Note that a range of measuring rods are supplied, so that the correct rod must be inserted to suit the diameter under test. The same principle is adopted when measuring the depths of slots or recesses; in these cases a depth micrometer is used as shown in Fig. 2.44A.

Hand Processes 63

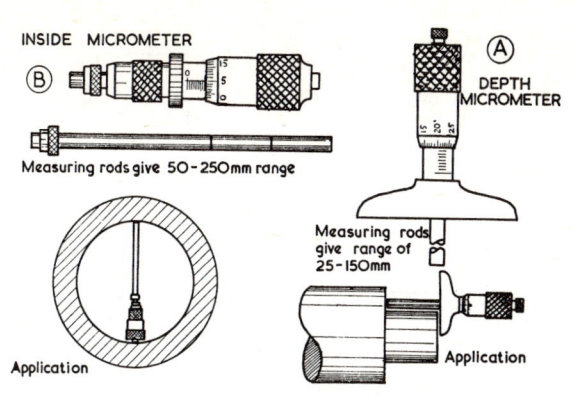

Fig. 2.44 Micrometer applications

Remember always, that all instruments using the micrometer principle are precision pieces of equipment and must be treated as such. When not in use, they must be stored carefully in boxes, and not left on a workbench, or on the worktable of a machine tool. Their accuracy must be checked at regular intervals especially if they have been accidentally misused, say dropped on the workshop floor.

The Vernier micrometer
The Vernier principle may be applied also to an external micrometer, increasing the accuracy of the reading. However, as we have seen, the scales on both the barrels and sleeves of micrometers are line standards, with the thickness of the engraved lines at least 0·3 mm. It follows therefore, that although the Vernier metric micrometer is capable of working to an accuracy of 0·001 mm, consistent results are difficult to achieve, unless the instrument is used by a highly skilled person. The Vernier principle is illustrated in Fig. 2.45 where it may be seen that the Vernier scale is engraved on the micrometer barrel as shown in the diagram. There are five divisions on this Vernier scale, and the total distance they take up is nine divisions on the thimble scale, hence one division on the Vernier scale is

$\frac{1}{5}$ of 0·09 mm = 0·018 mm

As two divisions on the thimble scale are equivalent to 0·02 mm between the anvils, then the difference between one Vernier division and two thimble divisions is

$$0·020 - 0·018 \text{ mm}$$
$$= 0·002 \text{ mm}$$

This is the accuracy possible using a Vernier micrometer, and it can be said that Metric Vernier micrometers find little practical use for the determination of linear dimensions to a high degree of accuracy. The

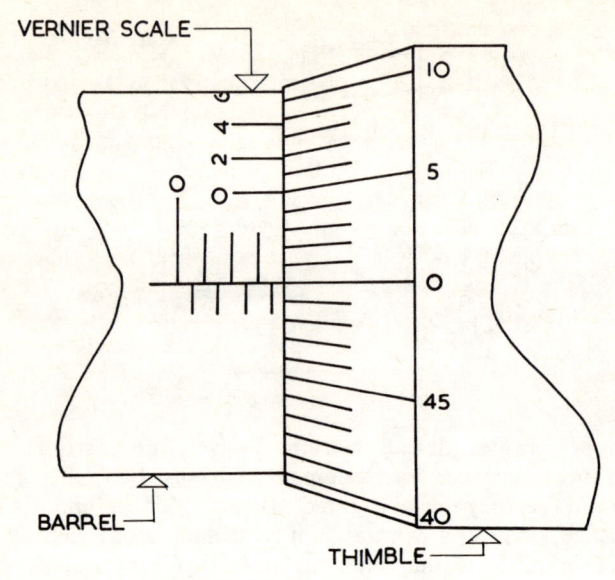

64   *Technician Workshop Processes and Materials*

Fig. 2.45 Vernier metric micrometer

best method to adopt is the principle of measurement by comparison, or the use of end standards.

2.3.6 Measurement by Comparison

The mass production which characterises so many branches of modern engineering manufacture would be impossible if component parts

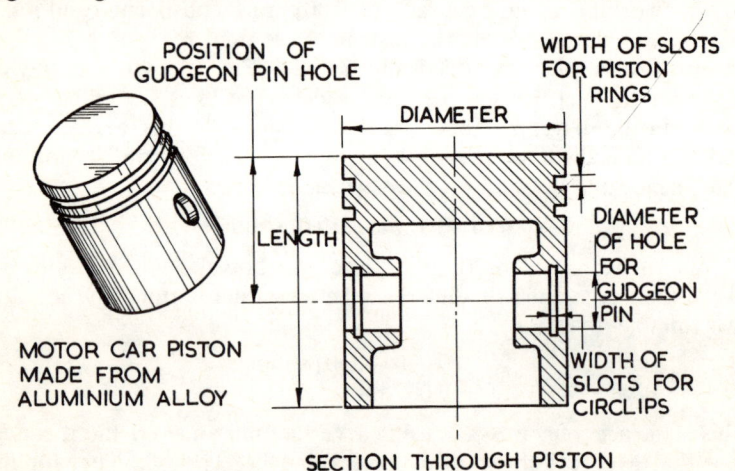

Fig. 2.46 Engineering component

Hand Processes 65

could not be produced to close dimensional tolerances. As we have seen, the use of line standards such as Vernier and micrometer calipers require a considerable degree of skill if consistent results are to be obtained. Consider the engineering component shown in Fig. 2.46. This is an aluminium piston for a motor car engine and may be considered as a typical example of the high degree of precision now demanded in the modern motor vehicle engine. Very large numbers are required, and this means that the piston must not only be mass-produced, but also all dimensions must be checked with the same kind of precision and speed as that used in their manufacture. Clearly the use of micrometers and Vernier calipers is not practical, for as we see in Fig. 2.46, there are many dimensions to be checked. If, however, the principle of measurement by comparison is adopted, say for the height of the piston, then the set-up would appear as shown in Fig. 2.47, and

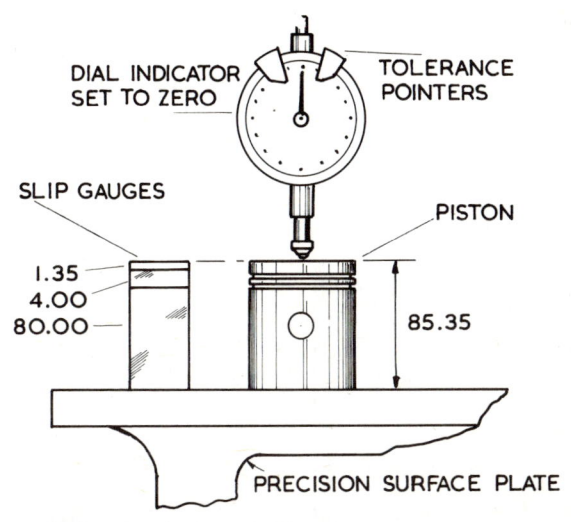

Fig. 2.47 Precise determination of an angular component

the determination of the height to a high degree of accuracy would take only a few seconds. Of greater importance, little or no skill is required from the operator and the consistency of the measuring operation would be of a high standard.

It is clear that two elements are involved in this system of dimensional control, and these are as follows:

1 Visual comparator,
2 end standards.

We see from the diagram that end standards, totalling 85·35 mm are set-up on a precision surface plate with the dial indicator pointer set at the zero position. If now, the end standards are removed, and replaced by the component, we are comparing the height of the piston against the known height of the end standards. Any difference in height will be shown by the amount the pointer differs from the zero setting. The comparator, therefore, is a magnifying device; the greater the magnification, the higher the degree of accuracy possible.

This magnification is not difficult to express in arithmetical terms; it is the ratio between the movement of the plunger and the resultant movement of the pointer. In other words, if M = magnification, p = plunger movement and P = pointer movement, then

$$M = \frac{P}{p}$$

The principle of plunger type comparators

The comparator shown in Fig. 2.47 is more commonly referred to as a *dial test indicator*, and this instrument is widely used in all workshops where accurate work is carried out. In more precise terms it is referred to as a mechanical comparator of the plunger type, the term mechanical derived from the fact that the magnification is obtained by mechanical devices such as gears and levers. Figure 2.48A shows the

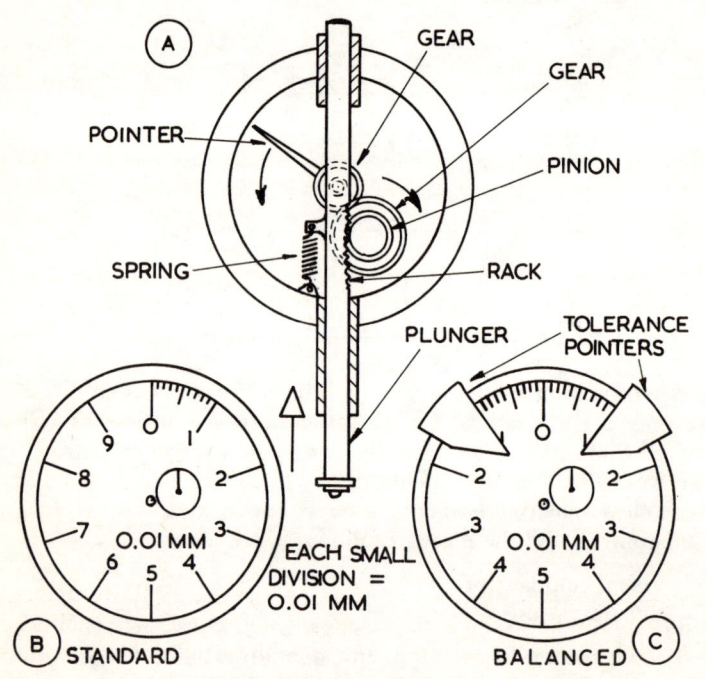

Fig. 2.48 Dial test indicator

Hand Processes

basic mechanism of a dial test indicator, the magnification obtained from a rack and pinion and gear systems. A suitable spring provides constant plunger pressure while hair springs are used to eliminate back lash; this is the name given to unwanted movement caused by slackness between mating parts. It is clear that if a dial test indicator is to provide faithful magnification of the plunger movement, the dimensional and functional elements of the gears, racks and pinions must possess a high degree of accuracy. Because of the compact, easily portable nature of dial test indicators designed for workshop use, dial diameters seldom exceed 60 mm, and this means that the sizes of the moving parts such as the gears, rack and pinion are, of necessity, quite small and hence exceedingly difficult to manufacture. This means that the accuracy possible from a mechanical dial test indicator is limited to about 0.002 mm. It will be found in practice, that dial test indicators reading to 0·002 mm, require very great care in use and handling; very often better results are possible using a dial test indicator with scale reading of 0·01 mm.

Types of scales
The two types of scales commonly used on dial test indicators are shown at Fig. 2.48B and C. The standard scale is calibrated in units of ten:

> One small division = 0·01 mm
> Ten small divisions = 0·01 mm (numbered division)
> Ten numbered divisions = 1·0 mm (one revolution of pointer)

The small dial gives the number of complete pointer revolutions, hence the total movement of the plunger can be determined. The maximum plunger movement, or the measuring range, is about 10 mm, therefore the pointer makes 10 complete revolutions when the plunger movement is at its maximum. It is seldom however, that a dial test indicator is used in order to obtain a linear dimension ranging from 0 to 10 mm. As we have seen, the mechanism comprises small gears, and it is an impossible task to ensure precise movement of this mechanism; the best use of the dial test indicator is to detect and magnify small variations from a known size, and for this purpose a *balanced scale* is used. This is shown in Fig. 2.48C, where it may be seen that the divisions are numbered each side of the zero. This allows immediate reading of any error plus or minus from the zero position.

Use of tolerance pointers
The use of a balanced dial, as we have just seen, allows immediate reading of any variation from a zero setting. If we refer back to Fig. 2.47, which shows the set-up to measure the height of a piston, then because the tolerance is known, tolerance pointers can be set at the

required positions as shown in Fig. 2.48C. The determination of the piston height is now greatly simplified; provided that when the piston is gently placed under the plunger the pointer does not move outside the tolerance pointers, the piston is within its limits of size. In this way, rapid and accurate checking of piston heights is easily achieved without the use of skilled and trained operators. This precise and consistent method of maintaining *dimensional control* would be impossible if line standards such as Vernier calipers and Vernier height gauges were used, and even the use of micrometer calipers would demand the maximum concentration and expertise from skilled craftsmen.

The principle of lever-type comparators
This type of comparator is simply illustrated in Fig. 2.49A, and because of its design it is most suitable for use in restricted places such as

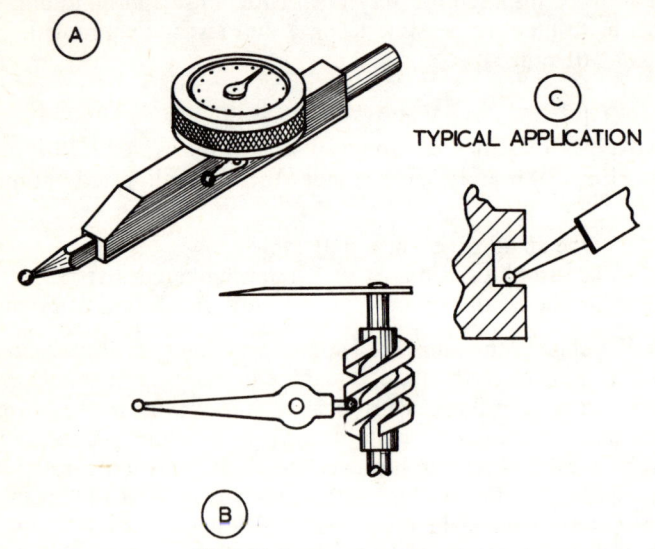

Fig. 2.49 Lever-type comparator

the mouths of small holes or bores, where the plunger-type comparator would be quite unable to enter. The principle underlying its operation may be seen in Fig. 2.49B; small displacements of the stylus giving rise to large movements of the pointer. Because their measuring range or stylus movement is relatively small, it is usual to find balanced dials fitted to lever-type dial test indicators only.

2.3.7 Use of end standards
Without end standards it would be virtually impossible to set a dial test indicator to zero. The end standards widely used in engineering

Hand Processes

workshops are more commonly known as *slip gauges*, and these were invented by a Swedish engineer, C. E. Johansson, early this century. When first introduced they were regarded as a novelty, and often referred to as 'Jo Blocks'. Today, however, their use is indispensable to all forms of engineering manufacture. Made from high grade cast steel to exceptionally close tolerances, slip gauges are available in several grades or qualities, and the choice of the correct grade is determined by the sort of work for which the slip gauges are needed. There are five grades available as follows:

Grade 2   This is the workshop grade. Typical uses include setting up machine tools, positioning milling cutters and checking machined widths.

Grade 1   Used for more precise work, such as that carried out in a good-class toolroom. Typical uses include setting up sine bars and sine tables, checking gap gauges and setting dial test indicators to zero.

Grade 0   This is more commonly known as the Inspection grade, and its use is confined to toolroom or machine shop inspection. This means that it is the Inspection Department only who have access to this grade of slips. In this way it is not possible for these slip gauges to be damaged or abused by the rougher usage to be expected on the shop floor.

Grade 00   This grade would be kept in the Standards Room and would be used for work of the highest precision only. A typical example would be the determination of any errors present in the workshop or Grade 2 slips, occasioned by rough or continual usage.

Calibration grade   This is a special grade, with the actual sizes of the slips stated or calibrated on a special chart supplied with the set. This chart must be consulted when making up a dimension, and because these slips are not made to specific or set tolerances, they are not as expensive as the Grade 00. It must be remembered that a slip gauge, like any other engineering component, cannot be made to an exact size. All slip gauges must have tolerances on the following elements:

1   length,
2   flatness,
3   parallelism of measuring faces.

Except for the calibration grade, all slip gauge sets are manufactured to within specified limits; the closer the limits the more expensive the slip gauges, but in the case of the calibration grade, greater tolerances on length are permissible. Because the actual lengths are known or recorded in the calibration chart, due allowance can be made when the slips are used.

Slip gauge sets
Metric slip gauges are available in a range of sets, and these sets are described according to the basic thickness of the slips and the number of slip gauges in the set. For example, a slip gauge set described as M46/2, is a metric set of 46 slips, 2-mm based, and made up as shown below:

9 slips	2·001	to	2·009	in steps of 0·001	mm
9 slips	2·01	to	2·09	in steps of 0·01	mm
9 slips	2·1	to	2·9	in steps of 0·1	mm
9 slips	1	to	9	in steps of 1·0	mm
10 slips	10	to	100	in steps of 10	mm

Protector slips
In order to prolong the life of slip gauges, protector slips may be used at each end of a slip pile. Such slips are usually made from tungsten carbide and identified by a letter P marked on one face, and allowance must be made for the thickness stamped on the protector slips. Always available as a pair these slips are best used for the workshop examples shown in Fig. 2.50A and B. Note that in each case, there is frictional contact on the end slips and this could lead to wear and consequent error of the slip size. The use of protector slips at each end of the slip pile will extend the accuracy and hence the useful life of the slip gauge set.

Choice and assembly of slip gauges
It is a well-established rule that the minimum number of slip gauges must be chosen in order to make up a given dimension, and the following practical example will help to make the principle clear. Referring to Fig. 2.50A we see that the gap under check has a width of 43·716 mm. When selecting the appropriate slips, adopt the technique shown below, assuming that an M46/2 grade 2 set is available.

Width required = 43·716
1. Select 2·006 and subtract from 43·716 leaving 41·710.
2. Select 2·01 and subtract from 41·710 leaving 39·7.
3. Select 2·7 and subtract from 39·7 leaving 37.
4. Select 7 and subtract from 37 leaving 30.
5. Select 30 leaving zero.

It is a wise plan to add the total selections together as a check, this total must be equal to the dimension required.

Hand Processes 71

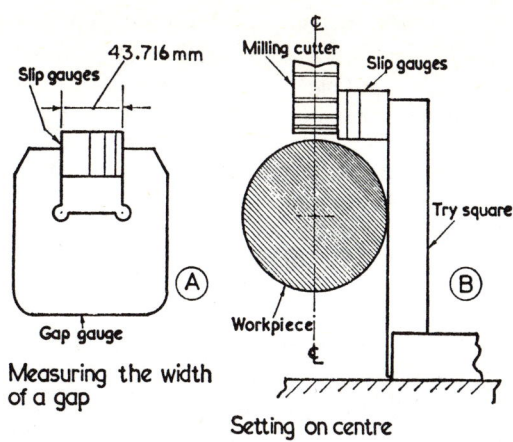

Fig. 2.50 Use of protector slips

2·006
2·01
2·7
7·0
30·0
―――
43·716 mm

If protector slips are to be used, the following procedure should be adopted, and we will assume that the size of each protector slip is 2·500 mm.

   Width required = 43·716
1 Select two protector slips.
     = 5·000 and subtract from 43·716 leaving 38·716.
2 Select 2·006 and subtract from 38·716 leaving 36·710.
3 Select 2·01 and subtract from 36·710 leaving 34·70.
4 Select 2·7 and subtract from 34·7 leaving 32·0.
5 Select 2·0 and subtract from 32·0 leaving 30.
6 Select 30 leaving zero.
Add to check:
     5·000 (two protectors at 2·50)
     2·006
     2·01
     2·7
     2·0
     30·0
     ―――
     43·716 mm

The importance of neatly laid out calculations cannot be over stressed, neither can the necessity for double checking at all times. It is a very easy matter to make a simple arithmetical error when selecting and calculating the slip gauges required, but the consequences can be most serious. It is essential also to check each slip on removal from the set. Although the slip size is indicated on the case opposite each slip, it is possible that the slips have been incorrectly replaced by the previous user, and it is vital that the selected size is checked by reference to the size clearly engraved on the slip face.

Wringing slip gauges
Slip gauges, when properly assembled or wrung together are held in firm contact by molecular adhesion, and a certain amount of force is needed to break this molecular bond. The correct procedure for both wringing and breaking a slip gauge assembly is simply illustrated in Fig. 2.51, and it may be noted that two distinct stages are involved.

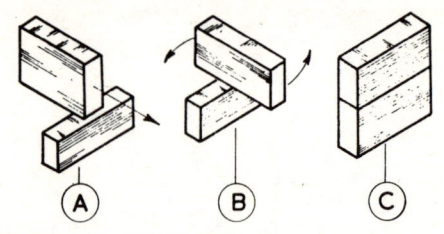

Fig. 2.51 Wringing slip gauges

The two slips are first placed at right angles as shown at A, then the top slip is slid across and rotated through 90°. This technique ensures that any dust or foreign matter on the slip faces is pushed off and not trapped between the slips causing abrasive wear. The reverse process is adopted when breaking the two slips; rotate as shown at B, then slide off as shown at A.

Care of slip gauge sets
Slip gauges are high precision measuring devices and must be treated as such. As soon as the correct slips have been chosen, the lid of the box must be closed to prevent any dust or swarf entering. Before wringing, the working faces may be gently wiped with a clean piece of soft cloth; after use the surfaces must again be carefully wiped with a thin smear of vaseline applied. Especial care must be taken if the slip gauges are to be used near a grinding machine, for the dust produced by the grinding operation is exceedingly abrasive, causing rapid and excessive wear of the slip gauge surfaces This is also true if slip gauges are used to check the accuracy of a component during a lapping operation. Once again, the lapping compound is highly abrasive, and every

Hand Processes

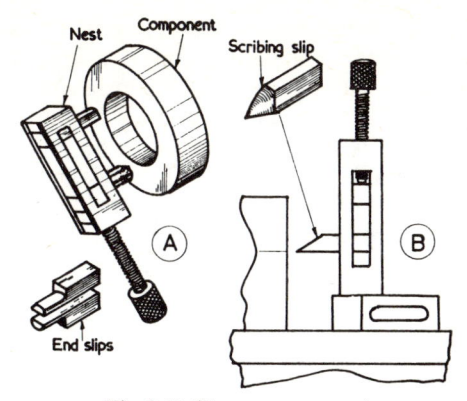

Fig. 2.52 Slip gauge accessories

care must be taken to ensure that no trace of this compound gets in contact with the slip gauges.

At Fig. 2.52B we see how a suitable accessory called a marking-out block, allows the use of slip gauges for very precise marking-out; a special marking-out slip is used to scribe the line. While this technique makes possible the scribing of lines, it must be remembered that the scribed lines will possess the linear dimension of thickness, dependent on the pressure used. In this way we have the unusual circumstance of end standards used to produce lines and check interval diameters.

Alternative end standards

It must not be thought that the only end standards available are the familiar slip gauges of rectangular shape. While these slips are ideal for measuring or setting dial test indicators to zero in order to measure linear dimensions they are quite unsuitable for other operations, such

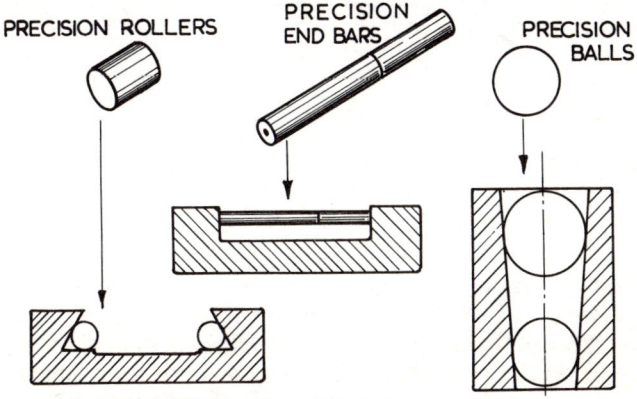

Fig. 2.53 Use of precision balls and rollers

as angular measurement. Figure 2.53 shows typical applications of the use of alternative end standards, such as rollers, balls and end bars, all of high precision quality. These are available in sets, manufactured to the same high degree of accuracy as that found in slip gauges; and in many cases a good working knowledge of basic geometry and trigonometry is an invaluable asset if full use is to be obtained from these alternative end standards. We shall see, in later volumes, how the use of simple trigonometrical formula, together with the use of the appropriate end standards, is used to resolve a wide variety of engineering problems associated with precise measurement.

The use of master gauges

Consider the component shown in Fig. 2.54A. It is a piston used in the hydraulic braking system of a motor car, and is manufactured to very

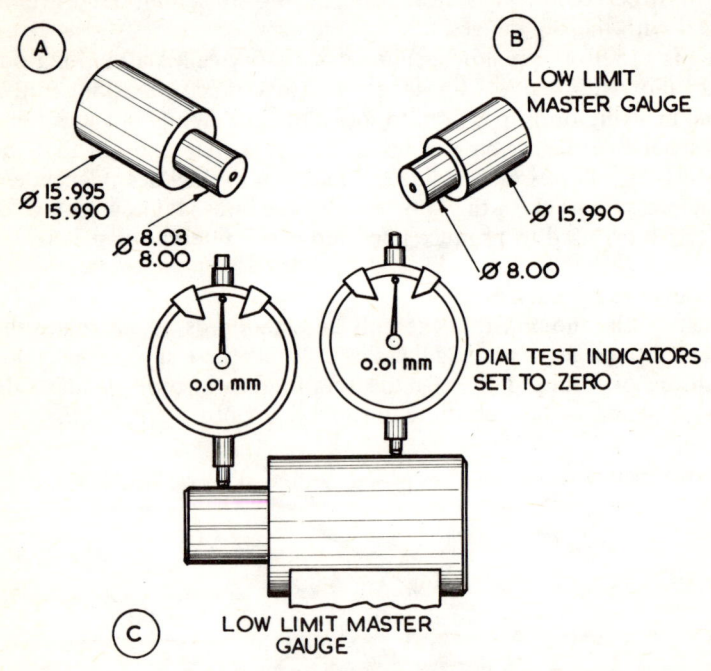

Fig. 2.54 Use of master gauges

close limits of size on the diameter, as may be seen from the dimensions given in the diagram. We have seen that a diameter such as this may be checked by using the technique of measurement by comparison, that is to say setting a suitable comparator to zero by means of slip gauges, and then noting any variation from zero when the component is placed under the comparator. A better method is to make *two* master gauges from hardened steel, one to *low limit* conditions, and the other to *high limit* conditions. The low limit master gauge is shown in Fig. 2.54B, and it is this gauge which is used to set the comparator. It is not difficult to check the two diameters at the same time as shown in Fig. 2.54C, where the low limit master gauge is being used to set the comparators to low limit. Note that in addition the set-up checks the non-linear function of concentricity of diameters and also, if provision is made for rotation of the job, the roundness of the diameters.

With all low limit dimensions set by adjustment of tolerance pointers, the low limit master gauge is replaced with the high limit master gauge, and the tolerance pointers adjusted to indicate the high limits. This technique is known as *combination gauging* allowing not only rapid and simple setting-up, but also equally rapid checking of the linear dimensions of diameter, and the non-linear functions of roundness and concentricity. It is possible, using master gauges, to set up and then check as many as twenty or more dimensions simultaneously, the component being located in a suitable receiving device and then being pushed into the checking position or station.

In this way it is not necessary for the operator to know the actual units or accuracy involved: provided the tolerance pointers are set to the low and high limit master gauges, the acceptability of the work piece is immediately determined by visual examination of the comparators.

2.3.8 Types of comparators

So far the only type of comparator discussed has been the dial test indicator of plunger or lever operation. These are essentially *mechanical* comparators, the necessary magnification obtained using levers and gear trains, but a wide range of comparators are available based on non-mechanical principles such as

> Electrical
> Optical
> Pneumatic
> Digital.

Irrespective of the type of comparator used however, the basic principle of operation remains the same: setting the comparator to zero using a known standard such as a slip gauge or master gauge, then noting any variation from the zero setting when the component is placed under test. In Fig. 2.55 we see, in a simple manner how the various comparators indicate variation from the zero settings. At Fig. 2.55A we see a typical large-dial mechanical bench comparator, with

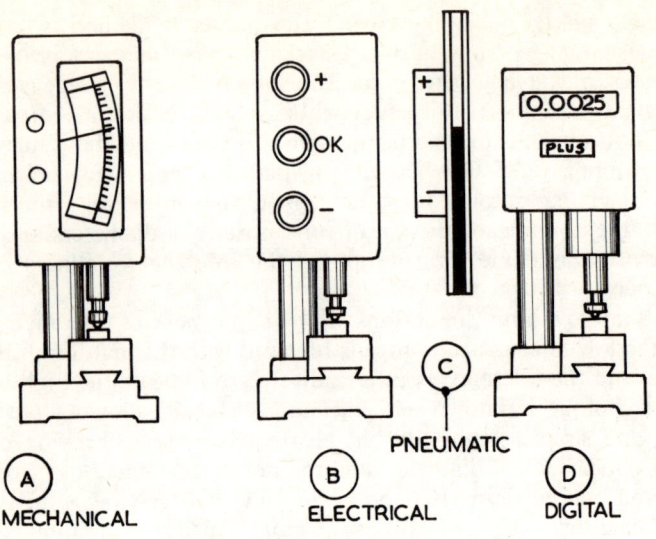

76 *Technician Workshop Processes and Materials*

Fig. 2.55 Modern comparators

two adjustable tolerance pointers, and a large easily read scale, each small division equal to a plunger movement of 0·002 mm.

Figure 2.55B shows an electrical visual gauging head having three coloured signal lamps; if the red lamp lights up the dimension is oversize, a yellow signal indicates an undersize dimension while a dimension within the limits and acceptable, results in a green signal. A pneumatic visual device is shown in Fig. 2.55C, consisting of a glass tube containing a column of coloured liquid. When the component is placed in the test position, the height of this liquid column must not move outside the limits shown. Figure 2.55D shows perhaps the ultimate in visual comparator scales, for the digital comparator gives the actual variation of dimension from the known setting. It needs to be remembered however, that in all cases, the comparators must be set with care and precision, and these techniques will be discussed in a later volume, together with a closer examination of the principles employed to obtain the required magnification.

2.4 The drilling machine

The purpose of a drilling machine is to produce internal cylindrical surfaces, in other words to drill, ream, countersink or counterbore. Drilling holes is an essential process in engineering manufacture, and there are very few engineering components that are without holes of one kind or another. Of all the metal removing operations carried out in engineering manufacture, the production of holes is perhaps the most difficult. It is possible to produce a hole at the bench using hand tools such as a cold chisel and file, but this can be done only if the metal is relatively thin. To produce several holes, 15 mm diameter in 30-mm-thick steel plate is an almost impossible task unless some kind of drilling machine is used. We have seen that powered hand tools are available, but tools of this kind are intended as an aid to hand or bench work. Before we consider the purpose and use of drilling machines it is vital to appreciate that all surfaces used in engineering manufacture must, of necessity, be *geometrical*, and the machines used to produce these geometrical surfaces are called machine tools. We may define a machine tool as a power-driven device designed to produce a given geometrical surface, and Table 2.3 shows the common names used for geometrical surfaces:

Table 2.3 Engineering surfaces

Geometrial surfaces	Engineering term
External Cylindrical	Round
Internal Cylindrical	Hole or Bore
Plane	Flat
Conical	Taper
Helical	Thread or Spiral

The essential geometric conditions to produce an internal cylindrical surface are relatively simple, namely that the centre-line of the drill be at 90° to the plane of the workpiece or the surface upon which the workpiece rests; this relationship to be maintained during the feed of the drill, and for any position of the drill or worktable. This is simply illustrated in Fig. 2.56, and the following features will be required:

(i) device for holding the work,
(ii) device for holding the drill,
(iii) rotation of the drill,
(iv) arrangements for feed of drill,
(v) adjustment of worktable.

The above conditions apply irrespective of the type of drilling machine, for there are several different types in use, although the actual difference is mainly one of size or capacity; that is to say the largest diameter of hole that can be drilled.

2.4.1 Tool holding on the drilling machine

The taper principle is used to ensure precise positioning of the drill, and this technique of drill location is illustrated in Fig. 2.57. Taper-shank drills are located directly in the spindle of the drilling machine,

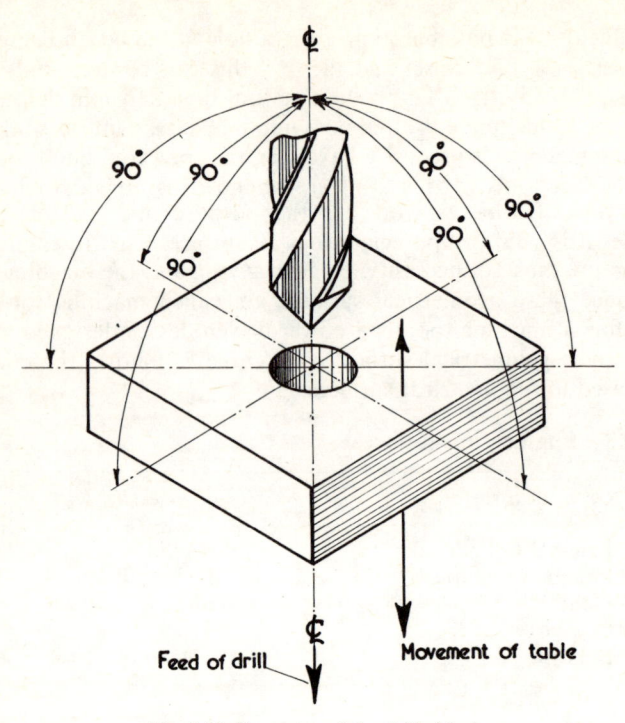

Fig. 2.56 Geometry of the drilled hole

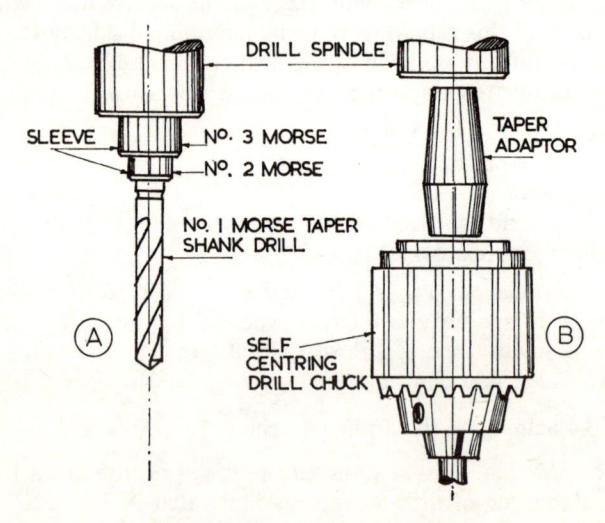

Fig. 2.57 Principles of drill holding

Hand Processes

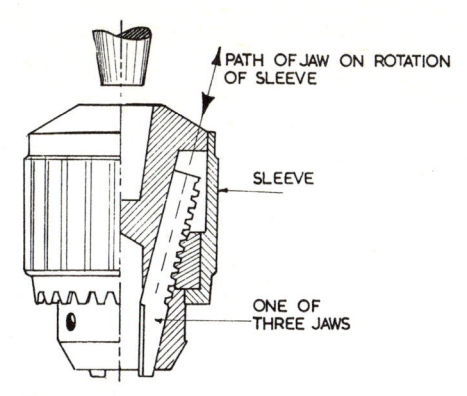

Fig. 2.58 Principles of self-centring drill chucks

and a set of Morse taper sleeves allows the use of relatively small diameter drills on a fairly large capacity machine. Parallel-shank drills are held in a self-centring drill chuck, that is to say the chuck is so designed that the centre line of the twist drill gripped in it, is automatically brought co-axial with the axis or centre-line of the drilling machine spindle. The principle of a self-centring drill chuck is simply illustrated in Fig. 2.58.

2.4.2 Work holding on the drilling machine

All drilling machines must have a worktable or reference plane. Holding work on this flat surface always presents problems leading to an undesirable tendency to hold the work by hand while the drilling operation takes place. This is an unnecessary and dangerous practice making the drilling machine a prolific source of injury and accident. Small work must be held in a hand vice and large work clamped to the

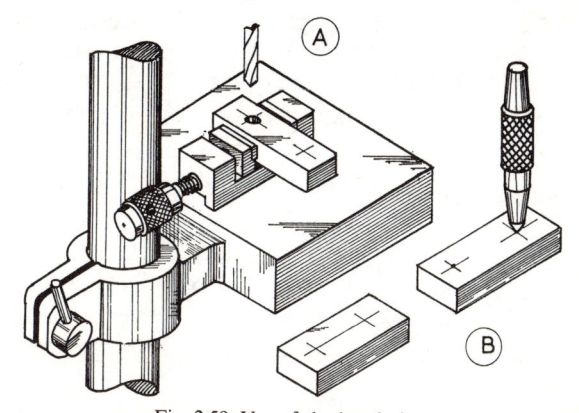

Fig. 2.59 Use of the hand vice

table of the drilling machine. A typical example of the use of a hand vice is shown in Fig. 2.59; note the use of a 90° centre punch to provide a good start for the twist drill, thus ensuring that the hole is drilled in the correct position.

Especial care needs to be taken when drilling fairly large diameter holes in thin metal plate. This can prove to be an especially dangerous undertaking, due to the fact that a twist drill tends to screw itself through the partly drilled hole, with the result that the work will be revolved and certain to cause serious injury to the operator if he is holding the component by hand. The proper technique involves

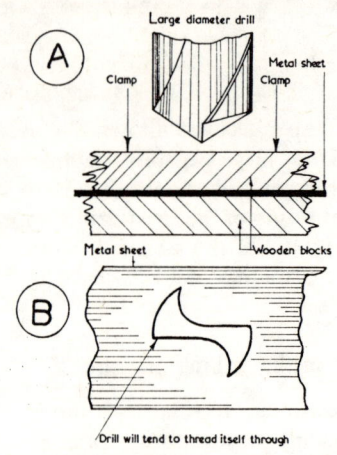

Fig. 2.60 Drilling thin plate

clamping the thin plate between two flat wooden blocks as shown in Fig. 2.60A, while B shows the profile of a partly drilled hole through which the drill is able to screw or thread itself. The opening out of large diameter holes is also dangerous to the operator, due to the relatively small amount of metal left for the opening-out drill. There is a real tendency for the drill to snatch or bite into the previously drilled hole; this effect is due to the right-hand spiral possessed by the twist drill.

Choice of drilling machines

For ordinary work the choice of a drilling machine is mainly determined by the size or diameter of the holes to be drilled. Although special purpose drilling machines are available, *three* types of machines are to be found in engineering workshops and these are as follows:

(i) sensitive,
(ii) pillar,
(iii) radial.

Hand Processes

For all these drilling machines, the machining technique is the same; positive and secure holding of the work and choice of correct drill and spindle speed.

2.4.3 Sensitive bench drilling machines

A sensitive drilling machine may be described as a compact structure built around a sleeve within which a spindle rotates. Changes in spindle speeds are effected by changing the position of a driving belt on a three-diameter or coned pulley. The feeding or downward movement of the drill is motivated by hand, permitting a nice sense of touch or feel which is very necessary if breakage of small diameter drills is to be avoided; hence the name, sensitive driller. This type of machine is ideal for drilling small diameter holes, and this means that the spindle needs to be revolved at high speeds; for the smaller the drill diameter, the higher the revolutions needed. Figure 2.61 shows a front and side view of a simple bench-type sensitive drilling machine capable of taking twist drills up to 6 mm diameter. Note the pulleys commonly known as *coned* pulleys; in the position shown the belt will drive the spindle at its highest speed, and this speed is likely to be around 3000 revolutions

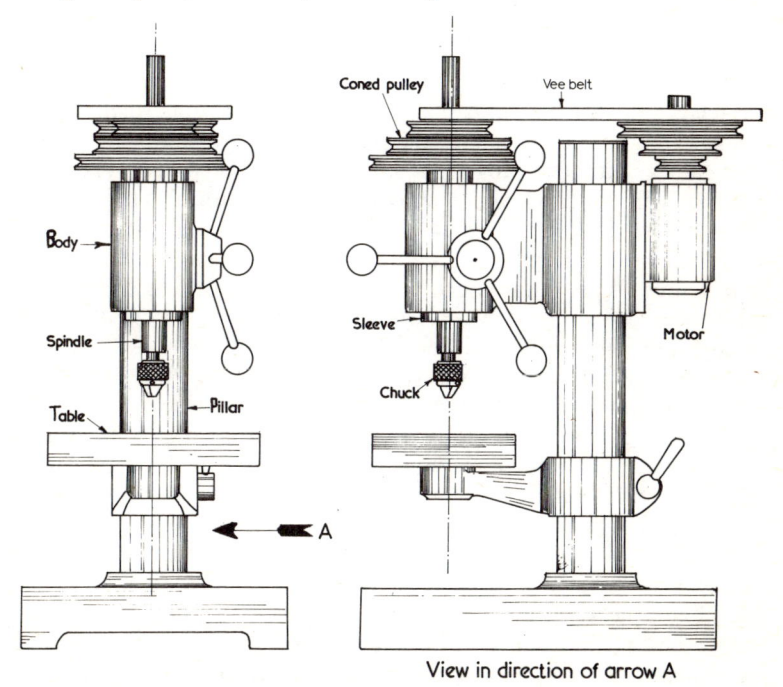

Fig. 2.61 Sensitive drilling machine

per minute. With the belt on the lower pulley, the spindle speed will be approximately 600 revolutions per minute, and this means that the operations of counterboring, countersinking and reaming are not suited for the sensitive driller because they require a low spindle speed for best results.

Figure 2.62 shows in a simple manner the essential movements of a sensitive drilling machine, and it needs to be remembered that the

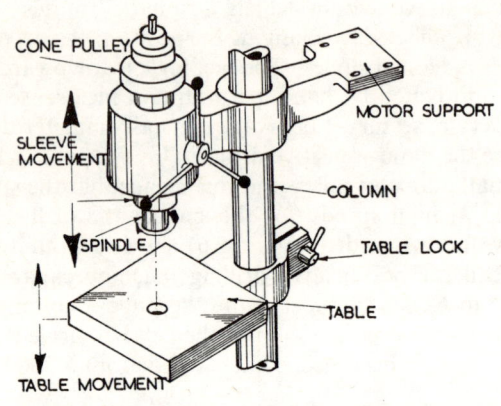

Fig. 2.62 Essential movements of the drilling machine

worktable may be swung to one side in order to drill work clamped to the base. As we have seen, the primary consideration when drilling holes is that the centre-line of the hole is at 90° to the surface of the work, and the importance of this basic geometrical condition is simply illustrated in Fig. 2.62, which shows a simple pictorial representation of a sensitive drilling machine assembly. Note that because the sleeve, worktable and base must maintain a 90° relationship to the pillar, a high degree of precision is needed when these respective components are machined. It may be stated that there are very few drilling machines in existence capable of drilling a hole truly at 90° to the surface of the work, and when holes are required to have an alignment accuracy to within say 0·002 mm, special-purpose machine tools called *jig borers* must be used. However, for most workshop jobs, drilling machines are quite adequate, and provided the machine is given the care and maintenance required by all machine tools, accurate and well-finished work can be produced by a drilling machine in the hands of a skilled operator.

Spindle and sleeve details
Figure 2.63 shows details of the spindle and sleeve of a sensitive drilling machine. The lower end of the spindle accommodates the work-holding device, usually a self-centring chuck, while the top end is

Hand Processes

keyed to the coned pulley. The purpose of the key is to transmit the turning movement or torque from the pulley to the spindle, while at the same time allowing the spindle and sleeve to slide vertically downwards as the hole is drilled.

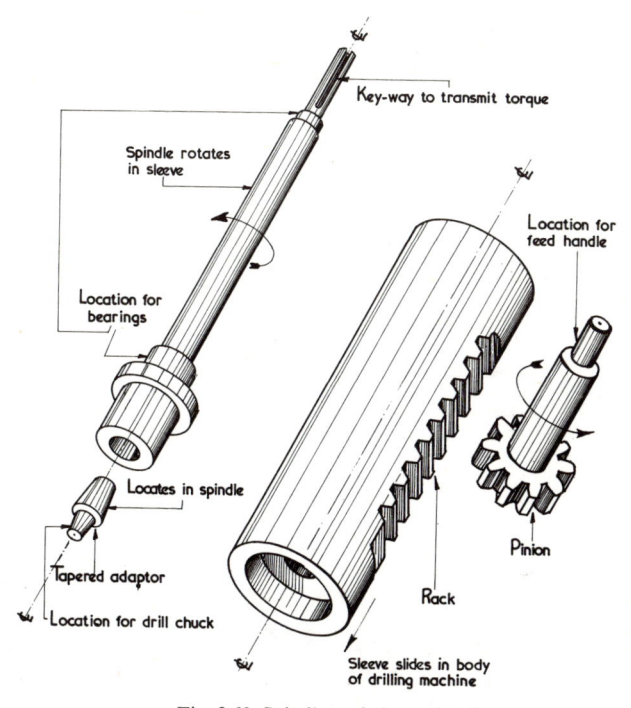

Fig. 2.63 Spindle and sleeve details

Note the provision of a taper at the lower end of the spindle. It is this taper which acts as a location for the tapered adaptor which, in turn, locates the self-centring chuck. The spindle rotates within the sleeve, supported by thrust roller bearings at the lower end, and by a ball race at the top. A rack is milled on the side of the sleeve, and the action of a constrained pinion meshing with this rack provides vertical movement of the sleeve under hand pressure only. This is shown in Fig. 2.63, and we now see that the sensitive drilling machine can be considered as a structure built around this compact unit, with an electric motor supplying the energy or power needed to drill the hole. Thus the operator is freed from excessive manual exertion or fatigue; all that is needed is to ensure that the work is securely held, and that light pressure is applied, this pressure to be eased off just as the drill breaks through.

Setting on the sensitive drilling machine

The amount of setting on a sensitive drilling machine is limited to control over the depth of a drilled hole, achieved by the use of a drill-stop. Many types of sensitive drillers however, are fitted with a friction dial graduated in millimetres, and when used in conjunction with the drill-stop, positive and automatic control over the depth of the drilled hole is easily obtained. Let us assume that the engineering component shown in Fig. 2.64A, requires four 5-mm-diameter holes drilled to a depth of 12 mm, as shown in the sectional view on the cutting plane

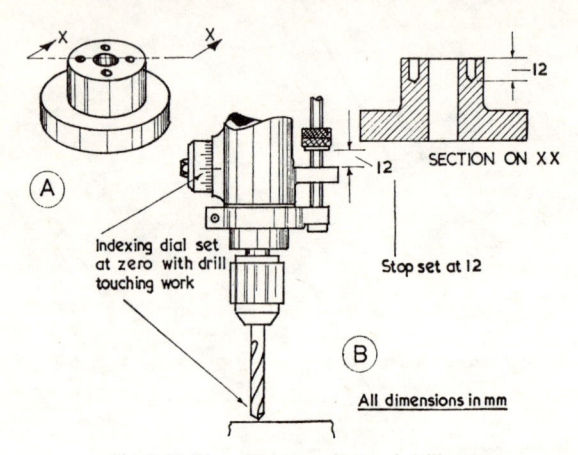

Fig. 2.64 Use of friction dial and drill-stop

indicated as xx. At Fig. 2.64B we see the correct procedure to adopt. With the drill point touching the work the chuck is securely tightened with a chuck key. The friction dial is now rotated until a zero reading is obtained; that is to say until the zero line on the dial coincides with the line engraved on the drilling machine body. While this setting to zero takes place, the drill point is kept in contact with the surface of the component to be drilled. Provided the four holes have been marked out and centre-dotted, drilling can commence, with a close watch being kept on the friction dial. When the dial reads 12 mm, the adjustable threaded drill-stops are screwed down so that the bottom stop is in contact with the stop arm. The purpose of the upper drill-stop is to lock the lower drill-stop on the threaded bar, otherwise the vibration set up during the drilling operation may lead to rotation of the drill-stop and errors in the depths of the drilled holes.

Sensitive-column drilling machines

A sensitive-column drilling machine is bolted or fixed to the floor of the workshop. Once again vertical feed of the spindle is actuated by

Hand Processes

the hand of the operator, permitting a nice sense of touch through the rack and pinion, which changes rotary motion to linear or straight-line motion. Column-type sensitive drilling machines are more rigid than bench drillers, and have capacities up to 12 mm; that is to say a hole of 12 mm diameter can be drilled using either a parallel or taper-shank twist drill. All the previous remarks with regard to the use of friction dials, and drill-stops when drilling blind holes to prescribed depths, apply equally to a column-type sensitive drilling machine.

2.4.4 Speeds and feeds

The Formula, cutting speed $= \dfrac{\Pi D N}{1000}$

may be transposed in order to calculate the approximate spindle speed when drilling holes. However, this is seldom carried out in practice, and the best procedure is to refer to the tables supplied by twist drill manufacturers, or rely on experience. In general, the smaller the drill the higher the spindle speed; attempting to drill small diameter holes at low spindle speeds is certain to result in drill breakage.

No hard and fast rule can be stated with regard to the rate of feed. Much depends on the composition of the metal to be drilled, the sharpness of the twist drill point, the amount and type of cutting fluid used, and the condition of the drilling machine. Excessive feed pressure, if it does not break the drill, will almost certainly cause it to deflect or bend, leading to an alignment error. In general, easy but firm

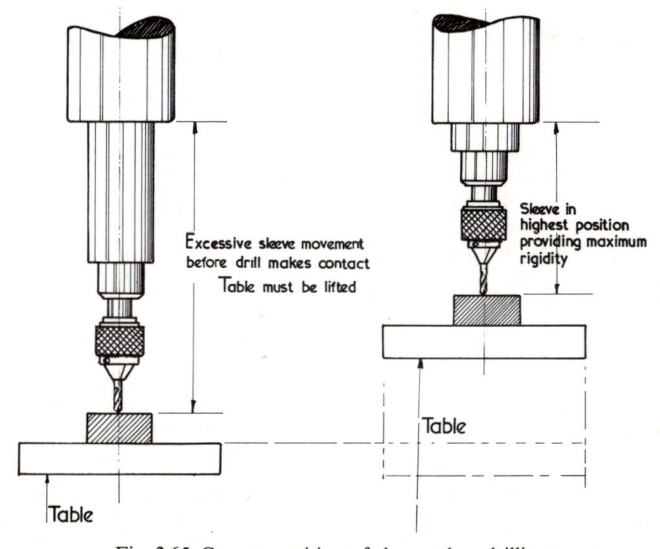

Fig. 2.65 Correct position of sleeve when drilling

pressure should be applied, care being taken to ensure that this pressure is eased off as the drill point breaks through. All drilling should be carried out with the sleeve in its highest position, thus ensuring maximum rigidity—the golden rule for all machining operations. This means that the worktable must be raised so that the surface of the work just allows the twist drill to be placed in the self-centring chuck with the sleeve in its highest position. This is simply illustrated in Fig. 2.65.

2.4.5 Pillar drilling machines

Both bench- and pillar-type sensitive drilling machines are suitable only for drilling fairly small diameter holes, and because of the range of high spindle speeds they are totally unsuitable for reaming, countersinking and counterboring. At the same time, the worktables are relatively small, and any attempt to drill holes in heavy castings would result in bending or deflection of the worktable. Pillar drilling machines however, are of very robust construction capable of heavy work or the drilling of large diameter holes at high speeds. Figure 2.66 shows the

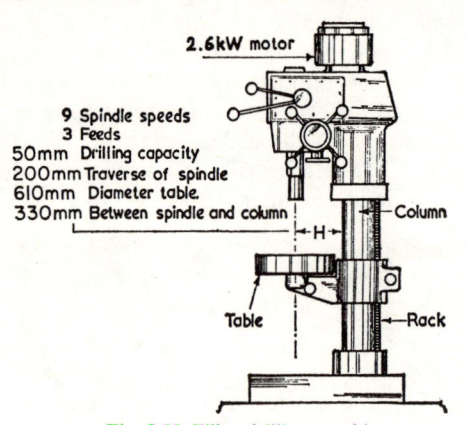

Fig. 2.66 Pillar drilling machines

outline of a pillar drilling machine with the main parts named. Although provision is made for hand feed of the drill, it is usually possible to employ automatic feed, and a range of feeds are available. The spindle speeds will have lower values than those of a sensitive driller, allowing the use of reamers, countersinks and spotfacing tools. A typical range of spindle speeds would be from 50 to 1200 rev/min, with about six to eight speeds provided; this would permit the drilling of holes from 5 mm to 35 mm diameter. Automatic feeding of the drill is arranged through suitable gearing, with feeding rates ranging from 0·07 mm to 0·4 mm per revolution of the spindle. The greatest of care needs to be exercised when using automatic feed on a pillar drilling

Hand Processes

machine. Absolute rigidity of the clamping set-up is essential, and it is a wise plan to use small diameter drills to start, opening out with larger diameter drills, reducing the feed per revolution as the diameter of the drills increases.

Tool holding on the pillar driller

Parallel-shank twist drills may be held in a self-centring chuck, while taper-shank drills are inserted directly into the spindle. All the tapers on the shanks of taper-shank drills are *Morse* tapers, as are the tapers machined in the spindles of the drilling machines. In order to accommodate a range of taper-shank drills, a set of Morse tapers are needed and the diameter of the drill determines the size of the Morse taper. Reference back to Fig. 2.57A shows the principle underlying the use of Morse taper sleeves, and in this way, relatively small diameter drills can be used in large capacity drilling machines. Table 2.4 gives the available sizes of Morse tapers together with the taper on diameter, and drill diameter range for each taper.

Table 2.4 Morse tapers

Morse taper No.	Drill diameter (mm)	Taper on diameter
1	Up to 14	0·049 88
2	14–24	0·049 95
3	24–30	0·050 19
4	30–50	0·051 93
5	Over 50	0·052 62

Precautions when tool holding

Although the drilling machine is one of the simplest of machine tools, it is the most prolific source of injury in the workshop. As we have seen, the basis of all machining is correct and positive holding of both work and tool, with a combination of movements or feeds in order to generate or form the required geometrical surface. Figure 2.67A shows the wrong method of inserting a taper-shank drill. A damaged drill point is certain to result, with the grave danger of a fragment of the drill point entering the eye of the operator. Figure 2.67B shows the correct method of inserting a taper-shank drill. The same precautions need to be taken when removing a taper-shank drill; never use a file as depicted in Fig. 2.67C, but use the correct drift, allowing the drill to fall on a piece of soft wood placed on the drilling machine table.

Although the insertion of parallel-shank drills into a self-centring chuck may seem a simple matter, it is essential to ensure that the drill is firmly held. Failure to grip the drill properly, possibly caused by the

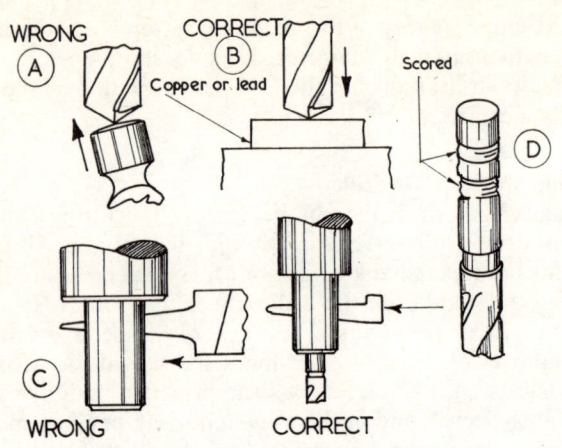

Fig. 2.67 Precautions when drill holding

use of a worn chuck key, results in a badly scored drill as show in Fig. 2.67D. Such a drill is certain to have its size removed by the score marks, and may not run true if the shank is badly scored. Always ensure that the correct chuck key is available for the appropriate chuck, and it is a wise plan to attach the chuck key to the drilling machine using a length of light section chain.

Work holding on the pillar driller
Not only does the pillar driller offer the advantages of automatic feed, together with the use of large diameter drills, but also the workholding capabilities are much greater. This means a wider and more accurate range of operations is possible when using a pillar driller.

Round table
This type of worktable is probably the most common of all drilling machine tables, and a typical table is simply illustrated in Fig. 2.68. Diameters range from about 500 to 600 mm and all have T slots to accommodate clamping bolts. The table is strongly ribbed on the underside giving added rigidity, which is very necessary because grey cast iron is the material used. Note that the table can be rotated about its axis, and also through 360° about the axis of the pillar or column, and locked in any position. The table is raised or lowered for vertical adjustment by means of the rack and pinion principle. The combination of these movements allows a considerable amount of accurate setting with respect to drilling, reaming, counterboring or countersinking.

Figure 2.69 shows the technique involved, and represents a plan view of the job, as set-up on the round table of a pillar drilling

Hand Processes

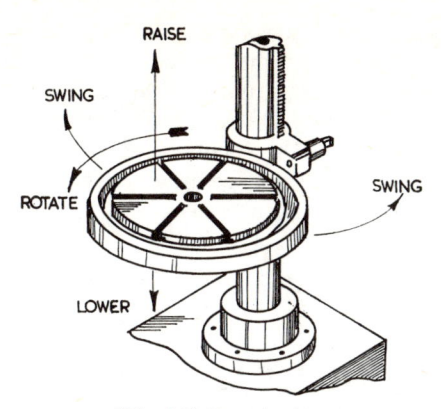

Fig. 2.68 Round table

machine. This job requires one large central hole and two smaller holes, and the best technique is to drill these holes in *one* clamping; that is to say, once the two plates are clamped together they are not separated until all the machining is completed. With the two plates clamped together after marking-out and centre-dotting the top plate, the centre-dot for the large central hole is brought to the centre-line of the drill spindle. This is easily achieved by a combination of table swinging and rotation, the centre-dot being picked up using a 60° pointed round bar held in the self-centring chuck. The table clamps are now tightened, the hole centre-drilled and opened out to the size required. Any further operations such as reaming, counterboring or countersinking, must be carried out immediately the hole is drilled. On completion of the machining for the first hole, the table lock is loosened and the table adjusted so that the second hole is brought to the centre-line of the drill spindle; the table locked and this hole machined. The same pro-

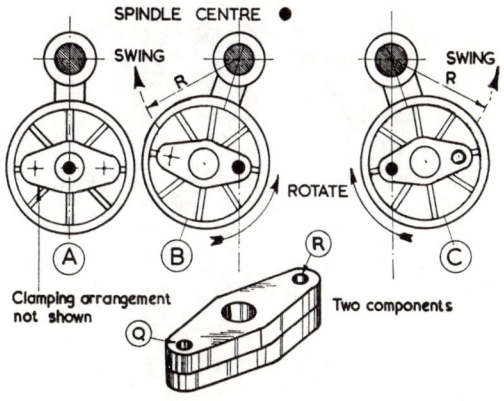

Fig. 2.69 Correct use of the round table

cedure is used for the third hole shown as Q in the diagram, and in this way the three holes may be drilled in the one clamping, ensuring that the alignment of the holes in the two components will be without error. The accuracy of the linear distances between the hole centres, will depend of course, on the accuracy of the marking-out, centre-dotting and picking-up of the centre-dots. Nevertheless, provided reasonable care has been taken, fairly consistent results are possible

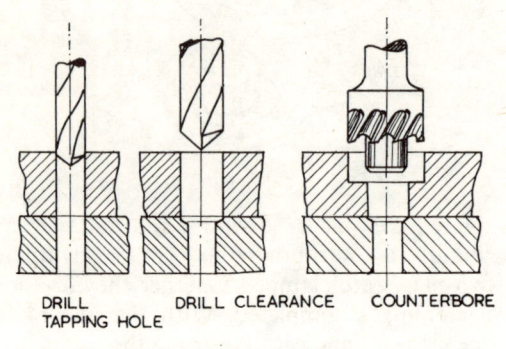

Fig. 2.70 Principle of the fixed centre-line

when using the round table of a pillar drilling machine, and the principle of the fixed centre-line, obtained by completing all machining during the one clamping should be adopted at all times. This principle is further illustrated in Fig. 2.70.

Rectangular table
The rectangular table is illustrated at Fig. 2.71. It is also made from grey cast iron, strongly ribbed on the underside and is capable of movements similar to those of the round table.

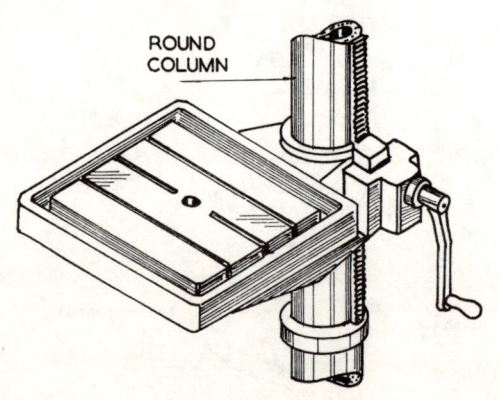

Fig. 2.71 Rectangular table

Hand Processes

Compound table
This table removes the main defect of the round and rectangular tables, namely the lack of control over the linear movement of the table, making it impossible to set the table at prescribed linear distances. A typical compound table for a pillar drilling machine is shown in Fig. 2.72, where it may be seen that the table is capable of both transverse

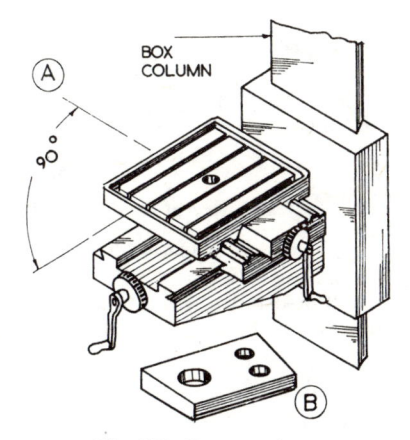

Fig. 2.72 Compound table

and longitudinal movement. The amount of movement is controlled by the indexing dials, and means that the centre distance between holes may be readily obtained to a high degree of accuracy without the need for marking-out. We may regard a compound table pillar drilling machine as a less accurate type of jig-boring machine.

2.4.6 Radial drilling machine

Referring back to Fig. 2.66, which shows a typical pillar drilling machine, it may be noted that the distance between the spindle centre-line and the column limits the size of work that can be accommodated on this type of machine. For example, it would not be possible to drill a hole in the centre of a round steel plate the radius of which exceeded the distance H shown in Fig. 2.66. At the same time it will be noted that the table is not properly designed to take heavy weights, and is likely to deflect or bend if unduly heavy castings are placed on it. Let us assume that the grey cast iron casting illustrated in Fig. 2.73A is to have the six 75-mm diameter holes drilled in the top face. With the overall dimensions of the component as shown, it is clear that even the largest capacity pillar drilling machine is unable to cope with a casting of this size and mass. We need, in effect a drilling machine with a

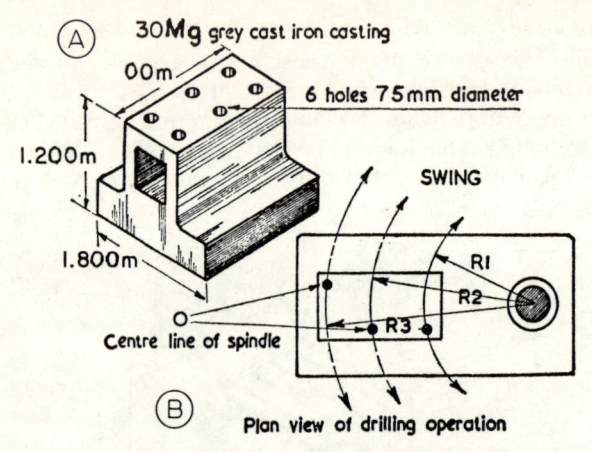

Fig. 2.73 Large casting requiring drilled holes

movable spindle; in other words the ability to take the spindle to the hole centre-line, as opposed to taking the hole centre-line to the fixed spindle when using sensitive or pillar drilling machines.

This principle is simply illustrated in Fig. 2.73B, which shows a plan view of the drilling operation. Let O represent the centre-line of the drilling spindle. If now we are able to vary this distance as shown in the diagram (where R1, R2 and R3 represent different positions of the spindle centre-line), the drilling of large castings will present little difficulty. It will be necessary to remove the need for a movable table, that is to say a table which can be raised or lowered, for it is not practical to raise or lower heavy castings. It is easier to raise or lower

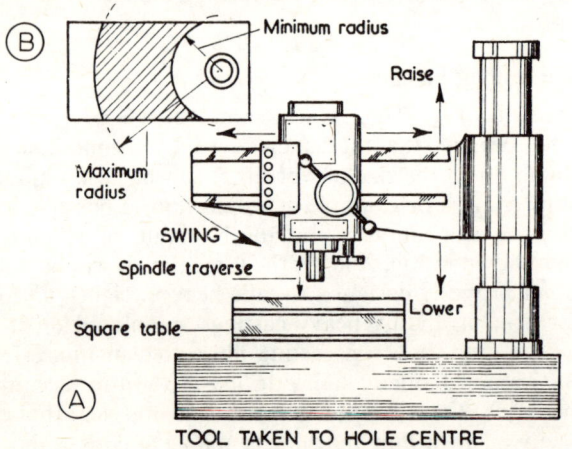

Fig. 2.74 Radial drilling machine

Hand Processes 93

the drilling head. The radial drilling machine is shown in Fig. 2.74A together with its essential movements. Note the large area that can be covered by the movable drill spindle as shown in the simple plan view at B. The base of the machine is a worktable with T slots to provide for work clamping, while an additional piece of equipment is an *auxiliary* or square table which may be bolted or clamped to the base. The capacity of radial drillers is very great. Small diameter drills are held in a self-centring chuck while a 50-mm-diameter twist drill would be located using the taper principle. Automatic feed is provided for the downwards movement of the spindle, with a range of spindle feeds and speeds. Once again, the greatest care must be exercised when drilling large diameter holes using automatic feed. The arm must be in its lowest position and securely locked before drilling commences, with the work adequately and securely clamped. The raising or lowering of the arm would entail considerable manual effort on the part of the operator, and this is avoided by the provision of an automatic lifting and lowering device.

2.4.7 Twist drills

We may consider a twist drill as a spade drill given a permanent twist. This principle is shown in Fig. 2.75. It is difficult to twist a flat piece of metal as shown in the diagram and retain cylindrical accuracy, together with providing a means of holding this twisted strip so that it

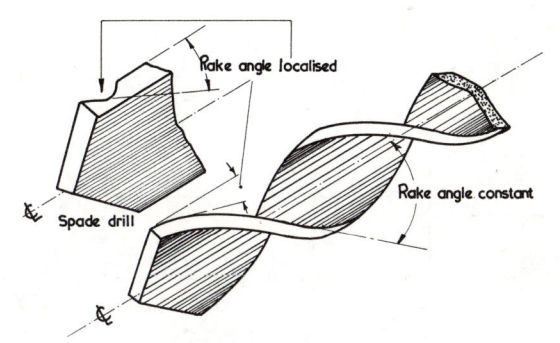

Fig. 2.75 Development of the twist drill

revolves about its centre-line. Because of this, twist drills are made from cylindrical steel bars, and the twist or spiral obtained by milling two helical flutes. In order to reduce the friction on the bearing surface or outside diameter of the drill, a small land is left behind each cutting edge, and most twist drills have a slight taper towards their shanks.

Figure 2.76 shows the cutting point of a typical twist drill together with a cross-sectional, and end view. Note that the rake angle is equivalent to the helical angle of the flutes, and a comparison is made

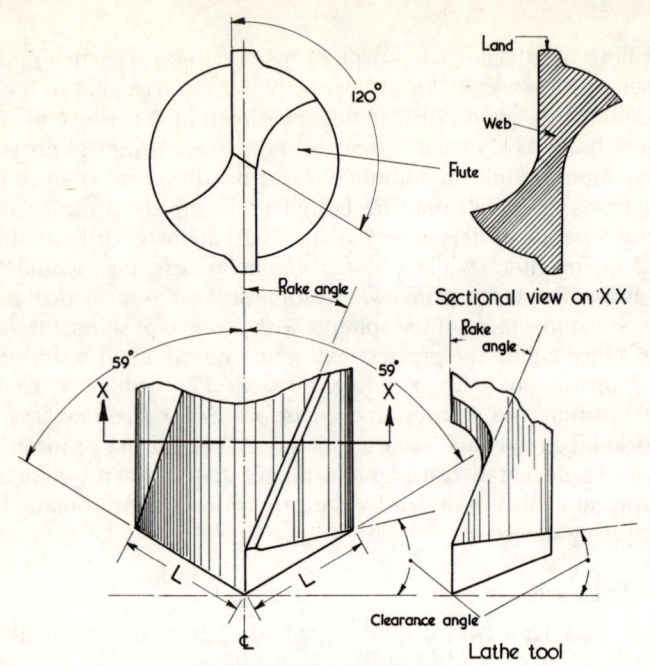

Fig. 2.76 Twist drill details

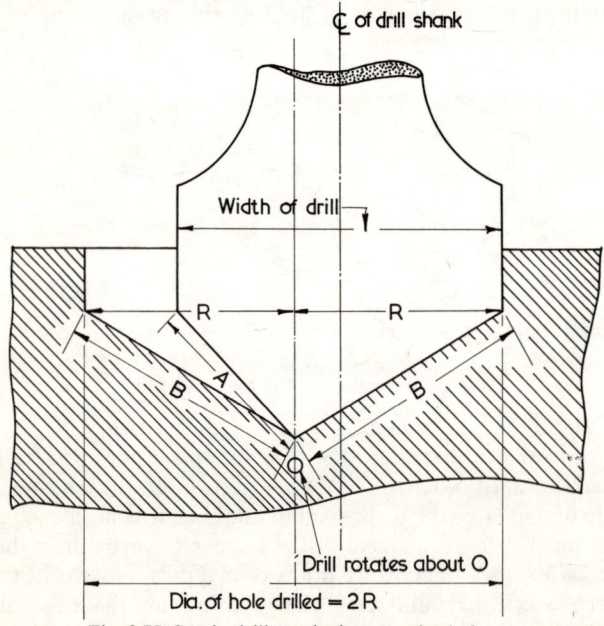

Fig. 2.77 Spade drill producing oversize hole

to a lathe tool. These helical flutes provide not only a rake angle which is unaffected by continual sharpening, but also a path for swarf exit; for the turning movement of the twist drill tends to move the swarf up the flutes and out of the hole. The correct angles for sharpening a twist drill are also shown in Fig. 2.76, and it is essential that these angles be correct, together with the lengths of the cutting edges. Incorrect grinding or sharpening of a twist drill point will produce oversize holes, together with excessive wear of the drill point, and excessive strain on the drilling machine spindle. This is simply illustrated in Fig. 2.77, where we see a simple spade drill with the cutting edge B longer than the cutting edge A. Although the centre-line of the spade drill will be co-axial with the centre-line of the spindle of the drilling machine, on rotation the spade drill will tend to rotate about its point which is off centre due to incorrect grinding. Thus the spade drill will tend to produce a hole of diameter 2R, considerably in excess of the true diameter of the drill, and subjecting the spindle bearings to very great stresses, with the possibility of permanent damage to the drilling machine spindle.

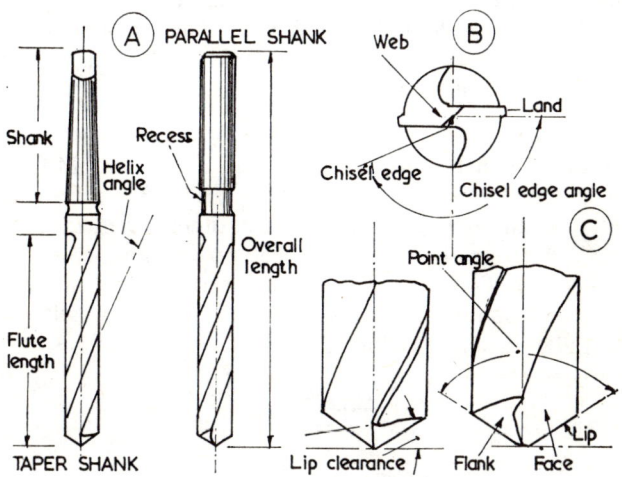

Fig. 2.78 Elements of the twist drill

Twist drill nomenclature
Figure 2.78A shows the names or nomenclature of the main parts of both taper- and parallel-shank twist drills. At B we see an end view of the cutting point with the main parts named, while the two views at Fig. 2.78C show the important cutting angles and faces. The names given are those recommended in British Standard 328.

Parallel-shank jobber series twist drills
These drills have two helical flutes with a parallel shank approximately

the same diameter as the cutting end. The smallest drill is 0·2 mm diameter and has an overall length of 19 mm. The greater the diameter, the longer the drill, for example a 6-mm drill has a length of 93 mm. The largest drill in this series is 16 mm diameter, and there are a total number of 191 drills available. This series is very popular, widely used in engineering workshops, and especially suited for sensitive and pillar type drilling machines.

Stub drills
A stub drill may be defined as a shortened form of parallel-shank jobber series twist drill, the reduction in length applicable to the flute only; the shanks being the same length as jobber series drills. Starting at 0·5 mm diameter and finishing at 25 mm diameter, there are 72 drills available. This series of drills is best used for portable drilling machines, or for drilling harder materials; the shortened length of flute appreciably reducing the possibility of drill breakage.

Parallel-shank long series twist drills
These drills have increased flute length and are used for deep holes or for holes close up to shoulders or flanges. The smallest drill has a diameter of 1 mm and the largest a diameter of 25 mm, with a total number of 175 drills in the series.

Morse taper-shank series drills
These are drills with two helical flutes and a standard Morse taper shank for holding and driving. The smallest drill is 3 mm diameter and the largest 100 mm diameter. As with the other series, the drill diameters increase in small increments together with an increase in flute and overall length. This Morse taper-shank series provide a wide range of drills, a total number of 204 being available, and widely used on pillar and radial drilling machines.

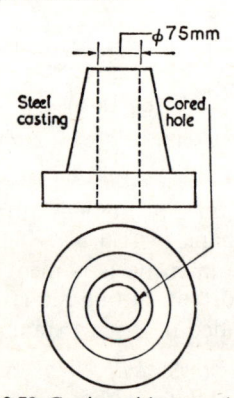

Fig. 2.79 Casting with a cored hole

Hand Processes

Morse taper-shank core drills

Figure 2.79 shows a casting with a cored hole, that is to say the hole is produced by the casting process. This hole is to be drilled or opened out to 86 mm diameter, and it would be unwise to choose a drill from the Morse taper shank series drills for the following reasons:

1. The drill will tend to wander or deflect,
2. the drill will tend to snatch or grab,
3. the drill is very liable to break.

All these faults are due to the relatively small cross-sectional area of a two-fluted twist drill as shown in Fig. 2.80. If, however, the flutes are

Weak, and easily bent or broken under drilling loads.

Fig. 2.80 Section of two-fluted twist drill

increased in number to three or four, they can be made less deep, giving a much stronger section. Figure 2.81A shows a three-fluted core drill with spiral flutes, very suitable for opening out cored holes in mild steel castings or forgings. At B we see a four-fluted core drill with straight flutes, and much used for opening out holes in grey cast iron or brass castings. At Figure 2.81C we see a four-fluted core drill with spiral flutes, very suitable for heavy duty work. The sizes of these core drills are identical in all respects to those available in Morse taper-shank series drills.

4 Fluted drill with straight flutes

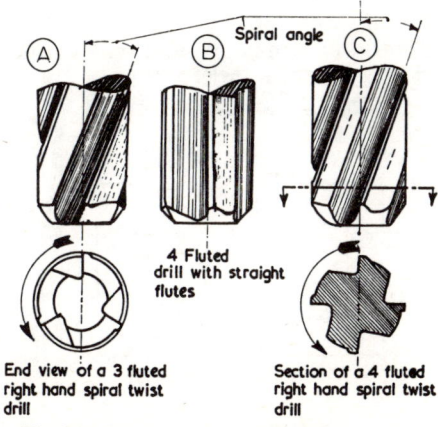

End view of a 3 fluted right hand spiral twist drill

Section of a 4 fluted right hand spiral twist drill

Fig. 2.81 Three and four-fluted drills

It needs to be remembered that a three- or four-fluted twist drill is quite unable to start its own hole, as it is impossible to drill a point on one of these drills. The ability to drill its own hole from scratch, is an important feature of the two-fluted twist drill, but much care is needed when sharpening the point if an accurate well-finished hole is desired. Under no circumstances should a large diameter twist drill be sharpened by hand at a pedestal grinding machine. As previously mentioned, an incorrectly sharpened drill point may cause serious damage to the spindle of a drilling machine and this is especially so when the drill is of large diameter. This means that a drill grinding device or jig is an essential piece of equipment, and should be present in any engineering workshop.

Reamers

The purpose of a reamer is to produce a well-finished and accurate diameter hole. This means that primarily the reamer is a *finishing* tool; the removal of metal is a secondary feature. In general, the less metal left for the reamer the better, but this is conditional on the drill leaving a reasonably well-finished hole free from score marks (which the reamer will be unable to remove). About 5% of the finished diameter is a reasonable allowance for reaming, thus the amount left for reaming in a 40-mm-diameter hole would be 5% of 40, or 2 mm, and a 20-mm hole would have 1 mm left for reaming.

Hand reamers

A parallel hand reamer is shown in Fig. 2.82A. Intended for right-hand rotation, the reamer shown has left-hand flutes, although straight flutes are available if required. These reamers are given a slight taper to facilitate concentricity with the drilled hole, a tap wrench is used to provide the necessary turning movement or torque to the reamer.

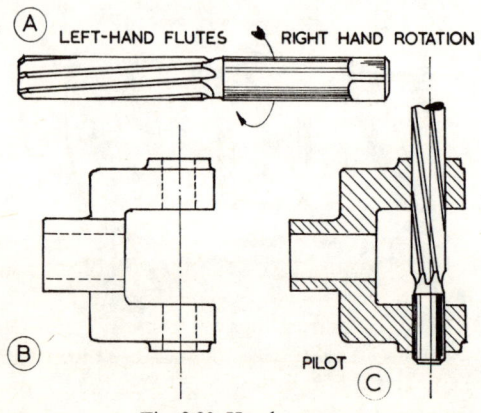

Fig. 2.82 Hand reamer

Hand Processes

Hand reamers with pilots
It is no easy matter to ensure that a hand reamer follows the axis of a drilled hole. Figure 2.82B shows a medium carbon steel forging having two holes requiring reaming, and the advantage of a pilot reamer for a job such as this is clearly indicated in Fig. 2.82C. Here we see that the pilot portion of the reamer locates in the drilled hole, acting as a guide to the reamer proper. Any attempt to ream the forging with an ordinary hand reamer is certain to result in a misaligned hole. It is of course important that the pilot be a reasonably good fit in the drilled hole.

Machine reamers
Figure 2.83 shows a parallel machine reamer with left-hand helical flutes. This reamer has a Morse taper shank, although parallel-shank machine reamers are available. Note that the flutes are left-handed, as a machine reamer with right-hand flutes has an undesirable tendency to screw itself into a drilled hole. The forces acting on both types of reamers are simply illustrated in Fig. 2.83B where it may be seen that while the reamer with right-hand spiral tends to pull itself into the hole, the left-hand spiral reamer tends to be pushed out.

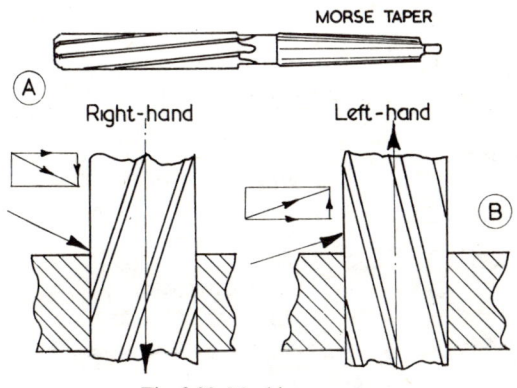

Fig. 2.83 Machine reamer

Reaming speeds and feeds
In general, when reaming, the number of spindle revolutions is considerably smaller than that used when drilling a hole of equivalent size. On the other hand, the feed or vertical movement of the reamer is in excess of that employed when drilling. A general rule is to reduce the spindle speed by one half and double the feed used for drilling the hole. A suitable lubricant or coolant greatly prolongs the life of a reamer, in addition to improving the surface finish of the reamed hole.

Countersinking

Figure 2.84A shows the cover for a small aluminium alloy junction box used in electrical installation, and it may be seen that the cover is secured with four set-screws. At B we see a section through the cover, and it is clear that a special tool is needed to machine these countersinks. Figure 2.84C shows a simple high speed steel countersinking

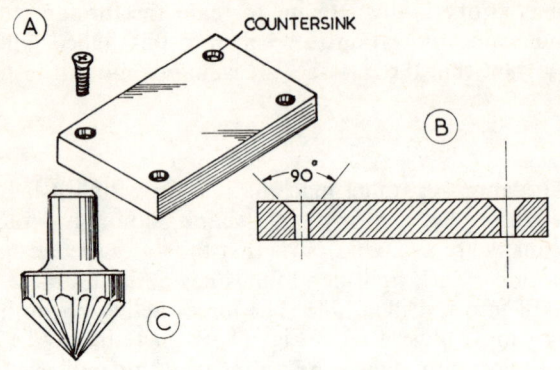

Fig. 2.84 Countersinking

tool, and it is important that this tool is used at fairly low spindle speeds. Once again, a suitable lubricant greatly improves the surface finish of the countersink, and if the drill-stop is correctly used, (as previously described) to ensure that the countersinks are machined to a constant depth, a neat and workmanlike job results.

Counterboring

Counterboring consists of enlarging a previously drilled hole to a predetermined depth, usually to accommodate a socket-head screw. This

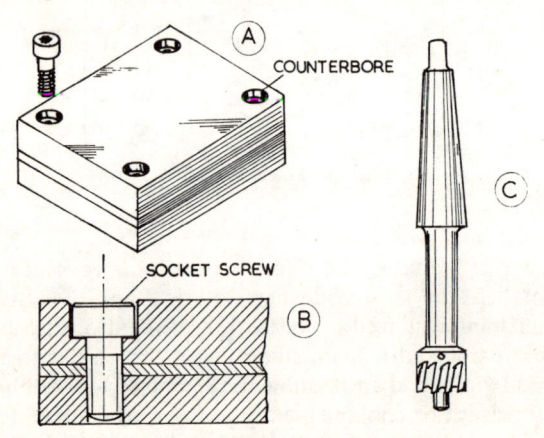

Fig. 2.85 Counterboring

Hand Processes

principle is illustrated at Fig. 2.85A, where it may be seen that a counterbore is simply a cylindrical countersink. A typical taper-shank counterboring tool is illustrated at Fig. 2.85C; the pilot or small peg at the cutting end enables the counterbore to be accurately located, machining the bore concentric with the drilled hole. Small-size counterbores are available with parallel shanks, and a set of interchangeable pilots may be obtained. When several holes need to be counterbored, as in the top bolster of a large press tool, a neat and pleasing effect is obtained by proper use of the drill-stop, thus ensuring that all counterbores are of uniform depth.

Spotfacing

Figure 2.86A shows an aluminium alloy head for a motor cycle, with a sectional view shown in B. The need for spotfacing now becomes clear; it is to provide a flat seating for the washer and nut as shown. A typical spotfacing tool is shown in Fig. 2.86B, and it is clear that a spotface

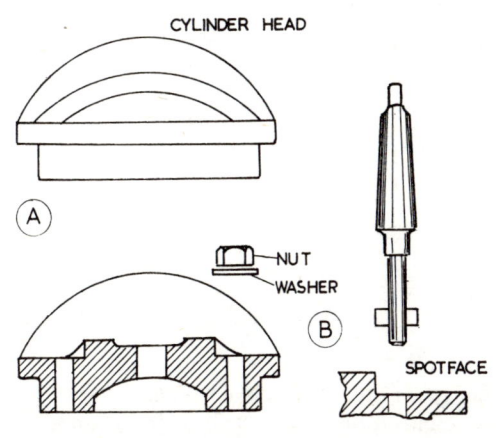

Fig. 2.86 Spotfacing

may be considered as a shallow countersink. Because of this, it is permissible to use a counterboring tool when a spotfacing tool is not readily available.

2.4.8 Drilling plastics

As a general rule, drilling holes in plastics materials is similar to drilling holes in brass. This is because most plastics have a low shear strength, and are best cut with a small rake angle and at relatively high spindle speeds and feeds. However, due to the large and ever-increasing range of plastics materials now available, it is almost impossible to lay down hard and fast rules with regard to the machining of plastics.

Plastics materials have the following properties:

1 Soft and yielding,
2 abrasive,
3 may chip or break.

Much depends whether the plastic is likely to be softened by heat, and in this case best results are obtained if the drill is kept as sharp as possible and revolved at a fairly high speed. Little time should be lost in feeding the drill to the required depth, and the drill needs to be frequently backed-out to free the chips. Figure 2.87 shows a drill point

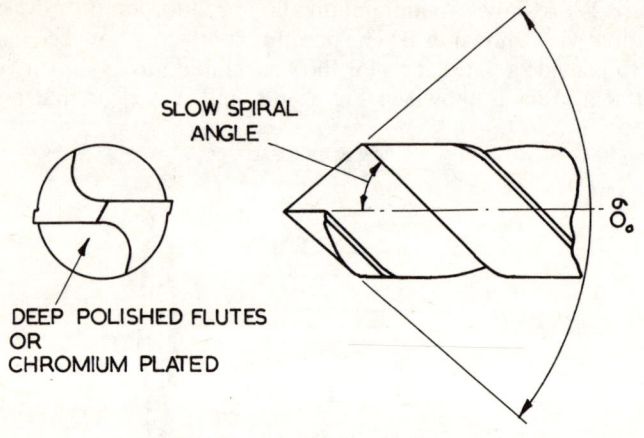

Fig. 2.87 Twist drill for plastics

suitable for most plastics; note the point angle is 60°, with a slow spiral, equivalent to a reduced effective rake angle. The flutes are wider and deeper and highly polished to reduce friction and facilitate chip flow from the cutting point.

Very small holes in plastics materials call for very high spindle speeds, ranging from 4000 to 6000 revolutions per minute, and it is clear that a bench sensitive drilling machine is required for this kind of work. Soapy water may be used as a lubricant, with very light feeds employed if drill breakage is to be kept to a minimum.

2.5 Sheet metalwork

It is a great mistake to regard sheet metalwork as the art or craft associated with the making or manufacture of simple utensils such as funnels, boxes or trays. Figure 2.88 shows some commonplace manufactured components, all of which are made from sheet metal and are

Hand Processes

therefore products of sheet metalworking. The motor car body or shell shown in Fig. 2.88A is an excellent example of a component made up of an assembly of sheet metal parts. So also is the deep freezer shown at B, together with the perambulator and beer can, shown at C and D.

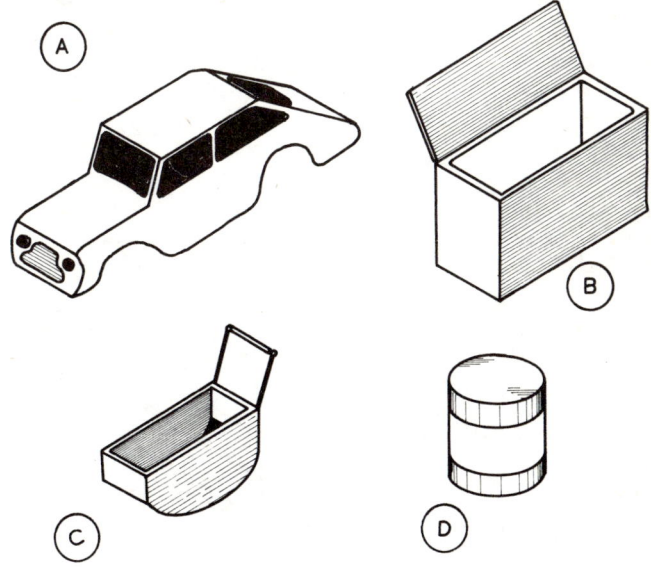

Fig. 2.88 Sheet metal components

The reason why such products are not generally associated with sheet metalwork is that the components shown in Fig. 2.88 are essential commodities in modern civilisation and required in very large numbers. While it may be possible for skilled craftsmen to produce any of the components shown in Fig. 2.88, using hand tools, both the time taken and the ultimate cost would be excessive in the extreme, and only the very rich could afford to buy them. It is certain that the sheet metal parts required for the items shown in Fig. 2.88, are produced by *press tools*, which are capable of cutting, bending, piercing, blanking, forming and many other metalworking operations. These operations are carried out at high speeds and with great accuracy, thus ensuring the mass production of sheet metal components at very reasonable cost. We shall see, in a later volume, the basic principles and techniques underlying the design and application of Press Tools. However, it needs to be remembered that all the operations mentioned above can be carried out in a small sheet metal workshop, and it will be of interest to consider them further. Figure 2.89A shows a simple sheet metal component which is to be made from a piece of mild steel 1 mm thick. The stages, together with the practical techniques are outlined below.

Development
A development of a sheet metal component may be defined as the shape to which a piece of metal must be cut in order to produce the job required. The development of the simple metal box shown in Fig. 2.89A, is shown at B, and it should be clear that a good working

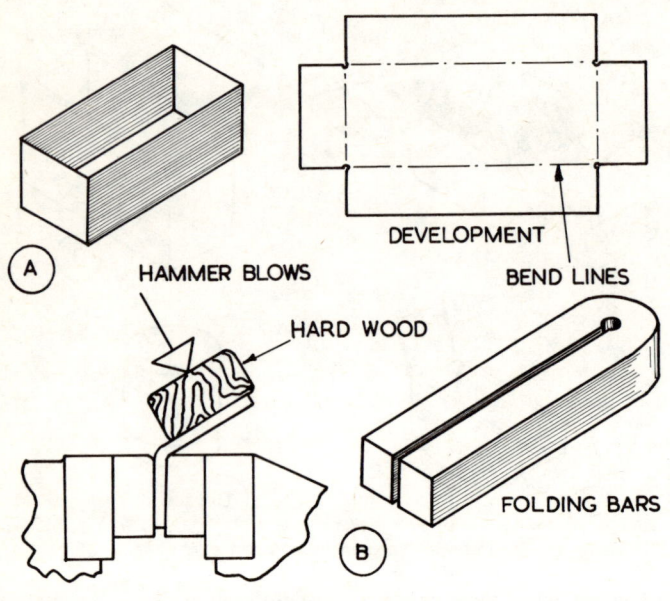

Fig. 2.89 Developing and bending

knowledge of workshop arithmetic and geometry is needed, if correct developments of awkwardly shaped jobs are to be achieved. Such awkward shapes include cylinders, cones and pyramids, and in addition to the basic developments of these geometric shapes, additional metal may be required in order to strengthen or make the job water- or airtight.

Bending
Bending is an essential part of all sheet metal manufacture, and the basic principle is simply illustrated in Fig. 2.89B. Best results are obtained using the *folding bars* shown, together with a piece of wood to ensure that the metal is brought down evenly and neatly. Bending machines are available which make the operation much quicker, the operator merely moving a lever. Long lengths are easily bent in a suitable bending machine, while small parts are bent in suitable press tools. The principle of a bending machine is shown in Fig. 2.90A. Figure 2.90B shows a typical component that would require the use of

Hand Processes

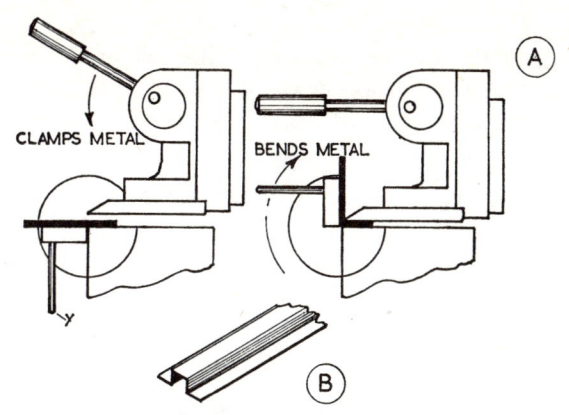

Fig. 2.90 Bending machines

a bending machine. It is a metal strip used to protect electric wires running down a wall, and it is subsequently plastered over.

Bending allowance

It is not often that the bending allowance is calculated for sheet metal components, but when the metal is of reasonable thickness and accurate work is required, it may be necessary to make the appropriate calculations. Let us assume that the simple bracket shown in Fig. 2.91A is to be made to a reasonable degree of accuracy. Because of the tendency of metal to distort when bent, such distortions will be at a minimum along what we may term as the neutral plane, that is to say the centre of the metal. Reference to Fig. 2.91A shows that the develop-

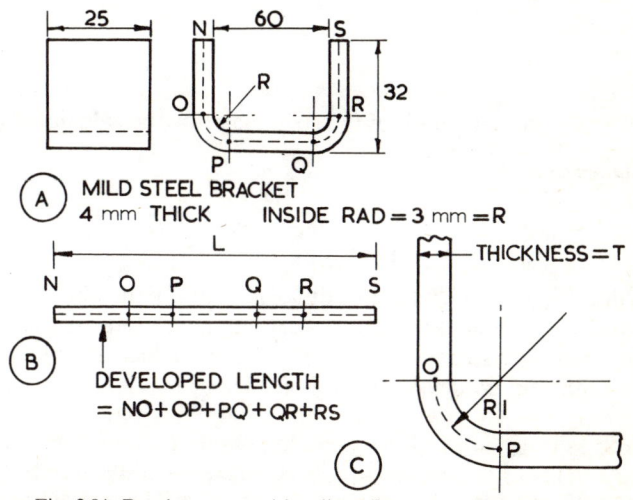

Fig. 2.91 Development and bending allowance

ment of the bracket will involve the calculation of the distance L as shown in Fig. 2.91B, otherwise known as the *developed length*. We see that this developed length will be equal to

$$NO + OP + PQ + QR + RS$$

NO, PQ and RS are straight lengths, while OP and QR are the actual bending allowances. The straight lengths are calculated as follows:

$$\begin{aligned} NO = RS &= 32 - (R + t) \\ &= 32 - (3 + 4) \\ &= 32 - 7 \\ &= 25 \text{ mm} \end{aligned}$$

$$\begin{aligned} PQ &= 60 - 2R \\ &= 60 - 6 \\ &= 54 \text{ mm} \end{aligned}$$

The method of calculating bending allowance is shown in Fig. 2.91C where it may be seen that OP is one quarter of the circumference of the dotted circle. Expressed mathematically

$$OP = \frac{2\pi R1}{4}$$

where R1 = inside Radius R plus half the thickness of the metal.

Hence
$$OP = \frac{\pi(R + \tfrac{1}{2}t)}{2}$$

$$= \frac{\pi(3 + 2)}{2}$$

$$= \frac{5\pi}{2} = 7 \cdot 85 \text{ mm}.$$

The developed length of the strip may now be calculated as follows:

Developed length L
$$\begin{aligned} &= NO + PO + PQ + QR + RS \\ &= 25 + 7 \cdot 85 + 54 + 7 \cdot 85 + 25 \\ &= 119 \cdot 7 \text{ mm}. \end{aligned}$$

With a little practice it is not difficult to apply the above technique to many types of bending jobs and developments, but the greatest of care must be exercised in all the calculations, which should be checked before the actual development is made.

Metal cutting
Provided the metal is not too thick, straight and bent snips are used to bring the sheet metal to the profile required. The technique underlying

Hand Processes

the use of straight snips is not unlike that adopted when using a hacksaw, and is simply illustrated in Fig. 2.92A. Straight cuts only are taken, so chosen as to leave the minimum amount of metal for filing,

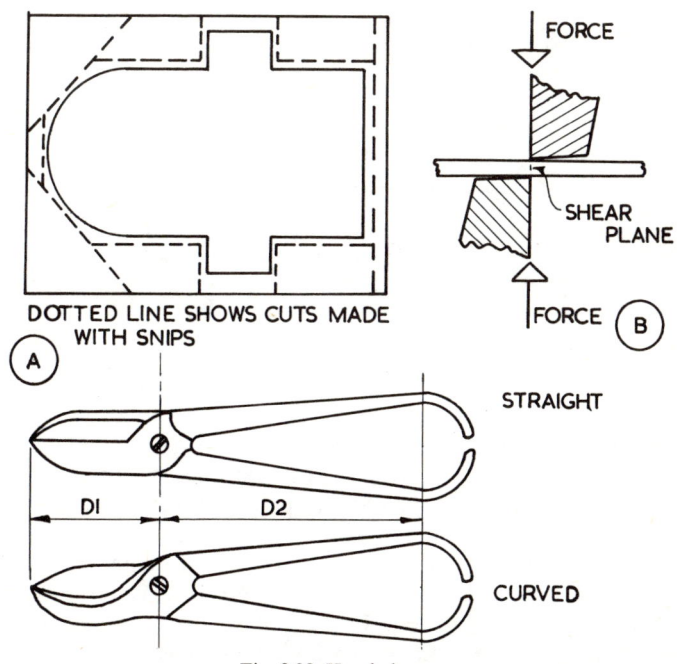

Fig. 2.92 Hand shears

and it is as well to appreciate at this point, that the sheet metal is *sheared* and not cut. This is clearly shown in Fig. 2.92B. The force exerted by the blades causes a shear force to act upon the sheet metal; this force is in excess of that applied at the handles because of the leverage provided by the difference in length between D1 and D2. This is a good example of the use of the lever principle in order to increase the value of a force provided by the muscular efforts of the operator. It is possible to cut a curved line using bent snips but some practice is needed before satisfactory results can be expected. The correct name for these hand cutting tools is *hand shears* but they are commonly referred to as *snips*.

Bench shearing machines

For cutting small quantities of sheet metal not thicker than about 5 mm, a bench shearing machine is invaluable. A typical machine is illustrated in Fig. 2.93, and it may be noted that the shearing principle is identical to that of hand shears, except that a much greater force may be applied to the blades because of the leverage provided by the long handle. Large sheets of metal will require power from an electric motor, and such metal-cutting machines are referred to as *guillotines*.

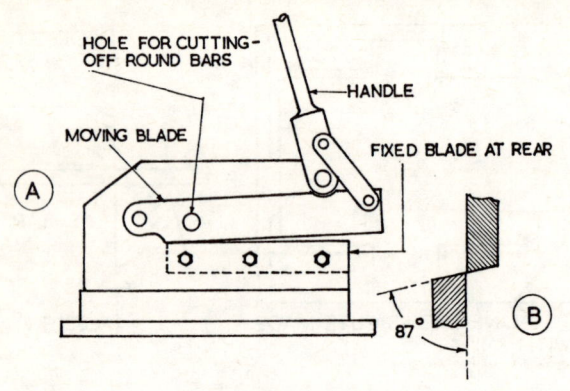

Fig. 2.93 Bench shears

Safety precautions in sheet metalwork

Perhaps the greatest hazard present when working sheet metal, is the possibility of cutting oneself on the sharp edges that are always to be found on thin metal. If handling metal sheets, gloves must be worn, but it is not practical to wear gloves when carrying out many sheet metal operations, and it is necessary that the sheet metalworker adopts at all times, a safe method of working. It is essential also, to ensure that the finished product is free from sharp and dangerous edges, and it is

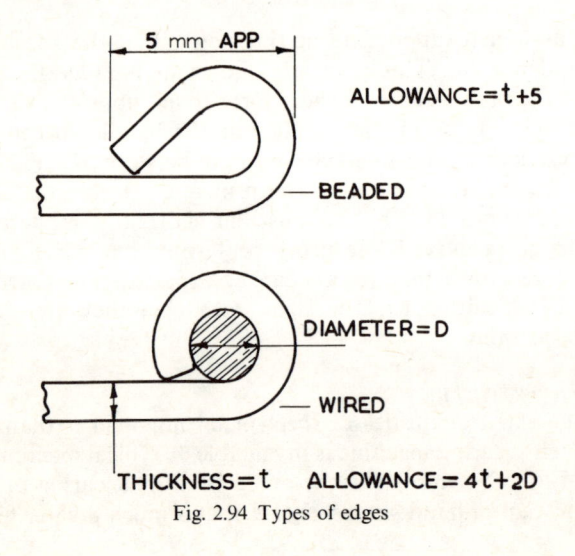

Fig. 2.94 Types of edges

Hand Processes

standard practice to *edge* all sheet metal components. Figure 2.94 shows the usual edges provided, and it may be noted that these edges not only make the component safer to handle, but provide also additional strength. At the same time it will be worthwhile having a look at the more common methods of joining sheet metal components, and

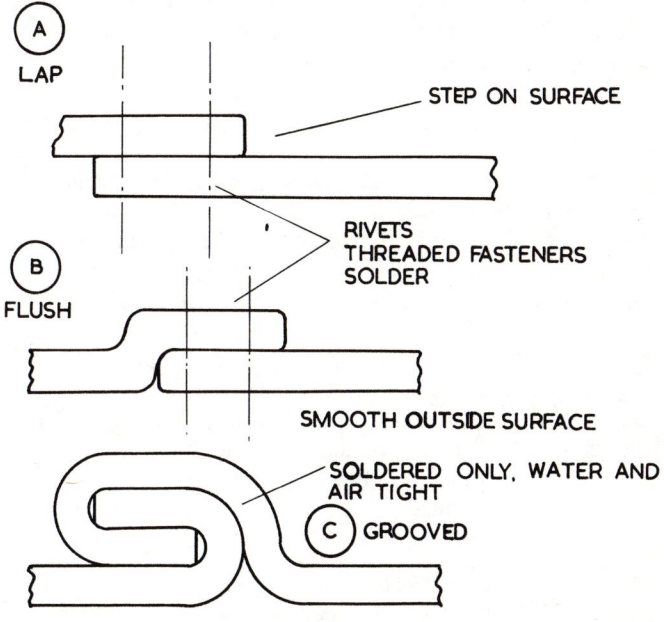

Fig. 2.95 Sheet metal joints

these are simply illustrated in Fig. 2.95. We shall see, later on in this volume, the correct techniques to apply when making the joints shown in the diagram.

Chapter 2 Assimilation Exercises

1. Make neat sketches of four marking-out tools used in an engineering workshop, and state the care that should be exercised in their use and storage.
2. Explain the importance of the following additional equipment used when marking out:
    (i) surface plate,
    (ii) pair of Vee blocks,
    (iii) scribing block.
3. Explain why several types of hacksaw blades are needed.

4. Make neat sketches, showing typical applications of *three* different types of hand files, and state the material from which the files are made.
5. Outline the safety precautions to be taken when filing and cold chiselling.
6. Sketch the tools required to cut an external thread on a steel pipe of 50-mm outside diameter at the bench.
7. Name three different hand powered tools available, sketching typical examples of their use.
8. Outline the precautions to be taken when using electrical hand drillers in a large workshop.
9. Outline the *two* main techniques adopted when marking-out engineering components.
10. Show three methods of determining a linear dimension, using line standards.
11. Explain why it is difficult to work to close limits of size using line standards.
12. Make a simple sketch of a Vernier caliper reading 15·48 mm.
13. Make a neat sketch of a Vernier height gauge reading 48·84 mm.
14. Make a neat sketch of a Vernier protractor reading 27° 45′.
15. State the pitch of a metric micrometer leadscrew, the number of divisions on the thimble, and sketch a reading of 21·87 mm.
16. Sketch micrometer instruments that can be used to measure the following:

    (i) large diameter bores,
    (ii) depths of holes.

17. Explain the difference between measurement by comparison and direct measurement.
18. State the difference between a plunger and lever-type comparator.
19. Explain what is meant by a 'balanced' dial.
20. Explain why several grades of slip gauge sets are necessary in engineering manufacture.
21. Sketch typical drilling jobs for each of the following machines:

    (i) Sensitive driller,
    (ii) pillar driller,
    (iii) radial driller.

22. Outline the precautions to be taken when drilling large diameter holes in thin metal sheet.
23. Outline the correct technique when inserting and removing a 25-mm-diameter drill at a pillar driller.
24. Sketch the point of a standard twist drill, showing the main features.

Hand Processes

25 Outline the technique to be adopted when reaming 20-mm-diameter holes.
26 A 12-mm-diameter hole is to be drilled in an aluminium casting at a cutting speed of 70 m/min. Calculate the rev/min for this operation.
27 Sketch typical examples of the following machining operations:

    (i) spotfacing,
    (ii) countersinking,
    (iii) counterboring.

28 Outline the precautions to be taken when drilling plastics materials.
29 Explain the circumstances that make it necessary to calculate the bending allowance when making a component from sheet metal.
30 Sketch four typical tools used for sheet metalworking operations.

3
MACHINE TOOLS

Objectives—The principles and applications of:
1. Machine tools
2. Generating and forming
3. The main features of a centre lathe
4. Methods of workholding and toolholding
5. Turning techniques
6. Screw cutting
7. Taper turning
8. Safety precautions
9. Main features of the shaping machine
10. Shaping techniques
11. Cutting tool theory
12. Single point cutting tools
13. Cutting fluids

3.1 The need for machine tools

A machine tool can be described as a powerdriven device which will produce a desired surface. It is important to appreciate that all engineering components are of geometrical shape or a combination of geometrical surfaces. This is illustrated in Fig. 3.1, which shows a typical engineering component. If we breakdown this component into separate parts, we will find that in each case the part is of geometrical shape, having a geometrical surface. For example part A has a round or cylindrical external surface, together with a bore or hole which is an internal cylindrical surface. Part B has plane surfaces, together with part cylindrical surfaces which are the internal radii of the slots shown as E. The part C also has flat or plane surfaces, two of which are at an angle. It would be possible to prefabricate this component at the bench, that is to say make it by joining the separate parts by welding A to C and then welding C to B. Clearly this would be a difficult, time-consuming and costly process, with the finished job unlikely to possess

Machine Tools 113

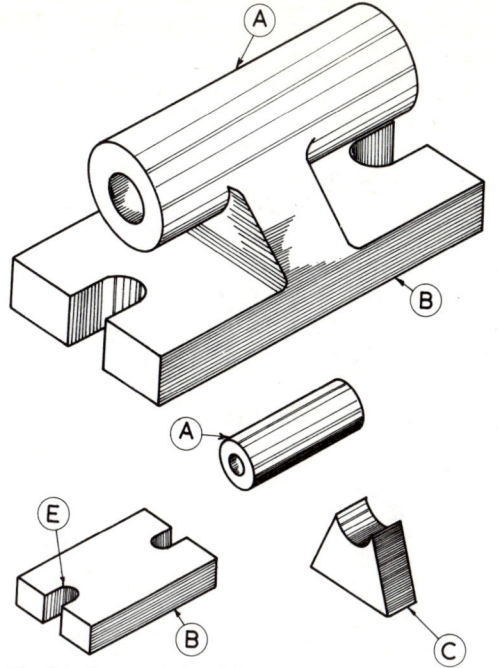

Fig. 3.1 Geometric breakdown of engineering component

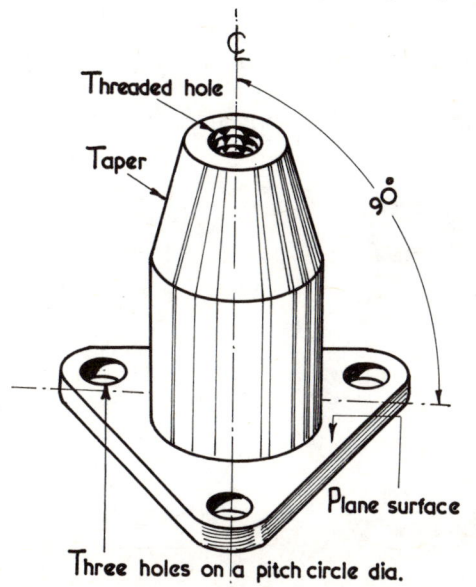

Fig. 3.2 Dimensions controlled by centre-line

accurate alignment or true geometric surfaces. For example it would be a most difficult task to file the base to a reasonable degree of flatness; what is needed is a machine tool, capable of rapidly and accurately producing the geometrical surface required; in this case a *shaping machine*. There are also other geometrical surfaces very difficult to produce by hand methods. Figure 3.2 shows an engineering component comprising plane, cylindrical, conical and helical surfaces. Conical surfaces are more commonly known as *tapers*, while a helical surface or form, is evidently, a screw thread. All these surfaces are extremely difficult to produce using hand methods, with the exception perhaps of the screw thread, which can be produced using a set of taps, provided the diameter is not excessive.

3.2 Generating and forming

It is true to state that most of the linear dimensions of the component shown in Fig. 3.2 evolve around the centre-line of the job. For example the three holes must be equidistant from this centre-line, as must the outside diameters of the cylindrical and tapered surfaces. The axis or centre-line of the threaded hole must be that of the centre-line of the job, which can be considered as an imaginary datum. It is difficult enough to work from a datum face, as we have seen in the previous chapter, but to expect accuracy when working from a non-existent or imaginary line is asking the impossible, especially when the component is to be produced at the bench. What is needed here is a machine tool, capable of rapidly and accurately producing jobs using the centre-line as a datum, in other words a *centre lathe*.

Before considering the basic principles and use of a centre lathe, it

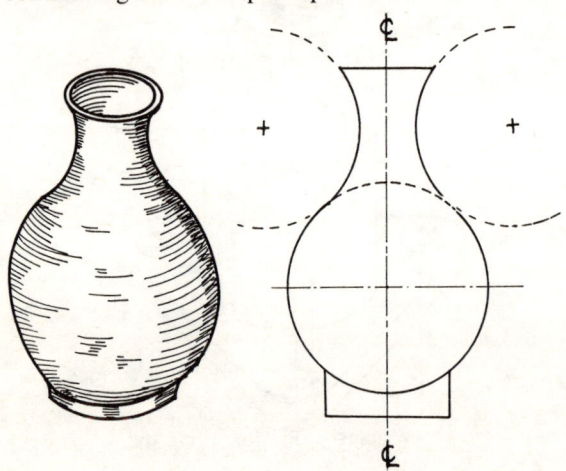

Fig. 3.3 Geometry of clay utensil

Machine Tools

may prove of interest to consider the problem that faced the potters who were concerned with the manufacture of clay utensils several thousands of years ago. Yet the shaping of clay is relatively easy, for it is a soft and pliable material, easily shaped by the potter's hand. We can regard the manufacture of pottery as one of the first engineering accomplishments of mankind, in which a primitive type of machine tool was used—the potter's wheel. Let us consider the problems involved when making or shaping the utensil or jug shown in Fig. 3.3. The shape is evidently geometrical, consisting of a combination of spherical and cylindrical surfaces, and the machining of this object from a solid block of aluminium, would be a good test of the machining skill of any engineering craftsmen. A very simple expedient, however, was adopted by the early potters, and this principle is still the basis of most machine-tool design and construction. This principle is known as *generating*, whereby a desired geometrical surface is achieved by the combination or summation of simple engineering movements or geometric motions. Perhaps the method used by the early potters will help to make this a little clearer.

The clay was first rolled into cylindrical rods between the palms of the hands. A series of circles was then made up, their diameters varying

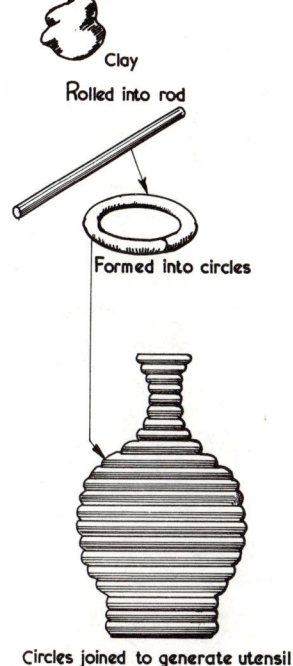

Fig. 3.4 Generating clay utensil by hand

according to the shape or profile of the utensil required. These circles were placed in position, smoothed over and the utensil baked or fired to provide the necessary hardness. This principle is shown in Fig. 3.4, and it may be appreciated that cylindrical, conical and helical forms can readily be produced in the manner shown. Engineers of the 16th century made wrought iron gun barrels using a similar basic process. The procedure involved wrapping wrought iron wire around a cylindrical former, and welding the wire at the outside, producing an iron cylinder. Screw threads were also produced in a similar manner. A length of soft wire would be wrapped around the bar to be threaded, a simple template used to ensure correct spacing of the wire, thus giving a constant pitch. A diamond pointed chisel would then be used alongside the wire, and the thread started in this manner.

Ingenious though these methods were in producing geometrical shapes or surfaces, they take no account of the fact that the accuracy of the form produced is governed by the relationship of the outside surface to the centre-line of the work. This important fact must have been appreciated by the early potters, for it led to the introduction of the potter's wheel. It has been said that the lathe is the father of all machine tools but this is not strictly true. The potter's wheel preceded the lathe, and is almost certainly the earliest example of generating by

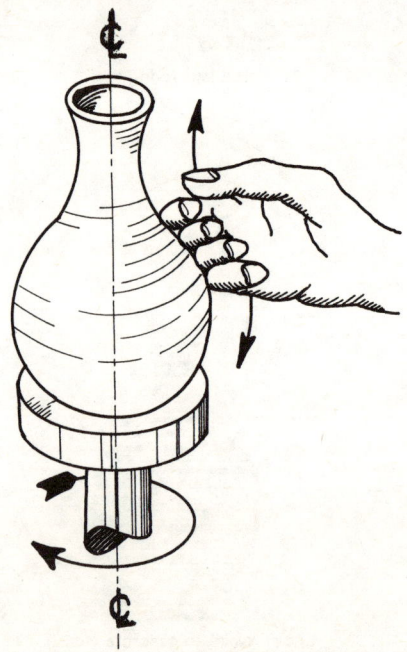

Fig. 3.5 Generating on a potter's wheel

Machine Tools

semi-mechanised methods. The principle of the potter's wheel is still embodied in most of our present-day machine tools, and is worthy of closer inspection. Figure 3.5 illustrates the principle involved. Note that the utensil revolves about its centre-line, and that the movement of the potter's hand, at the points of contact, produces a surface that is equidistant from the centre-line and thus geometrically accurate. It is essential, of course, that the wheel revolves, and that the centre-line of the pot is in alignment with the axis of rotation. These are precisely the conditions that are required in a centre lathe, except that the hand of the potter is replaced by a cutting tool that removes metal by a shearing action.

3.3 The centre lathe

The primary purpose of a centre lathe is the production of external and internal cylindrical surfaces, that is to say turning round bars and machining holes or bores. In order to machine a truly cylindrical sur-

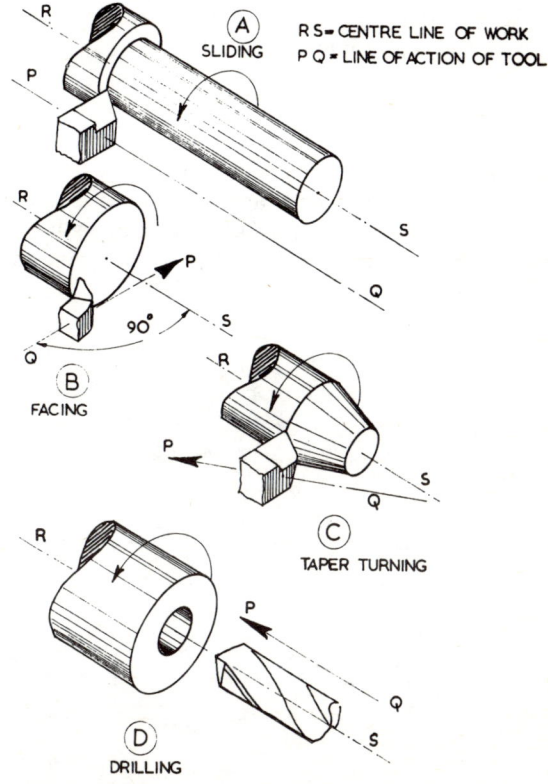

Fig. 3.6 Essential movements of the centre lathe

face it is necessary to arrange that the cutting tool moves along a path parallel to the rotating workpiece in both the horizontal and vertical planes. This is shown in Fig. 3.6A, and a standard straight-edged single-point lathe tool would be quite suitable for this operation. In the diagram, RS represents the centre-line of the work, while PQ is the path of the tool. Any deviation from parallelism of these two lines of action will result in a surface which is not truly cylindrical.

If the tool is able to move at precisely 90° to the centre-line of the work, a flat or plane surface is generated at the end of the cylinder or bar; thus with both ends machined a true cylinder results. Figure 3.6B shows the principle of *facing*, or machining a face at 90° to the centre-line. If now, a conical surface or taper is required, it is necessary to produce an out-of-parallelism effect between the lines RS and PQ, as shown in Fig. 3.6C, and this out-of-parallelism in the horizontal plane is achieved by changing the direction of the line PQ, or the path of tool. In practical terms this means moving or setting the compound side. At Fig. 3.6D we see the method of drilling holes in the workpiece, by feeding a twist drill into the rotating work; the drill centre-line coincident with the centre-line of the lathe.

Simple turning

Figure 3.7A shows the basic technique of simple turning, for example the production of a wooden leg for a table. The technique shown in the diagram is described as wood turning, and any cylindrical surface produced in like manner, is a *turned* surface, irrespective of the material or the type of machine used. Note that in the diagram, the path or line of

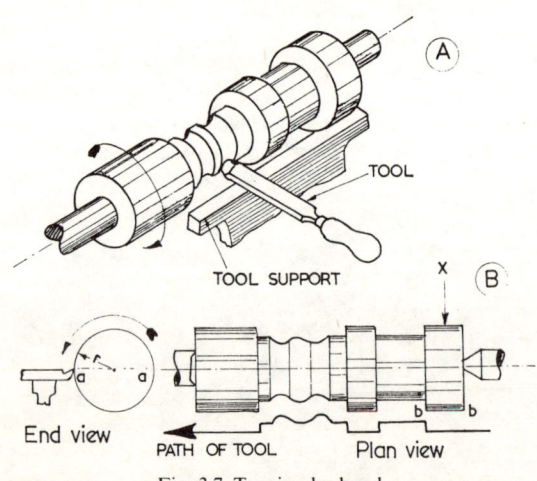

Fig. 3.7 Turning by hand

Machine Tools

action of the cutting tool, (in this case a wood chisel), is under the control of the craftsman. As long as the chisel rests on the support provided, the profile of the turned component is a replica of the profile of the chisel path. This is shown on the plan view in Fig. 3.7B. The geometrical accuracy of the turned profile must be correct, as can be seen from the end elevation, because the surface produced by the cutting action of the chisel is at a fixed distance from the centre-line of the work. In other words, if the chisel is held in one position, a circular shape or diameter of radius R is turned; thus movement of the chisel along the direction of the work centre generates a turned surface.

Let us now consider that portion of the turned table leg indicated as X in the diagram at Fig. 3.7B. If this part of the leg is to be truly cylindrical it is essential that the chisel point maintains a constant distance from the centre-line as the tool point traverses from *b* to *b*, as shown on the diagram. Clearly this accuracy is outside the ability of even the most highly skilled craftsman, and can be achieved only by some means of guiding the cutting tool.

Guiding the cutting tool
Figure 3.8 shows a view of a metal bar turned at a centre lathe, as seen from the tailstock. The centre of the work is shown by the dot O, and

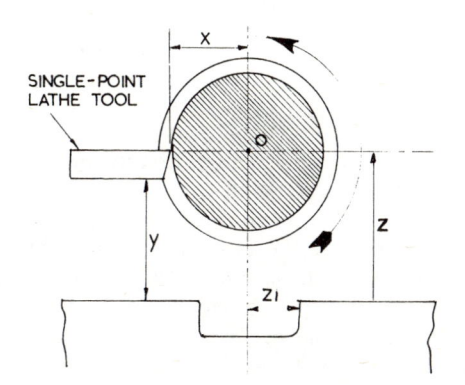

Fig. 3.8 Geometric requirements to generate a cylinder

is, therefore, the centre-line of the lathe. This centre-line is the datum of the lathe and is a fixed datum, that is to say, provided the workpiece is firmly held and rotated, there is no change in the position of its centre of rotation, hence the distances shown as Z and z_1 in the diagram are fixed or constant. If a truly cylindrical surface is to be generated as the tool traverses a path parallel to the lathe centre-line, the distance shown as X and Y in Fig. 3.8 must remain constant. This constancy, or parellelism of the cutting tool with the centre-line of the workpiece, in both the vertical and horizontal planes is achieved by the use of *guide-ways* and *bearing surfaces* machined on the body or bed of the centre lathe.

The lathe bed
The purpose of a lathe bed is to provide guideways and bearing surfaces to ensure that the cutting tool follows a straight and true path; also to give support for the main elements that make up a complete centre lathe. Figure 3.9A shows a simple pictorial view of the bed of a

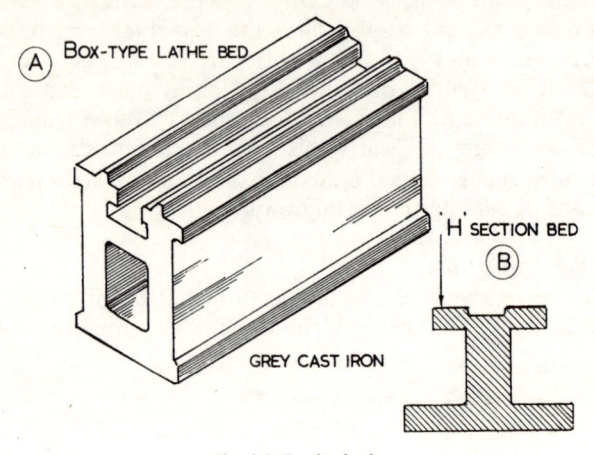

Fig. 3.9 Lathe bed

centre lathe. Note the box-like structure, first introduced by the eminent British engineer, Sir Joseph Whitworth; this innovation proved a great advance over the H section shown in Fig. 3.9B. The adoption of the box-like bed meant that the lathe was better able to resist the bending or deflection set up during the cutting action, especially when deep or heavy cuts need to be taken on hard or tough metals. A suitable material for the lathe bed would be grey cast iron. A large mass of metal is needed which can be cast in the fairly complicated shape of the box-like structure. A good even-wearing, low-friction sliding surface will be produced by the free graphite which is always present in the structure of grey cast iron. This graphite tends also to absorb vibrations and, added to the fact that grey cast iron is one of the cheapest metals available to engineers, it is clear that grey cast iron is ideally suited for the manufacture of lathe beds.

Guideways and bearing surfaces
The purpose of guideways and bearing surfaces is to provide straight and true paths for the saddle and tailstock. If we reconsider the action

Machine Tools 121

of generating a cylindrical surface on a centre lathe, we may recall that it is necessary to ensure that the tool moves along a path parallel in two planes to the centre-line of the lathe. This problem of sliding one part or machine tool element in accurate alignment with respect to another part, is common to most machine tools, and the method adopted is to use sliding faces or guideways. Figure 3.10 shows a section

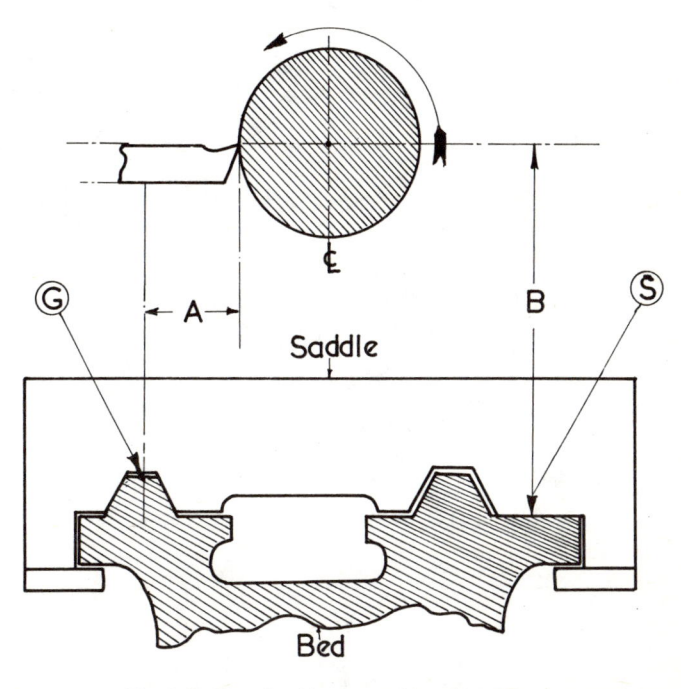

Fig. 3.10 Use of guideways and bearing surfaces

through a typical lathe bed, with the saddle in position. Note carefully that the saddle locates only on the guideway G and the bearing surface S; there is clearance everywhere else. We may consider the tool as part of the saddle, thus the guideway G controls the constancy of the dimension A, in other words parallelism of the path of the tool in the horizontal plane. Similarly the bearing surface shown as S in the diagram ensures that the tool follows a true path in the vertical plane. This principle is more clearly shown in the simple pictorial view in Fig. 3.11, which includes also the method of guiding the tailstock. Note that for both saddle and tailstock, the distance between guideway and tailstock is at a maximum, and this promotes greater stability. When cutting metal at a centre lathe the cutting force acts downwards and this is a further advantage when inverted V guideways are used. Figure 3.12 shows that

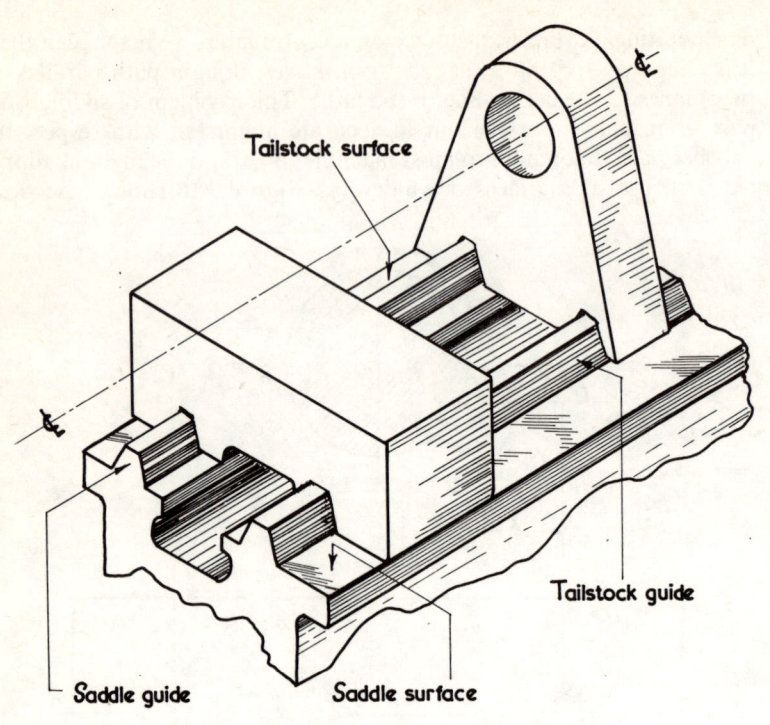

Fig. 3.11 Saddle and tailstock guiding

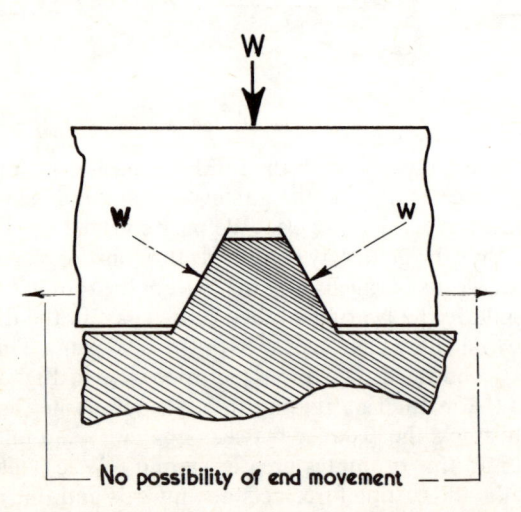

Fig. 3.12 Advantages of the inverted Vee guideway

Machine Tools 123

the force W has the effect of giving closer, and thus more accurate bearing contact on the guideway, together with the fact that irrespective of the amount of wear that may take place, no slackness or sideways movement is possible. Much the same principle is used for both the compound slide and cross slide. The purpose of the cross slide is to provide movement of the tool at precisely 90° to the lathe centre-line, allowing the ends of bars to be faced-off square, while the compound slide allows the tool to traverse at any required angle to the lathe centre-line, allowing short or fast tapers to be machined. The principle of the dovetail slide is simply shown in Fig. 3.13, where it

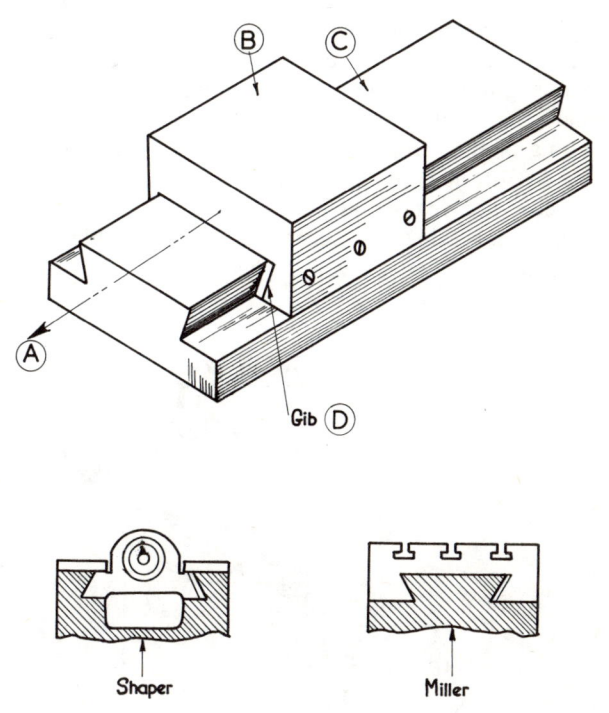

Fig. 3.13 Principle and application of the dovetail slide

may be seen that the only movement possible is along the direction of arrow A; because of the angled sides no sideways or upward movements are possible. The three adjusting screws at the side are used to exert light pressure on the gib strip shown as D, in order to remove any undue slackness and also take up any wear that may occur. This principle of the dovetail slide is widely used in machine-tool construction; examples are to be found in shaping, milling, planing and slotting machines.

Lead screw and indexing dial

Because the dovetail slide is widely used to bring the cutting tool to the required depth in most machine tools, it is important that this basic principle underlying dimensional control be fully understood. Reference to Fig. 3.14 will make the following clear. The sliding member receives its motion through the rotation of a lead screw which is constrained within the fixed member. Thus for *one* complete revolution of the lead screw, the sliding member will have a linear movement equal to the pitch of the lead-screw thread. Note that this thread is not used for tightening purposes, but for the transmission of motion, and will have a square or Acme form. If now a graduated dial is fixed to the lead screw as shown in the diagram, it is a simple matter to control the distanc moved by the sliding member, providing note is taken of the number of graduations moved in relation to a fixed line on the dovetail slide. The dial is usually referred to as an *indexing dial*, and represents one of the most useful and valuable devices present on a machine tool. Cutting-tool movements of less than two hundredths of a millimetre are readily achieved, as we shall see later on in this Unit, although whether a centre lathe is capable of removing such small amounts of metal will depend on the accuracy and condition of the centre lathe.

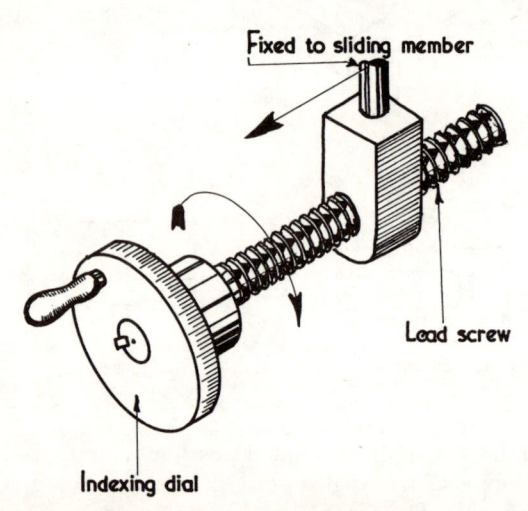

Fig. 3.14 Lead screw and indexing dial

One simple example will serve to illustrate the principle underlying the use of an indexing dial. Let us assume that the lead screw shown in Fig. 3.14 has a single-start square thread of 2 mm pitch, and an indexing dial with 100 divisions.

Machine Tools 125

Distance moved by tool for *one* complete turn = 2 mm

Distance moved by tool for *one* division = $\dfrac{2}{100}$

= 0·02 mm.

If therefore, it is required to move the cutting tool a distance of 1·8 mm, then the indexing, or number of divisions to be moved is calculated as follows:

Number of divisions = $\dfrac{\text{Distance required}}{0·02}$

= $\dfrac{1·8}{0·02}$ = 90 divisions.

Lathe toolpost
The purpose of a lathe toolpost is to position and clamp the cutting tool. It is important that the tool point coincides with the lathe centre-line, and thin strips of metal may be used to bring the tool to the correct height. There are several types of lathe toolpost, and Fig. 3.15

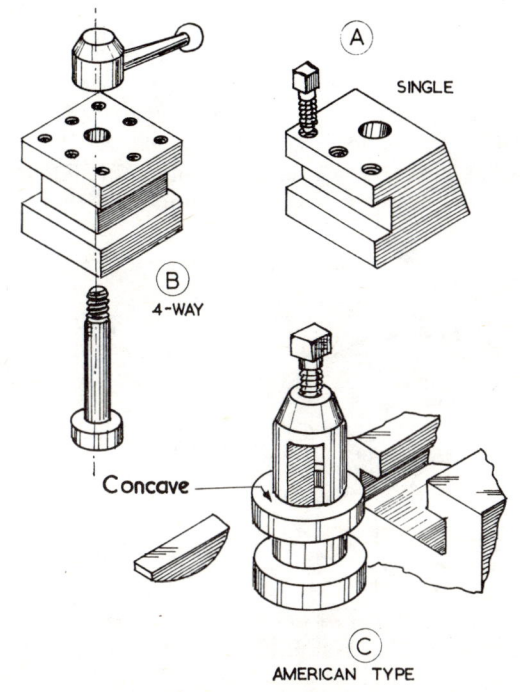

Fig. 3.15 Lathe toolposts

shows the three main examples in general use. At A we see a simple one-way toolpost, capable of holding one tool only; at B we see an improved version called a four-way toolpost. This toolpost saves considerable time, because four different tools may be held and presented to the work by the simple expedient of indexing or rotating the toolpost around the central pin, and then clamping it in the desired position. Figure 3.15C shows an American-type toolpost, more suitable for the smaller type lathe. Small variations in the height of the tool are achieved by tilting the tool through a small angle and tightening the locking screw, which holds both tool and toolpost in the required position. Irrespective of the type of toolpost used, it is vital that the minimum length of the tool protrudes from the toolpost; this promotes rigidity of the set-up, and makes possible a higher degree of accuracy with a better surface finish.

The compound slide
The compound slide serves two purposes:

(i) It provides support for the toolpost.
(ii) It makes possible movement of the cutting tool at an angle to the centre-line of the lathe.

We have seen that a truly cylindrical surface can be turned only when the tool traverses a path parallel with the centre-line of the lathe; thus if a conical or tapered surface is to be machined, it is necessary to alter the path of the cutting tool so that it is *not* parallel with the lathe centre-line. This is precisely what the compound slide does, and Fig. 3.16 shows a simple plan view of a compound slide set to turn a tapered component.

It may be seen from the diagram that the path of the tool lies along

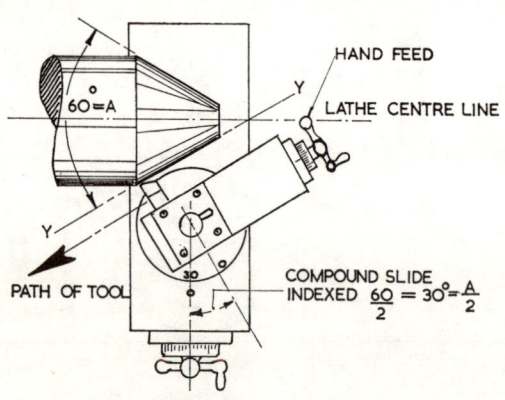

Fig. 3.16 Plan view of compound slide

Machine Tools

the line shown as YY in the diagram, movement being obtained by hand rotation of the handle adjacent to the indexing dial. The length of traverse is limited by the dovetail slide, and this means that fairly short tapers only can be machined by the technique of swivelling the compound slide. Note that the included angle of taper, shown as A in the diagram, is *twice* the angle through which the compound slide is swivelled; thus if an included angle of 60° is required on a component, the compound slide is swivelled through 30° and securely locked in position. A circular scale on the base, graduated in degrees allows easy setting, with two tightening nuts to lock the compound slide at the desired angle.

The cross slide
The purpose of the cross slide is to provide support for the compound slide, and movement of the cutting tool at right angles to the lathe centre-line. Figure 3.17 shows a simple plan view of the cross slide, the

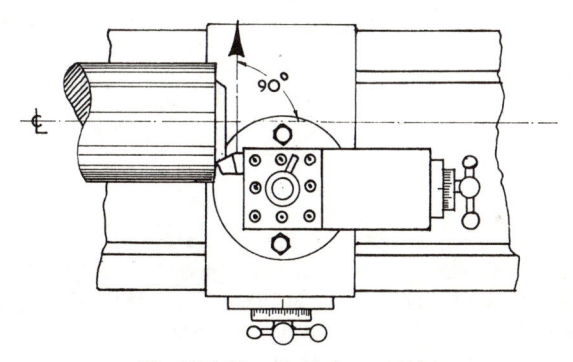

Fig. 3.17 Plan view of cross slide

operation in progress is the facing-off of a metal bar. Hand traverse may be used, although most lathes are fitted with a gearing arrangement by which the tool is automatically fed either towards or away from the centre of the rotating bar; these operations are called facing inwards and facing outwards.

The saddle
It may be seen from Fig. 3.11 that the saddle fits directly onto the lathe bed, thus the accuracy of the path of the tool is dependent on the alignment of the guiding and bearing surfaces machined on the underside of the saddle and the top of the lathe bed. The saddle provides support and guidance for the cross slide, and also accommodates all the gearing and mechanisms needed for automatic traverse of cross slide and saddle.

The tailstock
The tailstock of a lathe serves a dual purpose. Firstly it is used to support work which is being turned between centres; secondly it is used as a tool station for drills and reamers. It is essential that the centre-line of the tailstock be a continuation of the centre-line of the lathe at any position of the tailstock along the bed, and for this reason the tailstock is provided with its own guideways and bearing surfaces as may be seen by reference back to Fig. 3.11. Two views of a lathe tailstock are given in Fig. 3.18. Note that the distance shown as x

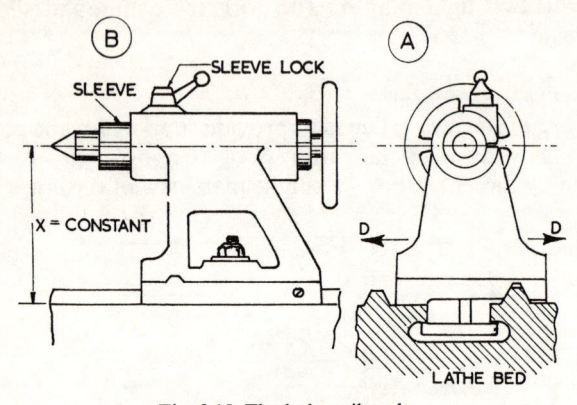

Fig. 3.18 The lathe tailstock

is constant—this is the height from the lathe bed to the tailstock centre-line—while small movements are possible in the direction of arrows D, as shown in the end elevation in Fig. 3.18B. This operation is called off-setting the tailstock, and has the effect of making the centre-line of a bar held between centres, non-parallel with the lathe centre. We shall see later on in this Unit how this technique of off-setting is used to machine slow tapers using automatic traverse of the saddle.

The taper principle is used for the location of drills or reamers in the tailstock, as may be seen in the sectional view shown in Fig. 3.19. In order to feed a twist drill into the revolving work a forward movement is needed, provided by rotation of the hand wheel; this transfers rotary motion into linear motion through a square thread as shown in the diagram. Note how rotation of the hand wheel in an anti-clockwise direction results in ejection of the centre.

The head stock
So far we have dealt with the principles and techniques adopted to ensure that the cutting tool moves along a precise path parallel to the lathe centre, which may be considered as an imaginary datum. We need now to consider the techniques adopted to achieve rotation of the

Machine Tools

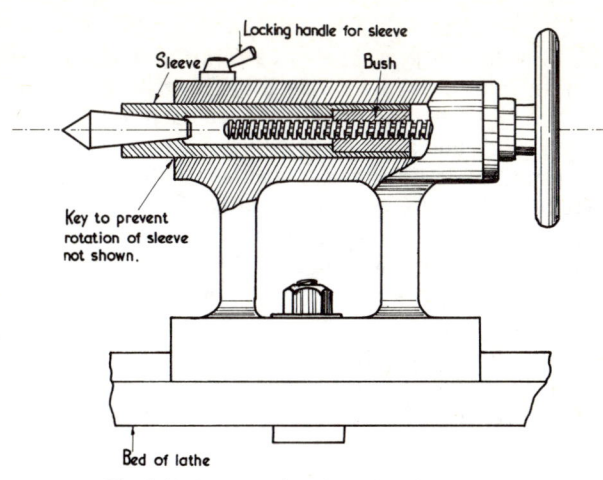

Fig. 3.19 Scrap section through lathe tailstock

workpiece, and the means of holding the various types of jobs likely to require machining at the centre lathe. We may consider the headstock as the support for the lathe spindle, and the means for ensuring true alignment of the rotating spindle at the number of revolutions required. Figure 3.20 shows an early-type open headstock with provision for lubrication of the phosphor-bronze bearings which support the spindle. Note the three-step coned pulley at the spindle centre, giving a choice of three different speeds by the simple expedient of changing the belt position. The spindle nose is threaded to take the work-holding

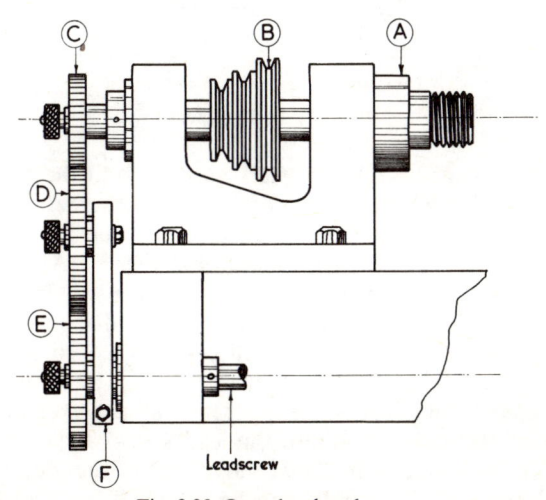

Fig. 3.20 Open headstock

device, while at the rear of the spindle provision is made for the interchanging of a driving gear, shown as C in the diagram. This gear meshes with an idling gear and thus transfers motion to the driven gear E, which is keyed to, and rotates the lead screw. These gears are known as spur gears, and a set would be supplied with the centre lathe, allowing a range of feeding rates together with a range of screw threads that could be cut on the centre lathe.

Although modern lathes no longer make use of open head stocks and a set of gears, the principle of gear changing using a *quadrant* is both important and interesting, and well worthy of further consideration. The quadrant shown as F in Fig. 3.20, can be rotated and locked at any position about the centre of line of the lead screw. The idler D can be set anywhere along the axis of the slot in the quadrant, which is more fully shown in Fig. 3.21. The procedure would be, having

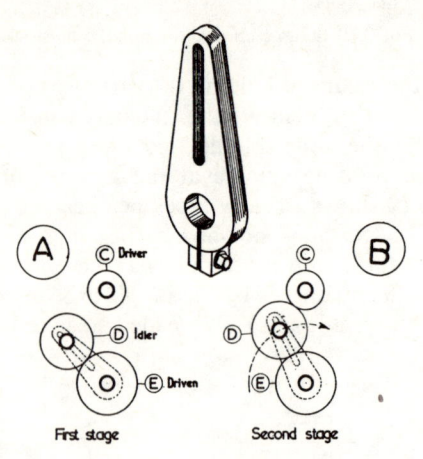

Fig. 3.21 Lathe quadrant

selected the driving and driven gears needed, to fit the driving gear to the lathe spindle, and the driven gear to the lead screw. An idler is then selected and set in the slot of the quadrant so that it may be brought into mesh with the driven gear as shown in Fig. 3.21A. This is the first stage of the operation, and the second stage consists in swivelling the quadrant in a clockwise direction so that the idler now meshes with the driving gear Fig. 3.21B. With all gears tightened in position, together with the tightening screw at the base of the quadrant, the drive is now complete from spindle to lead screw. In this way, motion is transmitted from the lathe spindle to the lead screw, and with the appropriate gears in position, the linear movement of the tool in relation to the revolutions of the spindle can be obtained, making possible the

Machine Tools

screw-cutting of a wide range of threads. It needs to be remembered however, that modern lathes have completely enclosed head stocks, and the required feeds or threads are directly obtained by the movement or positioning of external levers or handles.

3.4 Methods of work-holding

The method adopted for holding the workpiece depends largely on the type of job and the complexity of the turning. *Three-jaw chucks* are widely used because of the self-centring property they possess; that is to say their ability to bring the centre-line of the chucked workpiece co-axial with the lathe centre-line. Provided the workpiece is truly round, the three jaws will grip evenly with the job running true as shown in Fig. 3.22A. It is important to remember that bright mild steel

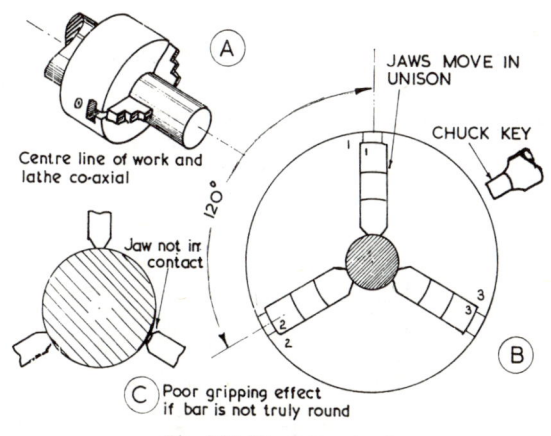

Fig. 3.22 Three-jaw chuck

or extruded bars only, should be held in a three-jaw chuck. As the jaws move in unison towards the centre, an out-of-round section prevents all the jaws from making proper contact, and a poor gripping action with risk of damage to the chuck jaws results. The effect of gripping a cast or forged bar is illustrated in Fig. 3.22C; both casting and forging are unlikely to produce bars having a truly round section. The workholding ability of a three-jaw chuck is greatly increased if a set of interchangeable jaws is utilised. Figure 3.23A shows a large diameter component held in a three-jaw chuck using standard jaws, the component held or gripped on its inside diameter. At Fig. 3.23B we see the same component gripped using a set of interchangeable jaws; in this instance the component is gripped on its outside diameter, an operation not possible using standard jaws.

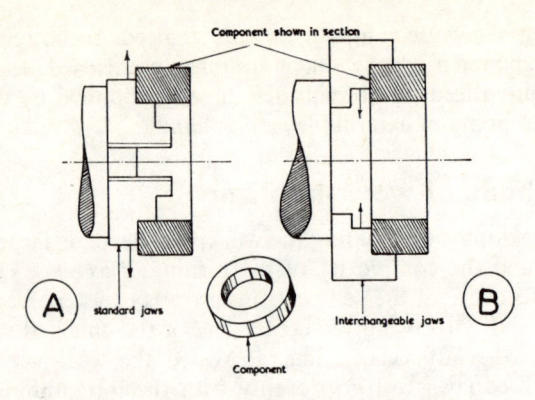

Fig. 3.23 Use of jaws for large diameters

Four-jaw chucks
Figure 3.24 shows a front view of a four-jaw chuck. These chucks are of much heavier construction than three-jaw chucks, and are very suitable for gripping forged or cast bars, rectangular or square sections, or most odd-shaped sections. The versatility of the four-jaw chuck is due to

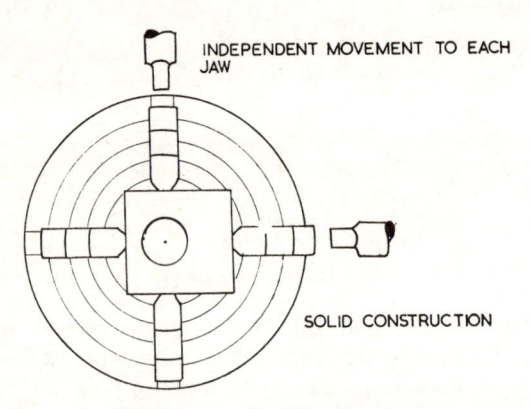

Fig. 3.24 Four-jaw chuck

the fact that each jaw has independent movement. A chuck key must be used to move each jaw in turn; thus a four-jaw chuck is *not* self-centring, and considerable skill is needed to adjust a round bar so that its centre is co-axial with that of the centre lathe. It must be remembered however, that because each jaw has an independent action, the gripping force is greatly superior to that of a three-jaw chuck, and care needs to be taken not to damage or distort workpieces which are held in a four-jaw chuck.

Machine Tools 133

Holding between centres

The use of centres overcomes the limitations of a three-jaw chuck in its inability to re-chuck a job accurately. The great advantage offered by the use of centres is the ability to remove and replace the workpiece without any loss of accuracy; that is to say, the centre-line of the workpiece always regains its correct alignment with the lathe centre. A further advantage is the fact that further operations can be carried out using the same location as that used when turning. Figure 3.25A shows

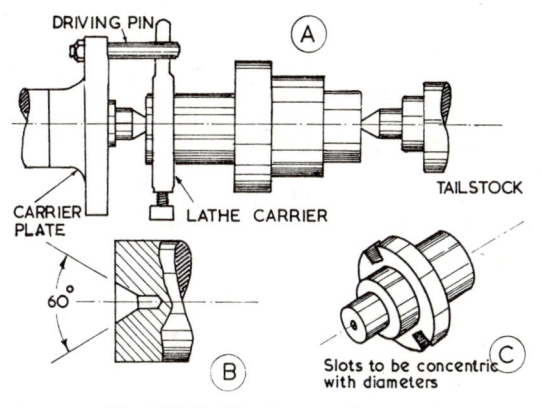

Fig. 3.25 Holding between centres

a front elevation of the technique of turning between centres, while at B we see a sectional view of the profile produced by a centre drill. At C we see a component requiring two milled slots which must be equidistant from the centre of the workpiece. With centres drilled at both ends, the component would be turned using these centres as locations,

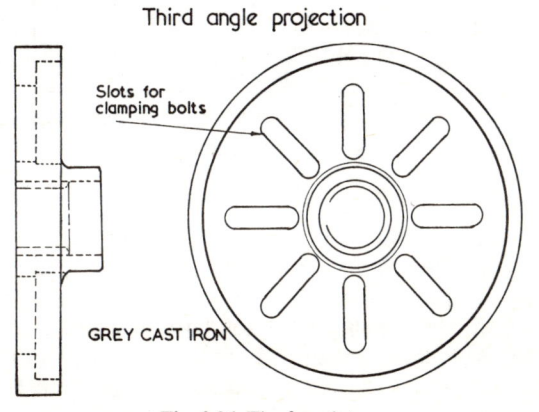

Fig. 3.26 The faceplate

and then passed to a milling machine where the two slots would be milled, using the centres as locations once again. In this way the concentricity of the milled slots with the turned diameters would be assured, because both operations have been carried out using the same datum or principle of location.

The face plate

The purpose of a face plate is to provide a datum face or reference plate at 90° to the lathe centre-line. This means that a component with a fairly large surface area can be clamped to the surface plate, allowing it to be turned or bored. A typical face plate is shown in Fig. 3.26, and as can be seen is made from grey cast iron.

3.4.1 The modern centre lathe

The modern centre lathe is a good example of engineering design and manufacturing skill. Figure 3.27 shows the outline of a typical medium duty centre lathe with the main features named. Note the totally en-

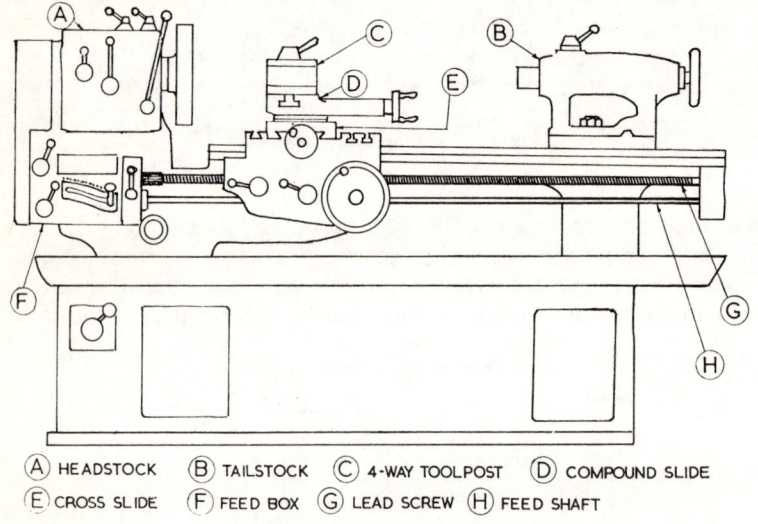

(A) HEADSTOCK (B) TAILSTOCK (C) 4-WAY TOOLPOST (D) COMPOUND SLIDE
(E) CROSS SLIDE (F) FEED BOX (G) LEAD SCREW (H) FEED SHAFT

Fig. 3.27 Modern centre lathe

closed headstock within which are heat-treated nickel–steel sliding gears providing eight spindle speeds. The quick-change feed box allows rapid selection of thirty-two different rates of feed (that is to say movement of the saddle along the bed of the lathe); there is also the instant choice of thirty-two different thread pitches. The bed of the lathe is of box section, diagonally braced to provide maximum strength; the guideways and bearing surfaces are precision ground, and may be hard-

Machine Tools

ened for longer life if required. Reference to the diagram shows that the lathe is fitted with a feed shaft, and this is used to provide movement of the saddle when taking sliding cuts along the lathe bed. The lead screw is used for thread cutting only, and is a good example of a precision-machined engineering component. The pitch error does not exceed 0·001 mm in any 100-mm length, and it is good practice to ensure that the lead screw is not allowed to rotate unless used for actual screw cutting.

Capacity of centre lathes
The capacity of a centre lathe refers to the size of work it can accommodate. It is a mistake to attempt large or heavy work on a small capacity centre lathe, and conversely it is wasteful to machine small components using a large capacity machine. The following items show how the capacity of a centre lathe is indicated.

Height of centres or swing over bed
This distance is shown as A in Fig. 3.28. The swing over bed is twice the height of centres, and indicates the maximum diameter of work that can be machined.

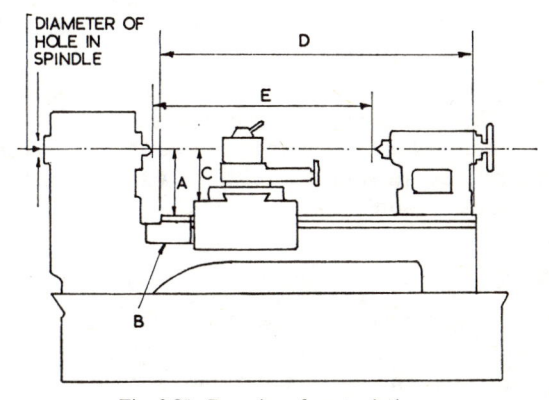

Fig. 3.28 Capacity of centre lathes

Gap bed
This indicates that the bed has a removable section, indicated as B in the diagram, and is just in front of the headstock. Removal of this section increases the swing of the lathe, permitting large diameter work to be machined.

Swing over saddle
This dimension is shown as C, and it may be seen that with the saddle in close proximity to the headstock, the effective swing of the lathe is

reduced, therefore the swing over the saddle is always less than the swing over bed.

Length of bed
This gives the total length of bed, shown in the diagram as distance D.

Distance between centres
Shown as E in the diagram, the distance between centres indicates the maximum length of work that can be machined between centres; this distance is always less than the length of bed.

Diameter of spindle hole
Hollow spindles allow long bars to have their ends machined by the simple expedient of passing the bar through the hollow spindle and gripping it with a three- or four-jaw chuck. The spindle diameter therefore, gives the maximum diameter of long bars that can be machined.

3.4.2 Spindle speeds and feeds

Lathes, like all machine tools, are very expensive pieces of equipment, and they are of practical value only when they are removing metal or producing work. The rate therefore, at which the metal is removed determines the output of the lathe, and it is a general rule that as much metal as possible will be removed in the minimum time. Large metal removal however, involves greater forces and increases the possibility of deflection of the workpiece, leading to dimensional inaccuracy and poor finish. For these reasons it is customary to divide most machining into separate operations:

1. **Roughing.**
2. **Finishing.**

The object of roughing is to achieve maximum metal removal in the minimum time, leaving a small amount of metal for the finishing operation. The object when finishing is to achieve the dimensional accuracy laid down, together with an acceptable surface finish; both operations more easily attained when small cuts only are taken, resulting in much reduced cutting forces. This principle, of course is identical to that adopted by the bench fitter, who uses a rough file for maximum metal removal, followed by a smooth file to bring the surface to the required accuracy and finish.

When turning at a centre lathe the rate of metal removal depends on,
   (i) the metal being machined,
  (ii) the revolutions of the workpiece,

Machine Tools 137

(iii) linear movement or feed of the tool,
(iv) depth of cut.

All metals have an optimum or ideal cutting speed which is usually stated in metres per minute. In the case of a round bar revolving in a centre lathe, the cutting speed is the speed of the metal in relation to

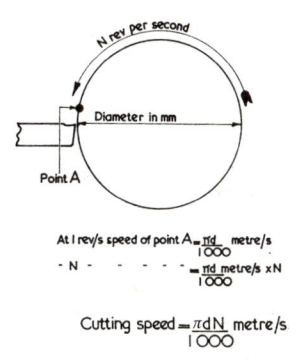

Fig. 3.29 Cutting speed of rotating bar

the cutting tool, and the principle and calculations are simply shown in Fig. 3.29. The cutting speed in metres per minute (m/min), can be expressed as:

$$CS = \frac{\pi dN}{1000}$$

where CS = cutting speed in metres per minute,
d = diameter of bar in millimetres,
N = revolutions per minute.

Provided then, that the cutting speed of a metal is known, it is not difficult to calculate the ideal or optimum revolutions at which to rotate the lathe spindle, using the formula:

$$N = \frac{1000\,CS}{\pi d}$$

Unfortunately, ideal conditions are seldom encountered when operating machine tools, and we have stated already that it is customary to separate all machining into roughing and finishing operations. A great deal of confusion exists because of this, and it is important for the apprentice technician to appreciate that the ideal cutting speed is applicable to the roughing operation only, when a fairly large amount of metal will be removed. Perhaps the following example will serve to illustrate the essential technique to adopt as shown in Fig. 3.30. A 50-mm bright mild steel bar is to have a portion of its length turned down

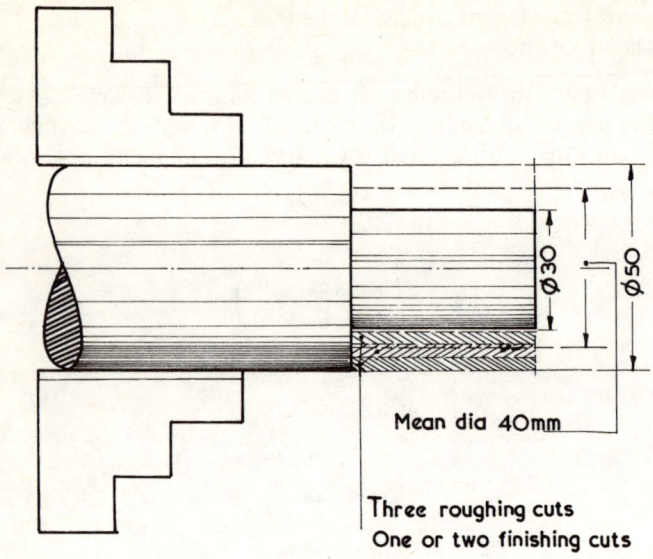

138 Technician Workshop Processes and Materials

Fig. 3.30 Turning techniques

to 30 mm, and this finished diameter must be within plus or minus 0·05 mm, the cutting speed given as 24 m/min. We see at once the difficulty when using our formula, for the diameter of the bar will reduce at each cut, and in theory, the revolutions per minute of the workpiece should change after each cut. A more practical method is to select the mean or middle diameter, which will be 40 mm, and use this figure as the d in our formula. The lathe revolutions can now be calculated:

$$N = \frac{1000 \, CS}{\pi d}$$

$$= \frac{1000 \times 24}{\frac{22}{7} \times 40}$$

$$= \frac{1000 \times 24 \times 7}{22 \times 90} = 192 \text{ rev/min.}$$

The lathe is now set to run at the nearest speed in excess of 192, for it is unlikely that the exact number of revolutions required will be available on the speed range of the lathe.

The problem that now remains is to determine the linear movement or feed of the tool per revolution of the work, and this is largely a matter of experience. In general, it should be about the maximum the lathe can take without overloading the motor or the driving arrangements. A medium-size centre lathe with a distance of 1·5 metres between centres will take a depth of cut of about 2·5 mm, at a feed of 0·25

Machine Tools

mm per revolution, with no undue strain. Reverting back to the example in Fig. 3.30, the total in-feed of the cutting tool is 10 mm, as 20 mm is to be removed and the following procedure is adopted.

1. Chuck the work using a three-jaw chuck, ensuring that it is running fairly true; because three-jaw chucks soon tend to lose their accuracy.
2. With the work stationary, feed in the tool so that it just touches the work.
3. Set the indexing dial to zero, removing all backlash.
4. Take the tool towards the tailstock until it clears the work and index in a distance of 2·5 mm, and take a cut. If coolant is available it should be applied in a steady stream. The feed should be approximately 0·25 mm per revolution of the workpiece, and at the end of the cut the new diameter of the bar will be 45 mm. This should be checked using a 25–50-mm external micrometer. This completes the first roughing cut.
5. Repeat with a second roughing cut of 2·5-mm depth, bringing the diameter down to 40 mm; once again checked with a micrometer.
6. Repeat with a third roughing cut of 2·5-mm depth, making the new diameter 35 mm.
7. The last roughing cut is now taken by indexing the tool a distance of 2 mm and at the end of this cut, 0·5 mm remains for finishing.

Finishing

By taking deep cuts and fairly coarse feeds the roughing operation is soon completed, with 0·5 mm remaining for the finishing technique, and we are now concerned with bringing the size within the allowable limits with a good surface finish. From now on, the amount of metal removed with each cut will be relatively small, so the speed of the lathe spindle can be *increased* and the feed *decreased*. It is generally permissible to double the spindle speed and reduce the feed by about one third. Hence the new spindle speed will be about 570 rev/min and the new feed about 0·01 mm per revolution. A finishing tool should replace the roughing tool and once again the tool is set to touch the workpiece with the indexing dial set at zero. A small cut, say 0·1 mm is taken and the resulting diameter read-off on a micrometer. Let us assume that 0·3 mm remains. Provided the machinist has confidence in the indexing dial, he now indexes the cutting tool a distance of 0·15 mm and this will bring the diameter just within the top limit of size. It is not good practice to come too close to finish size, because centre lathes of the highest quality only, are capable of taking very small cuts. This is due to the number of component parts between lathe bed and cutting tool, as can be seen by reference back to Fig. 3.27. It is evident that there is a certain amount of play or slackness between the saddle, lathe bed, cross slide, compound slide and toolpost, making very small cuts an impossibility. The importance of the roughing and finishing techniques cannot be over emphasized and they are applied to all types of machining, including milling, boring, shaping, planing and grinding.

3.5 Correct use of the centre lathe

In order to appreciate the versatility of the centre lathe we shall consider the essential techniques to adopt when turning various components. Firstly we need to keep in mind the following points regardless of the size of the component or the amount of machining required.

1 Correct choice of cutting tools.
2 Maximum support of the cutting tool.
3 Correct choice of the method of workholding.
4 Maximum support of the work.
5 Full use of all indexing dials.
6 Correct choice of spindle speeds and feeds.
7 Adoption of the roughing and finishing technique.
8 Due regard for operator safety and lathe maintenance.

Secondly we need to consider the basic methods of work holding and these may be listed as follows:

(*a*) One-setting jobs.
(*b*) Two-setting jobs.

3.5.1 One-setting jobs

By one-setting jobs we mean the completion of all machining during *one* setting of the work. In other words, once the job is securely held, it will not be removed until all machining is complete. In this way, the inherent geometry built into the centre lathe will be transmitted to the workpiece, ensuring that all diameters are concentric. Reference to the component shown in Fig. 3.31 will give an indication of the correct procedure to adopt. The drawing must be studied with great care and for some time, so that the completed workpiece can be visualised. This is what is meant by *reading a drawing*, and unless the technician can see, in his mind's eye as it were, the finished component, then he is unable to read a drawing. This calls for considerable skill and experience, but should the young technician be puzzled, some simple free-hand sketches may be made to help make a clear picture of the completed job. It is most unwise to proceed blindly with a turning job, and it can be said that once the machinist has a picture of the finished job then half of his problems are solved.

The component shown in Fig. 3.31 has a recess at one end, a drilled

Machine Tools 141

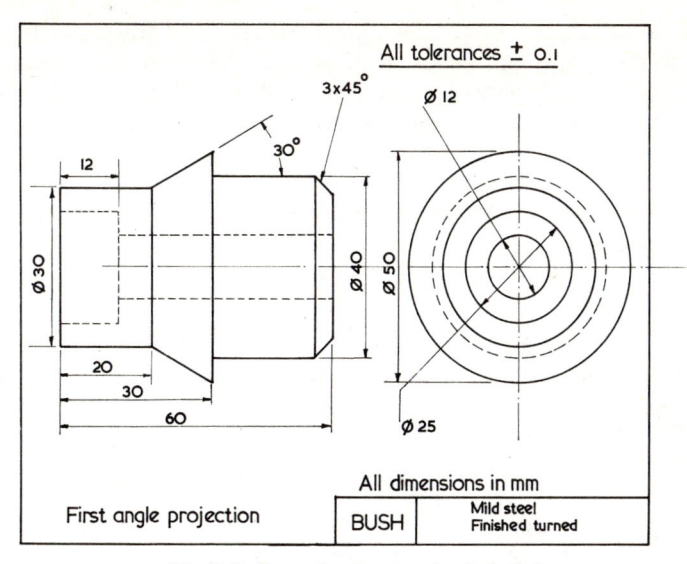

Fig. 3.31 Example of one-setting lathe job

hole and external diameters together with a tapered diameter. It is evident, that apart from the linear dimensions of the lengths and diameters, the important non-linear function of concentricity is involved,

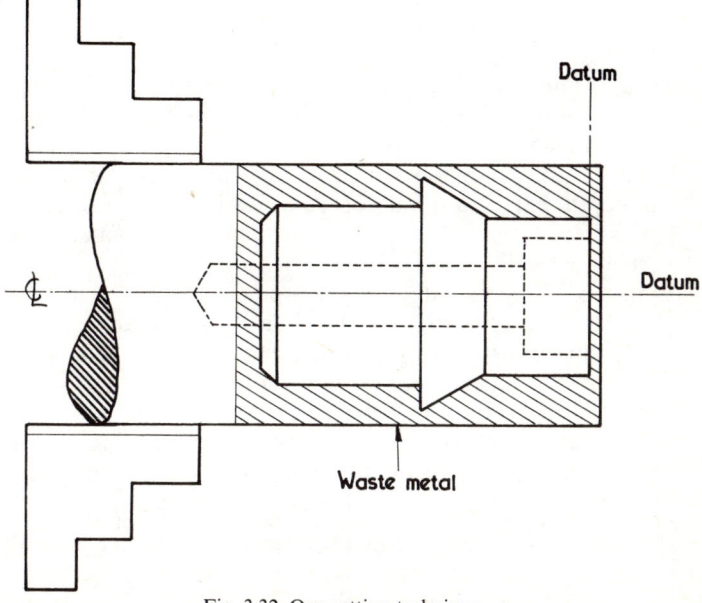

Fig. 3.32 One-setting technique

that is to say, the taper recess, external diameters and drilled hole must be concentric. The centre-line of the lathe is the datum that controls this concentricity, provided that the centre-line remains constant during the whole of the machining. In other words the job must be done in *one* setting, and this is fairly easily achieved by utilising the set-up shown in Fig. 3.32. Note that the job is reversed in order to bring the recess facing the tailstock, enabling it to be machined in the one setting. The skilled machinist, by virtue of his experience is able to plan a sequence of operations mentally, but it is not a bad idea if the young technician commits his sequence of operations to paper first. The following outline will give some indication of the correct approach.

1. Chuck job securely, holding on about 25-mm length and allowing about 12 mm in excess of job length.
2. Calculate and set spindle speed.
3. Face end, this will provide the second datum as shown in Fig. 3.32.
4. Centre end using tailstock and support with a running centre if possible. If a dead centre is used it must be well greased.
5. With odd-leg calipers and job rotating, mark-out 20-, 30- and 60-mm lengths.
6. Rough down 30 mm diameter, to within about 0·5 mm of line and finished diameter.
7. Rough down 50 mm diameter. Set compound slide to 30° and rough turn the taper using hand feed.
8. Rough down 40 mm diameter.
9. Remove tailstock centre; drill 6-mm-diameter hole and open out using 12-mm-diameter drill, ensuring feed shaft is disengaged and changing spindle speed to suit drill diameters.
10. Using 15- and 22-mm-diameter drills, drill recess just under 12-mm depth, reducing spindle speed as drill diameters increase.
11. Finish recess using a recessing tool, checking diameter with Vernier calipers.
12. With a finishing tool, doubling the spindle speed used for roughing and using a fine feed, finish off the outside diameters checking all dimensions with depth and external micrometers, using indexing dials to maintain precise control over amount of metal removed.
13. With a parting-off tool, feed in about 6 mm and bring the job to length. Set odd-leg calipers to 3 mm and mark line for chamfer. Machine chamfer and part-off, as shown in Fig. 3.33.

Note that the chamfering tool in the diagram produceds a tapered surface by *forming*, the accuracy of the 45° angle dependent on the grinding and alignment of the tool. It is customary also to remove all

Machine Tools 143

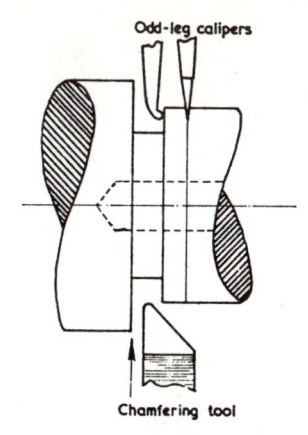

Fig. 3.33 Chamfering before parting-off

sharp edges on turned work, and a smooth file may be used for this purpose, or small chamfers can be machined using a suitable tool. The sequence of operations described above is for guidance only, but several simple rules emerge and they may be summarised as follows:

(i) Study the job carefully.
(ii) If possible machine in *one* setting.
(iii) Finish all roughing before starting the finishing operations.
(iv) Support the work whenever possible.
(v) Disengage feed shaft or lead screw when not in use.
(vi) Change the spindle speed to suit the operation at hand.
(vii) Provide maximum support for the work and tool.
(viii) Remove all sharp edges.

Use of the four-jaw chuck
The gripping power provided by the independent action of four separate jaws makes the four-jaw chuck ideal for holding work which may not be truly round, for example cast or forged bars. In addition the four-jaw chuck is capable of holding work of almost any section, or round bar in any position. Figure 3.34A shows a typical example of what may be termed as eccentric turning; this is a phosphor bronze pressure pad having two eccentric oil grooves. We see from Fig. 3.34B, that three different settings of the work are needed; at x an internal cylindrical surface or hole is to be machined on centre. At y, a circular, shallow depth groove is needed, a certain distance above centre, while at z we see that a similar groove is needed below centre. For each operation, therefore, the axis of rotation must be changed, and a four-jaw chuck is eminently suited for machining the two oil grooves. The central hole may be drilled with the work held in a three-jaw chuck,

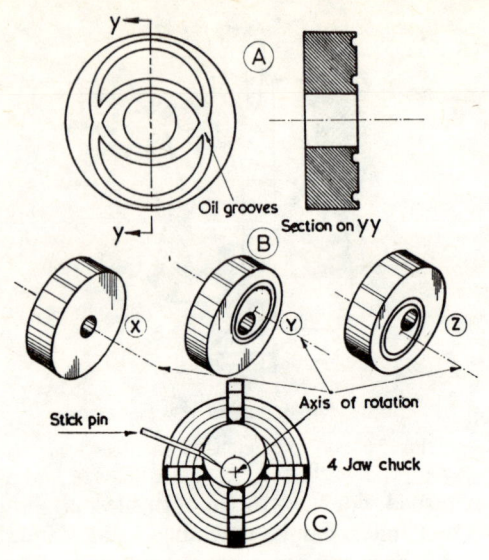

Fig. 3.34 Use of the four-jaw chuck

which must then be removed and replaced with a four-jaw chuck. This changing of chucks must be carried out with great care; the lathe first *isolated* from the mains supply and the bed protected by strong wooden boards. The circular oil grooves are marked-out and the scribed circles are picked up using a stick pin temporarily fixed to a toolpost. With the lathe isolated and in neutral gear, the chuck is rotated by hand, and the jaws adjusted so that the stick-pin point follows the scribed circle on the component. A front elevation of the set-up is shown in Fig. 3.34C where it may be seen that the centre of the circular oil groove is coincident with centre-line of the lathe or the axis of rotation. The same technique is adopted to machine the second circular oil groove and in this way, three separate set-ups are carried out in order to machine the central hole and the two oil grooves.

Turning between centres
Figure 3.35A shows a turned blank for a progressive plug gauge. The material is mild steel and the plug gauge is to be finish ground after case-hardening. Because the gauge is to be produced using *two* different kinds of machine tools, a centre lathe and cylindrical grinder, it is necessary that sufficient metal be left for grinding and that the same datums or holding methods be used for both turning and grinding. Figure 3.35B shows that there is very little difference between turning between centres and grinding between centres; in both cases the cylindrical surface is generated by traversing the tool when turning and

Machine Tools

PROGRESSIVE PLUG GAUGE

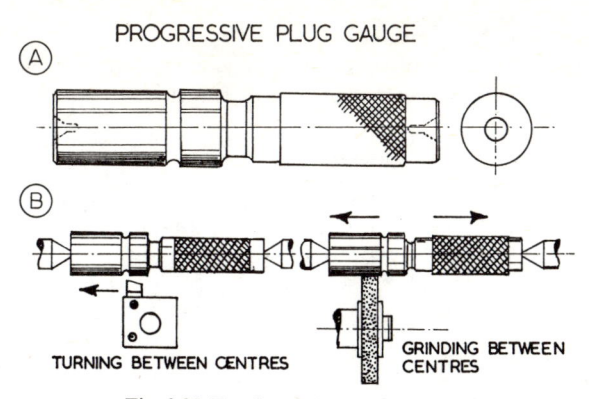

Fig. 3.35 Turning a plug gauge blank

traversing the work when grinding. A sequence of operations for turning the plug gauge between centres is shown in Fig. 3.36, and it needs to be remembered that the workpiece can be removed from the centres at any time, with complete confidence in the ability of the job to be relocated with no loss of accuracy.

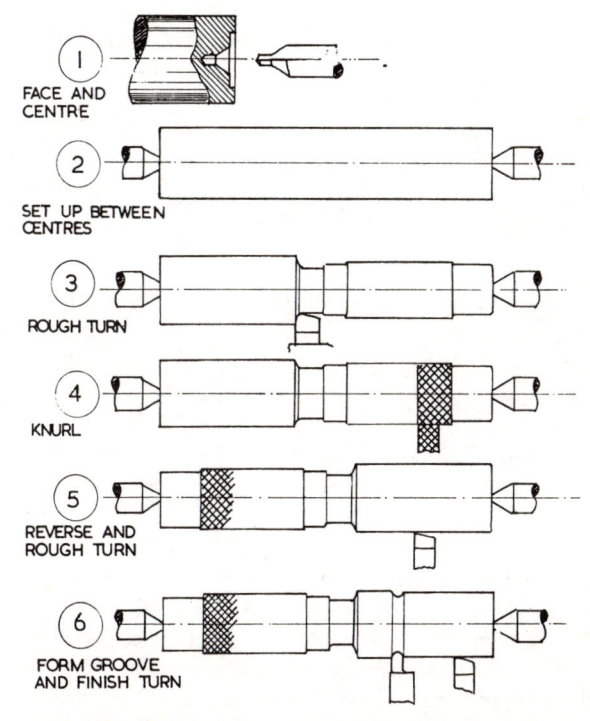

Fig. 3.36 Operational sequence for plug gauge

Use of collet chucks

Figure 3.37 shows a simple stud to be machined from extruded nylon bar of 19 mm diameter, supplied in 1-m lengths. Because of the high

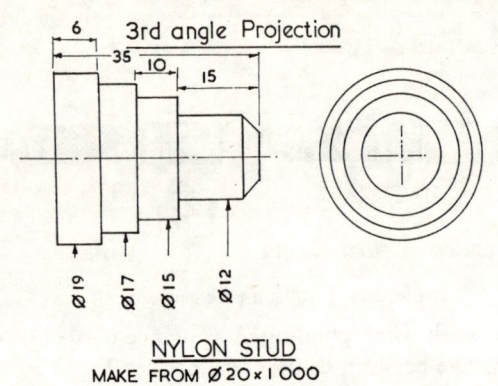

Fig. 3.37 Turned nylon stud

degree of accuracy together with the exceptional surface finish obtained from the extrusion process, the outside diameter of 19 mm has practically no variation in size or form. This means that a fixed-diameter work-holding device may be used to advantage, allowing the nylon rod to be moved forward on the completion of each component. For this kind of application a *collet chuck* is used and such a chuck is illustrated in Fig. 3.38. Note the relatively large contact area between

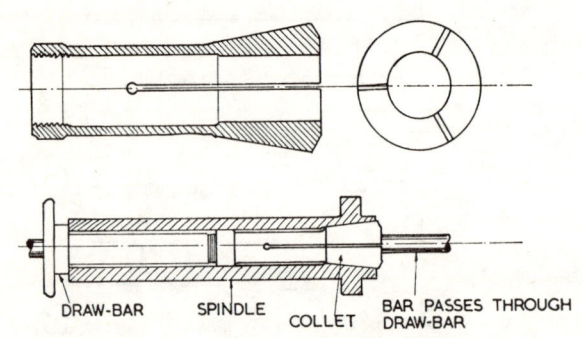

Fig. 3.38 Principle of the collet chuck

chuck jaws and workpiece, relieved only by the three narrow slots which permit the jaws to move inward when the collet is pulled up against the taper in the spindle nose of the centre lathe. At Fig. 3.39 we see an operational sequence for the nylon stud and once again it is clear that a concentric job results, because the nylon stud is machined at *one* setting. Because collet chucks are fixed-diameter work-holding

Machine Tools 147

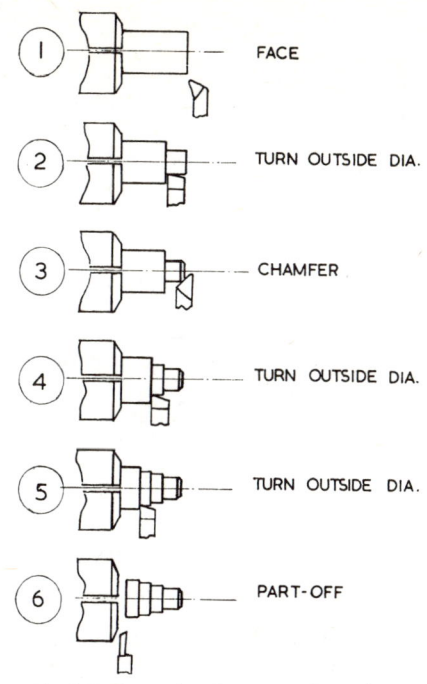

```
1 ─ FACE
2 ─ TURN OUTSIDE DIA.
3 ─ CHAMFER
4 ─ TURN OUTSIDE DIA.
5 ─ TURN OUTSIDE DIA.
6 ─ PART-OFF
```

Fig. 3.39 Operational sequence for nylon stud

devices, a set of them is needed to accommodate standard diameters, and the lathe must be equipped with a matching taper and hollow spindle. Absolute cleanliness of the collet diameter and taper, together with the mating taper in the lathe spindle nose is essential, the material rod or bar must be free also from any damage, swarf or burrs. The correct lathe cutting tools for nylon must be used, together with the appropriate spindle speeds and feeds, and if the indexing dials are marked with an indelible pencil, a type of semi-automatic turning will be achieved.

Use of the faceplate
As previously stated the purpose of a lathe faceplate is to provide a datum face at 90° to the lathe centre-line. The need for this kind of location may be appreciated by reference to Fig. 3.40, where we see a grey cast iron casting requiring a hole bored to 48 mm and a recess of 52 mm diameter. Clearly these are large non-standard sizes, thus a radial drilling machine would not be suitable for the machining operations of drilling and counterboring, and a centre lathe is the most suitable machine tool for machining this casting. It will be necessary to ensure that the centre of the hole coincides with the centre or axis of

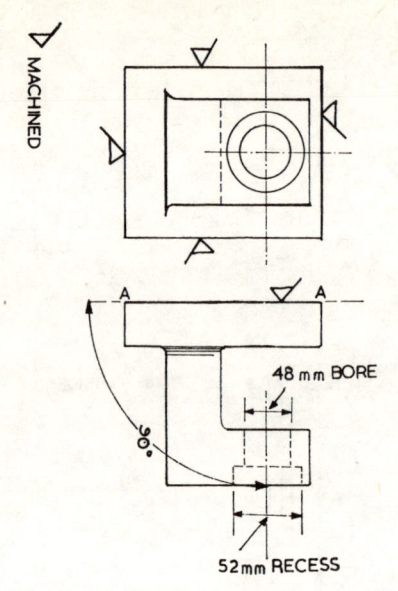

Fig. 3.40 Casting requiring bored and recessed hole

rotation of the lathe and whilst, as we have seen, this is possible using a four-jaw chuck, such a chuck will need to be of large capacity and extremely heavy as a result. A better and easier process, is to use a faceplate, and Fig. 3.41 shows both a front and end elevation of the set-up. Once again it is necessary to mark-out and scribe the hole circles, in order that a stick pin may be used to pick up this circle and so bring the centre of the hole in line with the axis of rotation. The clamps are lightly tightened while the setting-up proceeds; the lathe in neutral gear and the faceplate rotated by hand. When the circle runs true, relative to the fixed stick pin, the clamps are tightened to ensure that

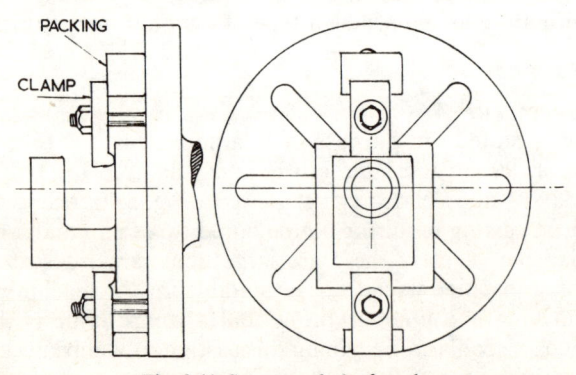

Fig. 3.41 Set-up on lathe faceplate

Machine Tools

the casting is firmly held to the faceplate. Under no circumstances must this faceplate be rotated at speed, for the addition of the casting and clamping arrangements are certain to lead to powerful out-of-balance forces caused by centrifugal force. This means that the faceplate set-up needs to be balanced before any rotation takes place, the procedure being as follows:

1. Place lathe in neutral gear and *isolate*.
2. Spin faceplate by hand and chalk-mark the faceplate at its lowest point after it comes to rest.
3. Add a balance weight at opposite position to chalk mark, using T slot to clamp the weight to the faceplate.
4. Re-spin faceplate by hand and if chalk mark still comes to lowest position, more weights are needed. If weighted end comes to lowest position, weight is excessive. It may be possible to move the weight either towards or away from the lathe centre.
5. The faceplate may be considered as balanced, when the chalk mark comes to different positions for each spin of the faceplate.

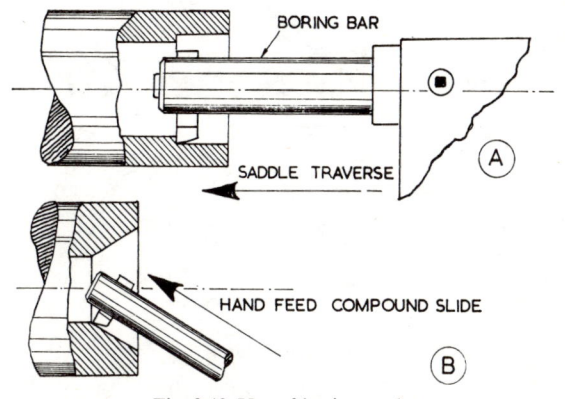

Fig. 3.42 Use of boring tools

With the faceplate balanced, the hole may be drilled from the tailstock after using a centre-drill to start the hole, and then bored to size using a suitable boring bar held in the toolpost. A typical boring bar is shown in Fig. 3.42A, and automatic feed can be used by engaging the saddle traverse. If, on the other hand, it is required to bore a taper as shown in Fig. 3.42B, it will be necessary to set the compound slide at *half* the included angle of taper and hand-feed only is possible.

3.6 Screw-cutting at the centre lathe

Cutting a thread using a centre lathe is more commonly known as screw-cutting. It is a mistake to think that the art of screw-cutting represents the ultimate in turning technique, and a great deal of unnecessary prominence is given to screw-cutting at a centre lathe. It needs to be clearly understood that every time a sliding cut is taken on a lathe, the tool advances a certain linear distance for each revolution of the workpiece, in other words a thread is cut. The external cylindrical surface generated by the interaction of tool path and work centre is, in effect, a screw thread of very fine pitch.

Screw thread essentials
Provided that the essential requirements of a screw thread are appreciated, screw-cutting at a centre lathe should present no greater problem than the machining of an ordinary round surface. There is, however, one important difference that must be appreciated; this is that screw-cutting is a combination of both *generating* and *forming* processes. Figure 3.43 will help to make this clear. At A we see a pictorial

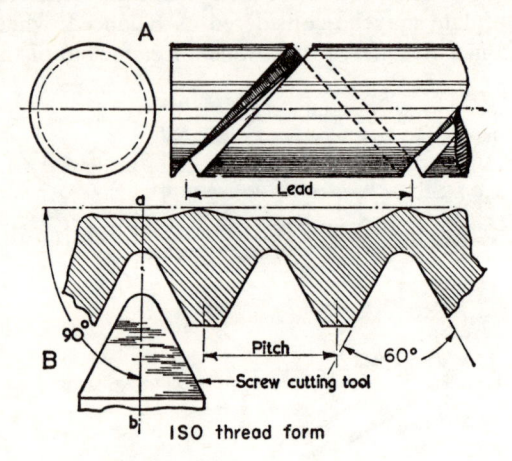

Fig. 3.43 Screw thread elements

representation of the *lead* of a thread. This is the distance a nut will move along the axis of the thread when given one complete turn. The lead is not necessarily the same thing as the *pitch* of a thread, which is the distance from one thread form to the next thread form as shown in Fig. 3.43B. Provided the thread is a *single start* thread, then the lead is equal to the pitch, but if the thread is a multi-start, then the lead will be equal to the pitch multiplied by the number of starts.

Thread form
It is clear that the accuracy of the thread form depends on the profile of the threading tool, because the tool transmits its profile to the workpiece, hence the thread form on the workpiece is an example of *form-*

Machine Tools

ing and not *generating*. At the same time the alignment of the screw-cutting tool is important; it is essential that the centre-line of the tool be at 90° to the axis of the work as shown in Fig. 3.43B. This means that the tool, in addition to having the correct profile of the thread to be cut, must be correctly positioned in the lathe toolpost. We see now that *three* essential factors are involved when screw-cutting at the centre lathe.

1. Pitch of the thread.
2. Accuracy of thread form.
3. Alignment of thread form.

Figure 3.44A shows an adaptor for a press tool, and it is required to screw-cut the thread indicated in the diagram. This thread is shown as M30 and this means that it has been chosen from the *metric coarse thread form series*. 'M' stands for Metric and the '30' stands for the diameter. Table 3.1 gives the basic sizes of the main thread elements for diameters 22 to 42 mm.

Table 3.1 ISO metric coarse series

Diameter	Pitch	Major diameter	Minor diameter
22	2·5	22	20·376
24	3·0	24	22·051
27	3·0	27	25·051
30	3·5	30	27·727
33	3·5	33	30·727
36	4·0	36	33·402
39	4·0	39	36·402
42	4·5	42	39·077

Referring to Table 3.1 we see that the pitch is 3.5 mm and the minor diameter 27·727 mm. The major diameter is the outside diameter of the thread, and this will be turned down to size as shown in Fig. 3.44B; note that the adaptor is to be completely machined, including the screw-cutting in *one* setting of the job, the last operation will be parting-off. The minor diameter allows us to calculate the depth of thread as follows:

$$\frac{(Major\ diameter - Minor\ diameter)}{2} = \text{depth of thread}$$

$$\frac{(30 \cdot 000 - 27 \cdot 727)}{2} = \frac{2 \cdot 273}{2} = 1 \cdot 136 \text{ mm}$$

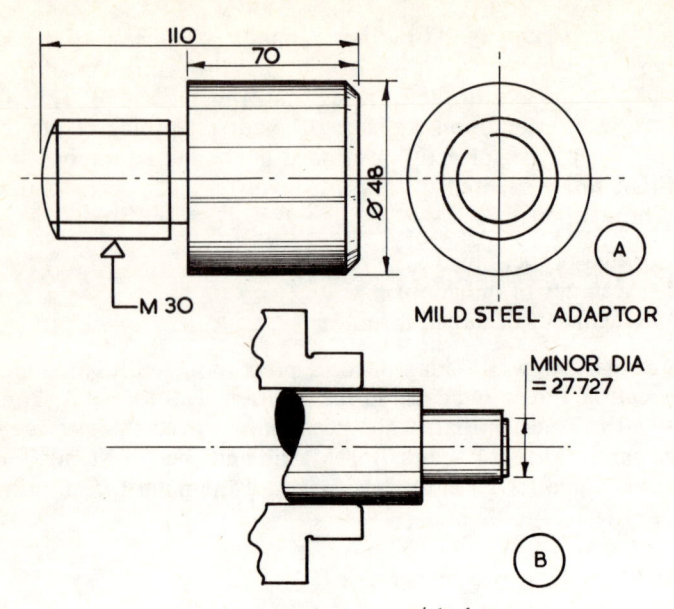

Fig. 3.44 Threaded press tool adaptor

This is the depth to which the screw-cutting tool is fed, and can be obtained by careful use of the indexing dial. Perhaps a better method is to turn down a small portion of the workpiece to the root diameter as shown in Fig. 3.44B; the infeed of the cutting tool to stop when the point just touches this diameter. When the screw-cutting operation is completed, this small step can be chamfered to give the adaptor a neat appearance.

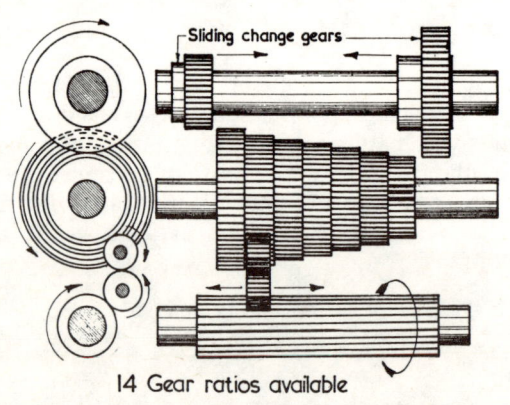

Fig. 3.45 Principle of Norton-type gear box

Machine Tools 153

Setting-up for the pitch
All modern lathes are fitted with *Norton-type* gear boxes and the setting for a wide range of both Metric and English pitches is easily and rapidly obtained merely by the movement of levers, much the same as selecting a gear when driving a modern motor car. The basic principle is illustrated in Fig. 3.45, while Fig. 3.46 shows a pictorial view which may assist in an understanding of the technique involved. It is, in

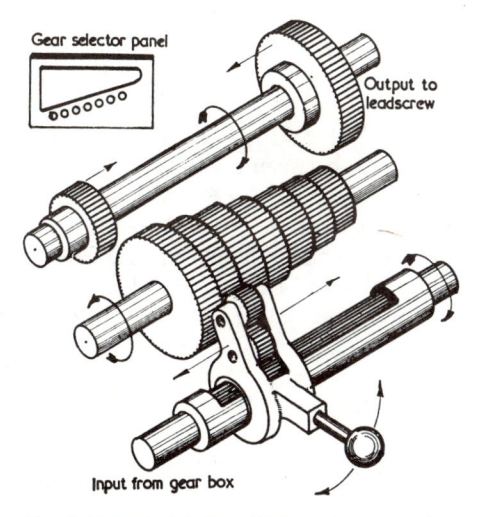

Fig. 3.46 Pictorial view of Norton-type gear box

effect, an automatic method of changing gears and perhaps the apprentice technician may now appreciate some of the high quality precision work necessary in the design and manufacture of machine tools. Provided the centre lathe is equipped with a Norton-type gear box, the positions for the pitch-changing levers is checked on a plate fixed to the headstock and the levers placed in their correct positions to screw cut a pitch of 3·5 mm. In the very rare event of having to select and change the gears manually, making use of a set of spur gears and the quadrant principle previously explained and illustrated, the following procedure and calculations are required, and it will be necessary to refer back to Figs. 3.20 and 3.21. These illustrations clearly identify the driving and driven gears and the following formula may be used to calculate the actual gears required.

$$\frac{\text{Driver}}{\text{Driven}} = \frac{\text{Pitch to be cut}}{\text{Pitch of lead screw}}$$

The pitch to be cut is 3·5 mm and for the purpose of solving this example, we are assuming that the centre lathe at our disposal has a lead screw with a pitch of 5 mm. Inserting these values, we have

$$\frac{\text{Driver}}{\text{Driven}} = \frac{3 \cdot 5}{5} = \frac{35}{50}$$

The set of gears available usually starts with a 20-tooth gear finishing at about 120 teeth, with increments of 5 teeth between each successive gear, hence for the screw-cutting of the thread shown on the adaptor in Fig. 3.44, a 35-tooth gear is placed on the spindle shaft, and the 50-tooth gear placed on the lead-screw shaft. With a suitable idling gear adjusted, the quadrant is swung and tightened in position when the idler meshes with the 35-tooth driving gear. This technique is made clear in Figs. 3.20 and 3.21, and it is worth noting that the idler gear has no effect on the *ratio* between driver and driven. Its purpose is to transmit motion and it does so by reversing the direction of rotation, as shown in Fig. 3.21, making the driven gear rotate in the same direction as the driving gear. This is the correct set-up for cutting a right-hand thread, but if a left-hand thread is needed, it will be necessary to insert *two* idlers between the driving and driven gears.

Simple and compound gear trains

The gear train used to cut the M30 thread on the adaptor is known as a simple gear train, consisting of only one driving gear and one driven gear, but if an awkward pitch is to be screw-cut on a centre lathe it may be necessary to use a compound gear train. Perhaps the following screw-cutting problem will help to show the principle involved. Let us assume that an M5 thread is to be cut on the same lathe. An M5 thread is a metric coarse series thread and reference to the coarse series table shows that the pitch is 0·8 mm and the diameter 5mm.

Using the screw-cutting formula

$$\frac{\text{Driver}}{\text{Driven}} = \frac{\text{Pitch to be cut}}{\text{Pitch of leadscrew}}$$

$$\frac{\text{Driver}}{\text{Driven}} = \frac{0 \cdot 8}{5} = \frac{8}{50} = \frac{4}{25}$$

The nearest gearing ratio would be

$$\frac{4}{25} \times \frac{5}{5} = \frac{20}{125} = \frac{\text{Driver}}{\text{Driven}}$$

but a 125-tooth gear is not available, therefore it is not possible to cut the thread using a simple gear train. The solution is to use a compound train in other words to make the gear ratio equal to the sum of two fractions.

Machine Tools

$$\frac{\text{Drivers}}{\text{Driven}} = \frac{4}{25} = \frac{A}{B} \times \frac{C}{D}$$

$$\frac{\text{Drivers}}{\text{Driven}} = \frac{4}{25} = \frac{4}{5} \times \frac{1}{5}$$

$$\frac{\text{Drivers}}{\text{Driven}} = \frac{40}{50} \times \frac{20}{100}$$

We now have two driving gears, 40 teeth and 20 teeth and two driven gears of 50 and 100 teeth. The set-up for these gears is shown in Fig. 3.47, where it may be noted that no idler is necessary when cutting a

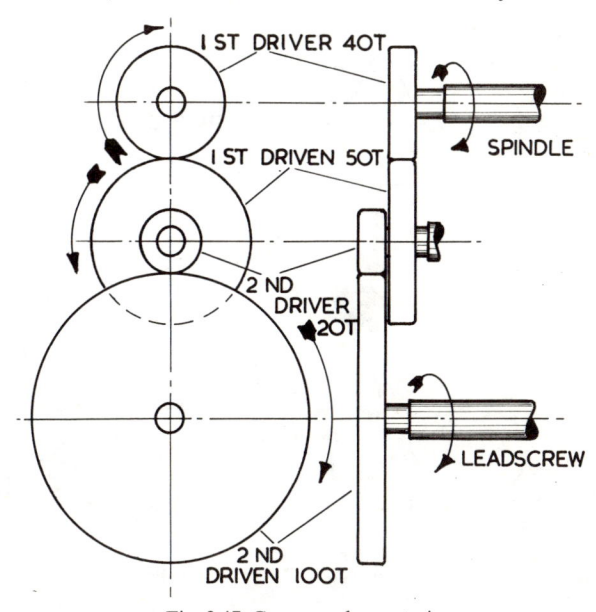

Fig. 3.47 Compound gear train

right-hand thread; due to the fact that the first driven and second driving gear are keyed or compounded to the same shaft which is free to rotate. In this way the compound gear made up of the 50-tooth driven and 20-tooth driver acts as an idler. Should, however, a left-hand thread require to be cut when using a compound gear train, it will be necessary to insert an idling gear. We see now the somewhat complicated procedure involved when changing gears by hand methods and it is very seldom that this technique is carried out in actual practice. We must remember that a centre lathe, like any other machine tool is of value only when it is actually removing metal. Any time spent on tool changing, gear changing or work setting may be regarded as unproductive time and must be kept to a minimum.

Setting the tool
Now that the lathe is geared so that for one revolution of the workpiece the saddle and hence the tool, advance a distance of the pitch (in this case 3·5 mm, the thread on the adaptor shown in Fig. 3.44) we may turn our attention to the correct alignment of the thread-cutting tool in relation to the centre-line or axis of the workpiece. It is certain that the profile of the thread form has been accurately ground and checked, as shown in Fig. 3.48A and the correct setting of this tool is also shown in Fig. 3.48B. With the setting shown, the tool is securely

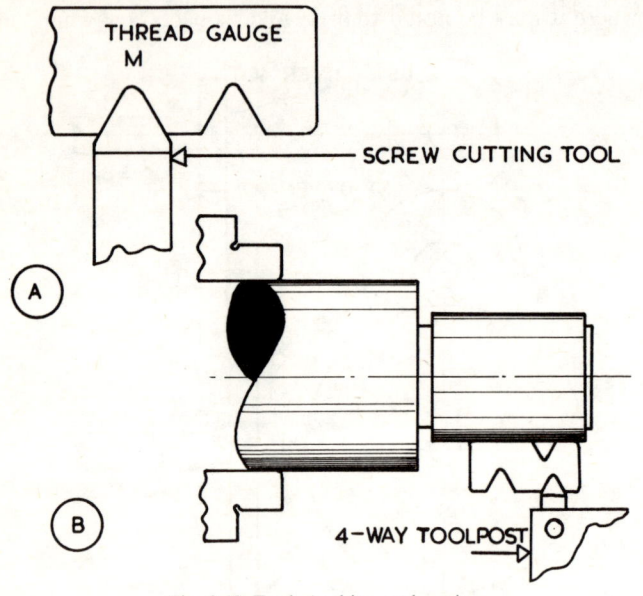

Fig. 3.48 Tool checking and setting

clamped and then re-checked for accuracy of alignment; the compound slide set at zero with backlash removed by feeding the slide by hand towards the headstock until the zero line on the indexing dial coincides with the fixed line on the slide. With the cross slide also fed towards the work and its indexing dial set to zero, when the point of the screw-cutting tool just touches the work, all is set for the actual screw-cutting operation.

Cutting the thread
The actual thread-cutting or metal removing operation is a matter of experience and confidence. Most turners prefer to feed directly into the work, reducing the depths of cut as the thread approaches its final size, and occasionally allowing the tool to run down the thread without

Machine Tools 157

taking a depth of cut. This cleans up the thread and ensures that there is no build-up of tension or spring in the tool. It is possible to set-over the compound slide at half the thread angle, the feed now being made along the compound slide. Perhaps the greatest problem is the vital necessity of ensuring that the thread-cutting tool picks up the thread exactly, at each successive cut and for this purpose, a chasing dial is fitted which can be engaged with the lead screw when desired. Figure 3.49 shows a plan view of a chasing dial, together with the rules that govern its use when screw-cutting.

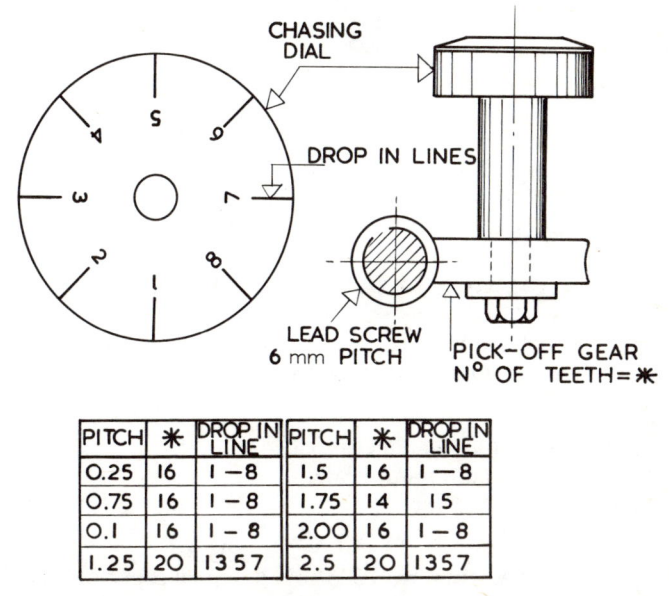

PITCH	✶	DROP IN LINE	PITCH	✶	DROP IN LINE
0.25	16	1—8	1.5	16	1—8
0.75	16	1—8	1.75	14	15
0.1	16	1—8	2.00	16	1—8
1.25	20	1 3 5 7	2.5	20	1 3 5 7

Fig. 3.49 Metric chasing dial

Use of chasing tools
It is possible to finish turn an ISO thread at a lathe using a single-point cutting tool, because flat crests are permissible. However, rounded crests are also allowed and this means that if rounded crests are required it is necessary to use a chasing tool to finish-off the thread. Figure 3.50A shows the ISO thread form; at B we see a simple hand-chasing tool. This tool is used by carefully running it down the thread with the lathe on a slow spindle speed, and keeping the chasing tool adequately supported by resting it on a bar clamped on a toolpost. It must be kept in mind that the sole purpose of the chasing tool is to produce the radii at the crests and great care needs to be taken not to remove excess metal and thus render the thread undersize.

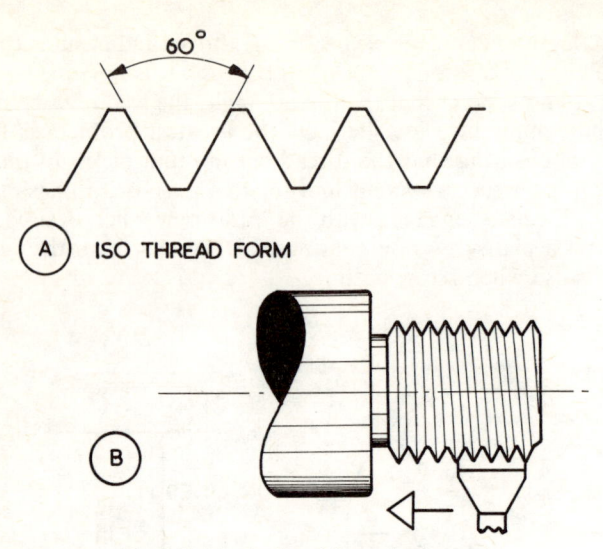

Fig. 3.50 Metric thread form chasing tool

3.7 Taper turning at the centre lathe

The production of a tapered surface is a common operation at a centre lathe and there are *four* main processes by which a tapered surface may be produced.

1. Using a form tool (chamfering).
2. Off-setting the compound slide.
3. Off-setting the tailstock.
4. Using a taper-turning attachment.

We have already seen in Fig. 3.33, how a simple taper or chamfer is produced using a form tool, while reference back to Fig. 3.16 shows how a bigger, or fast taper is produced by off-setting or indexing the compound slide at half the included angle of taper to be machined. On the occasions when a slow taper requires to be machined, that is to say a taper of small included angle, a better plan is to off-set the tailstock, allowing the taper to be machined using automatic traverse, a method not possible when off-setting the compound slide. Automatic feed will make the job less tedious and give a better finish to the machined taper. Figure 3.51A shows a component turned between centres with the tailstock centre co-axial with the lathe centre-line, while at B we see the effect when the tailstock is off-set towards the operator. If we turn back to Fig. 3.18, which shows two views of a lathe tailstock, we can see that the top part of the tailstock is capable of adjustment in the directions of the arrows shown as D, and then being locked in the

Machine Tools 159

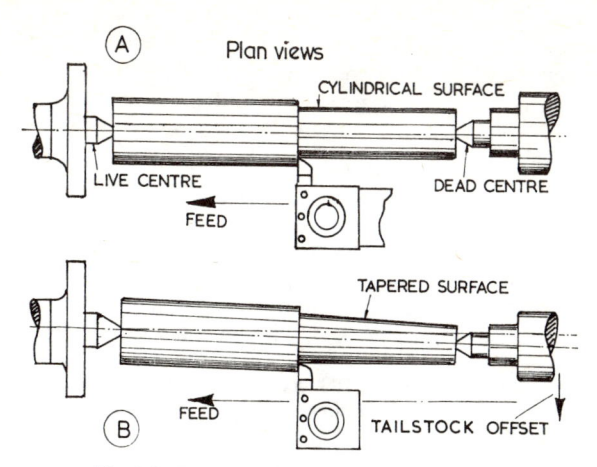

Fig. 3.51 Taper turning by off-setting tailstock

desired position. The amount of off-set is somewhat limited, not likely to exceed about 20 mm each side of true centre, so that it is not possible to machine a short fast taper by this technique. It is of the greatest importance to first check the alignment of the tailstock before any attempt is made to off-set it to produce a given taper. In other words the off-set is the distance moved from *true centre*, and it is therefore vital to ensure that the tailstock is first on true centre. The basic process for ensuring true centre may be seen from the plan view of a centre lathe shown in Fig. 3.52. The essential condition is that the dial test indicator maintains a zero reading on the test bar, while the

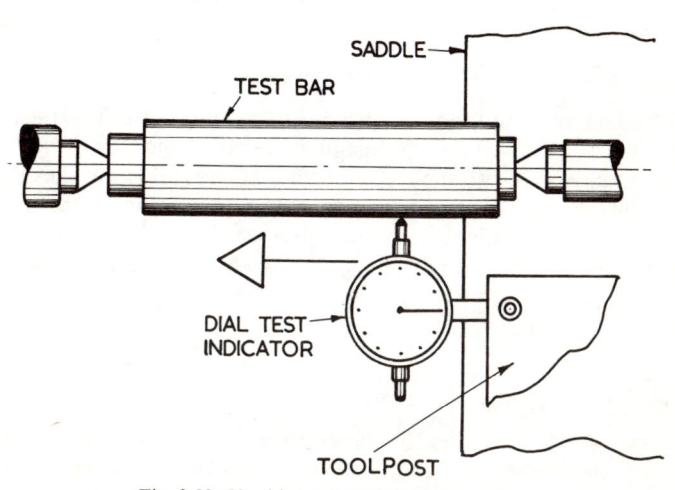

Fig. 3.52 Checking tailstock for true centre

saddle is hand traversed the length of the test bar. We are in effect, checking the parallelism of the line of action of the saddle with the lathe centre-line, in the horizontal plane. It is, of course, equally vital to ensure that the lathe tailstock is restored to its proper position on completion of the taper turning.

Taper turning attachments
These are fitted to the more expensive toolroom-type centre lathes and allow reasonably long but slow tapers to be turned using automatic traverse, but without the need for off-setting the tailstock. The basic principle is illustrated in Fig. 3.53, where it may be seen that the cross

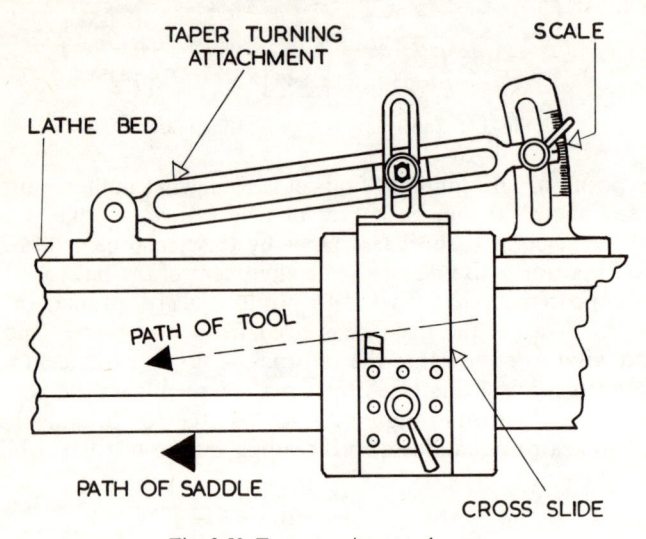

Fig. 3.53 Taper turning attachment

slide can be made to follow a path out-of-parallel with the lathe centre-line. This is achieved by disengaging the cross slide from its feeding arrangements at 90° to the lathe centre-line, thus allowing the off-set of the sliding bar arrangement to give sideways motion to the attached crosslide as the saddle or carriage traverses the bed of the lathe in the direction of the headstock. A suitable scale calibrated in degrees or unit run, makes it an easy matter to set the taper turning attachment to the taper required.

3.7.1 The designation of tapers

We have seen earlier, how the taper principle is utilised for the precise location of engineering components; for example the use of *Morse* taper-shank drills and reamers for insertion into the tapered spindles

Machine Tools

of pillar and radial drilling machines. It is very seldom that a taper is finish turned, and grinding is the standard method of producing most if not all tapered components, because of the high quality of the accuracy and finish obtained when grinding. It is important that the technician is able to understand and appreciate the methods by which a taper is described or designated and it is equally important that the method of calculating taper designation is fully understood. Perhaps one simple example will make the principle clear. Figure 3.54A shows

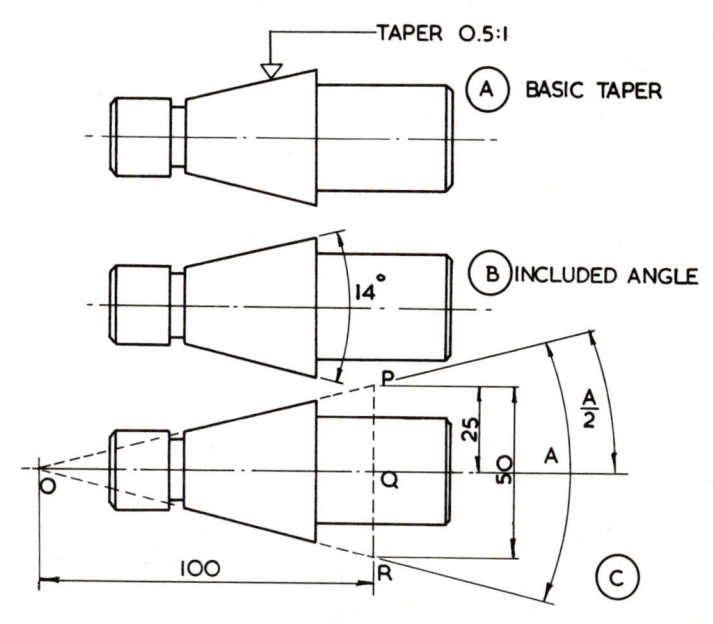

Fig. 3.54 Tapered component with basic taper

a turned blank, and it is required to rough turn the taper by off-setting the compound slide. We see from the diagram that the taper is dimensioned using a method known as *basic taper*; in the component shown at Fig. 3.54A it may be seen that the taper is indicated as follows:

$$0 \cdot 5 : 1$$

This means that for 1 unit length, the increase in diameter is 0·5 units and for 10 units of length the diameter increases by $0·5 \times 10 = 5$ units. This may be more clearly seen in Fig. 3.54C, where the distance OQ is given as 100 mm. At point O the diameter is zero and at PR the diameter will be 50, with PQ = 25. In triangle OPQ, half the included angle is shown as $\dfrac{A}{2}$ and

$$\text{Tangent } \frac{A}{2} = \frac{PQ}{OQ}$$

$$= \frac{25}{100}$$

$$= 0\cdot2500$$
$$= 14° \text{ (app.)}$$

The included angle is therefore 28°, but it needs to be remembered that the compound slide is set to *half* the included angle and this will be 14°. On the other hand, we may need to off-set the tailstock to turn a slow taper which has been designated as an included angle, as shown in Fig. 3.55A. We now need to calculate the amount by which the tailstock needs to be off-set, allowing us to turn the taper using automatic traverse. Reference to Fig. 3.54B shows that the following simple calculations are required.

In triangle OPQ

$$\text{Tan }\frac{A}{2} = \frac{PQ}{OP}$$

$$\text{Tan } 5° \, 30' = \frac{PQ}{150}$$

$$150 \times \text{Tan } 5° \, 30' = PQ$$
$$150 \times 0\cdot0962 = PQ$$
$$14\cdot443 = PQ$$

This is the amount the tailstock needs to be off-set and at Fig. 3.55C, we see a simple method of achieving this movement. Once again it is necessary to check that the tailstock is on true centre before any attempt is made to off-set it and a test bar and dial indicator are used for this purpose, as shown in Fig. 3.52.

3.8 Safety precautions on the centre lathe

There are several safety precautions which must be observed when using centre lathes. The first essential is to be able to stop the lathe quickly, should anything go wrong. Before starting work on a strange lathe the technician should make himself thoroughly familiar with all the controls. Secondly there is the danger of personal injury, for the centre lathe is a power-driven device capable of causing serious injury. The swarf produced by the cutting action of the lathe tool often possesses a razor-like edge, and must never be handled with bare hands. Loose clothing, and long hair are to be avoided at all costs and the greatest care is required when tool-changing. Four-jaw chucks, for example, are deceptively heavy and quite capable of causing severe

Machine Tools 163

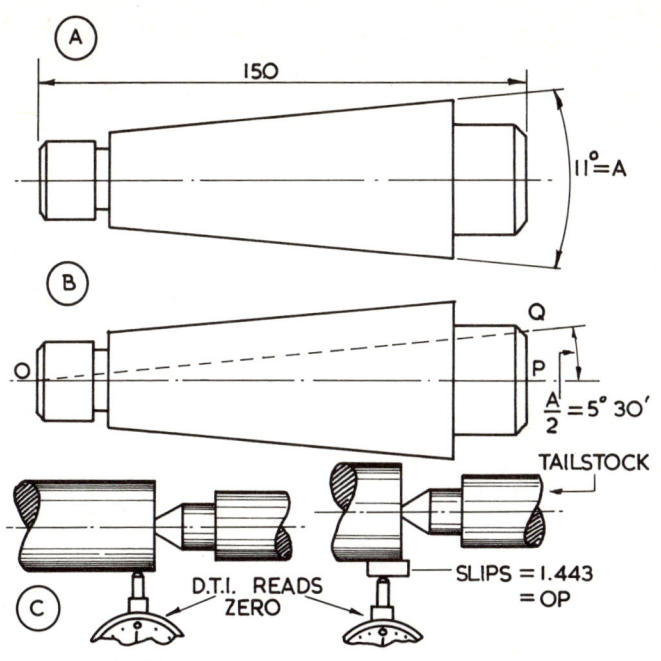

Fig. 3.55 Tapered component with included angle

damage to both operator and lathe bed. Thirdly there is the care and appreciation of the lathe, its accessories and measuring equipment used when turning. Micrometers and rules should never be left on the headstock or in the swarf pan. There is also no excuse for leaving spanners on the lathe bed whilst turning. The skilled machinist will ensure that his machine is free from all encumbrances, and there will be provision for the neat and safe storage of all accessories used.

Finally the lathe will be cleaned down at the end of the day's work and this will include the cleaning and checking of the accessories used. Regular and efficient maintenance of the centre lathe, in connection with oiling and greasing of moving parts is essential if the centre lathe is to continue to do its work efficiently and to the accuracy expected.

Production turning

The centre lathe as we have seen is used mainly for the manufacture of single components of cylindrical shape. Tapers, screw threads and bored holes are also produced, but considerable skill both in setting and machining are required if accurate well-finished components are to be machined. Much of the technician's time is taken up with work setting and tool changing, while linear accuracy of the machined component is dependent on skill and experience with regard to the use of indexing dials and measuring instruments. In general, an overall accuracy of about 0·02 mm is possible provided the lathe is in reasonable condition. It is evident that centre lathes would be most uneconomical for production turning, for each lathe would require a skilled operator, with more time taken on tool and work changing, than on the actual removal of metal.

Capstan lathe

We may consider a capstan lathe as a centre lathe modified for production turning and requiring operation by a semi-skilled machinist. It is similar in principle to a centre lathe but is fitted with a hexagonal turret or capstan which provides *six* additional tool positions. These additional tool positions allow a setter to tool-up the capstan lathe; that is to say to adjust or set the cutting tools so that the machinist is left with the function of *operating* the machine only. Dimensional accuracy is obtained by the use of adjustable *stops* as shown in Fig. 3.56B, so that the dimension shown as L is achieved by the simple

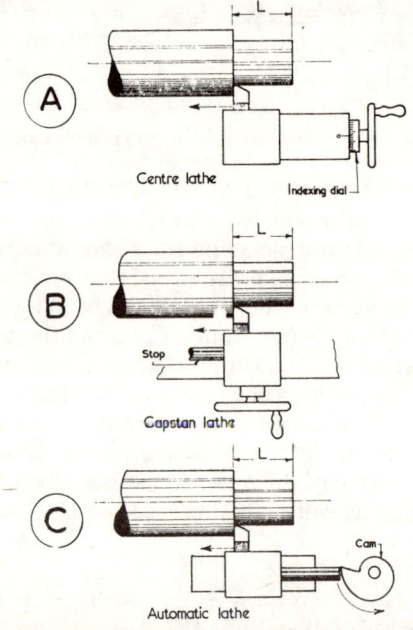

Fig. 3.56 Development of linear control

expedient of bringing the toolpost up against the stop. In this way a capstan setter can set the tool positions and stops and so, enable a semi-skilled machinist to produce the turned component by working to the stops. The main occupation of the capstan operator, therefore, is machining components, while the setter will have several capstan lathes to look after. This technique, of course, follows one of our stated machining principles, namely the completion of as much machining as possible in the one setting, while the principle of roughing and finishing is readily achieved due to the large number of tool positions available.

Generally speaking, linear accuracy of dimensions turned at a capstan lathe will be about 0·04 mm, according to the size of the component and condition of the capstan lathe, not to mention the skill and experience of both capstan setter and operator.

Automatic lathes

As the name suggests, these lathes are automatic in operation, except for loading bars into the headstock of the machine. While the use of stops in a capstan lathe allows a semi-skilled operator to machine components without the aid of the skill and experience necessary when using indexing dials and measuring instruments, the use of *cams* now replaces the operator. The automatic lathe is set by the auto-setter who adjusts the cams that control the distance traversed by the cutting tool, and this principle is simply illustrated in Fig. 3.56C. Provided the automatic lathe is supplied with material, usually in the form of long bars, it will produce turned components without any need for an operator. This is an important step forward in machine-tool operations. Human beings are subject to many shortcomings; fatigue, boredom and loss of concentration are all common factors that tend to reduce the efficiency of an operator and increase the possibility of the production of faulty or reject components. A well made, accurate, case-hardened mild steel cam will control the linear movement of a cutting tool with precision and unerring accuracy, and will continue to do so over a long period of time.

It is interesting to note that the use of magnetic tape now makes possible the precise setting of machine tools without the necessity for a skilled machine tool setter. Such a technique is known as *numerical control*. With this system, all the movements of both work-holding and tool-holding devices are independent of human control; all the relevant information is received from a magnetic tape.

3.9 The shaping machine

Experience suggests that the shaping machine does not always receive the consideration it deserves. It is often thought of as the handmaiden of the machine shop or toolroom; for whilst a craftsman may consider himself as a skilled turner or miller, skilled shapers are relatively rare. It is wrong to think that little or no skill is required to operate a shaping machine; this machine is designed to produce plane surfaces, and its correct and efficient use calls for a very high degree of skill indeed.

It may be admitted that the shaping machine is a relatively simple machine to operate, with regard to both work-holding and tool-holding, but the use of a machine tool to best advantage, involves a great deal more than just operating it. We shall see, later on in this Unit, the sort of work possible on a shaper, and it is hoped that the technician will form a new opinion of the machining ability of the shaping machine, and accord it its rightful place in the engineering workshop.

We have seen that the production of flat or plane surfaces at the bench using chisels, files or scrapers, is a slow and laborious process necessitating a high degree of skill from the craftsman or technician. If, however, we are able to guide a cutting tool along a prescribed path, and at the same time allow for movement of the work along another prescribed path, it is possible to generate a plane surface. This is precisely what a shaping machine does. Not only is the plane surface produced more accurate than that produced at the bench, but also the size or area of the work presents no problems provided it is within the capacity of the machine. The surface is produced in a fraction of the time required by hand methods, and the accuracy well beyond that produced by even the most skilled craftsman. The shaping machine, therefore, is a surface-producing machine tool, best suited to the production of fairly small surface areas.

Essential movements of the shaping machine

The essential movements required from a shaping machine are simply illustrated in Fig. 3.57A. Note that the single-point cutting tool removes metal on the forward cutting stroke only; the return stroke is

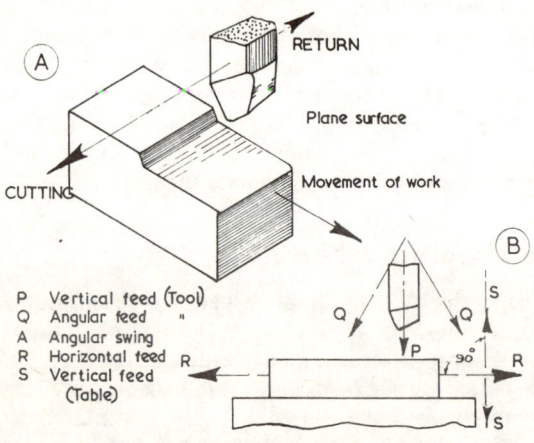

Fig. 3.57 Generating a plane surface

Machine Tools

unproductive and is usually carried out at a higher speed to increase the efficiency of the machine. In order that complete use is made of the shaper it is necessary to arrange for further work and tool movements. These are as follows:

  (i) Vertical feed of the cutting tool,
  (ii) angular feed of the cutting tool,
  (iii) horizontal feed of the table,
  (iv) vertical feed of the table.

Figure 3.57B shows a front view of a shaping-machine table, with arrows indicating the movements given above. It is evident that the design and construction of a shaping machine incorporating all the above movements, calls for the production of a rigid, compact and robust assembly of many machined parts.

The ram

The accuracy with which a shaper tool reciprocates along its horizontal path is determined by the precision of the ram assembly. A principle similar to that of the dovetail slide is adopted, the object being to constrain movement in one direction only. The method adopted is simply illustrated in Fig. 3.58A, where it may be seen that the sliding

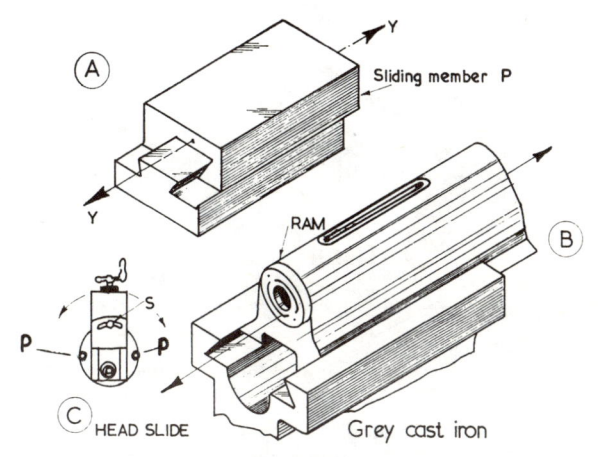

Fig. 3.58 Shaping machine ram

member P is able to move only along the path indicated by the arrow Y; movement in any other direction is not possible. This is also true for the simple shaping machine ram assembly shown in Fig. 3.58B; note that the ram is the sliding member, reciprocating in the body of the machine. The material used for both ram and body is certain to be grey cast iron, a cheap and easily cast metal, having a low friction bearing surface.

The head slide
This is shown in Fig. 3.58C, its function is to hold and support the single-point cutting tool, producing movement in both vertical and angular directions, with an indexing dial giving control over the distance moved. Linear motion of the slide is achieved through rotation of the handle, and slackening of the screw shown as s in the diagram allows the head to be tilted to the left or right, through about 60° of arc. This enables the tool to lift off the work when shaping a vertical or inclined surface. Slackening the two screws shown as P allows the whole head slide to be indexed through about 90° and then locked in position. In this way, angular or inclined faces can be shaped using hand feed of the head slide. This principle is identical in all respects to that used when machining tapered surfaces at the centre lathe, and another quick glance at Fig. 3.58C will serve to remind the technician of a centre lathe compound slide.

The worktable
The worktable of a shaping machine is usually a box-like structure, with T slots machined on both top and side faces as shown in Fig. 3.59. The workpiece is usually held in a large capacity machine vice

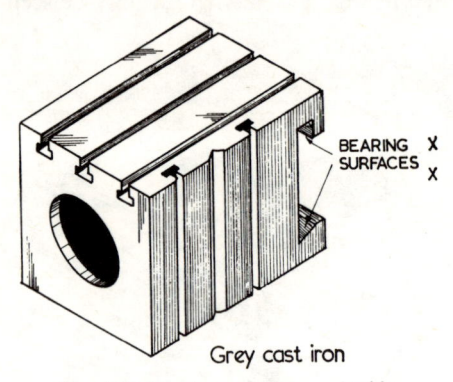

Fig. 3.59 Shaping machine worktable

bolted to the top of the table with suitable T bolts, and T slots at the side of the table permit the direct clamping of large castings. Once again grey cast iron is the metal chosen for the shaping machine worktable.

The table support
We have seen in Fig. 3.57B that the worktable of a shaping machine needs to be moved in both horizontal and vertical directions; these movements are made possible by the table support as illustrated in Fig. 3.60. The bearing surfaces shown as x in Fig. 3.60, locate, and

Machine Tools 169

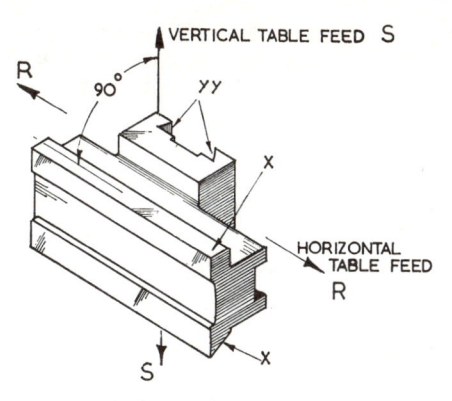

Fig. 3.60 Table support

are guided along the surfaces shown as x in Fig. 3.59, and in this way precise horizontal movement of the table is obtained in the direction of arrows R as indicated in Fig. 3.60. At the rear of the table support we see the bearing surfaces shown as Y in the diagram, and it is of the greatest importance to appreciate that these bearing surfaces are machined at precisely 90° to the bearing surfaces shown as x. If now we turn back to Fig. 3.57B, we may see the importance of this 90° relationship, for the accuracy of the 90° indicated between the movement S (vertical feed) and R (horizontal feed) depends on the precision with which the bearing surfaces on the table support have been machined.

The shaping machine body
This is a casting in grey cast iron, heavily supported or cross-ribbed on the inside to make it rigid and robust, and well able to support the ram assembly, table support and worktable. A typical shaping machine body is illustrated in Fig. 3.61; note the bearing surfaces machined to take ram and table support. The ram fits and slides on the bearing

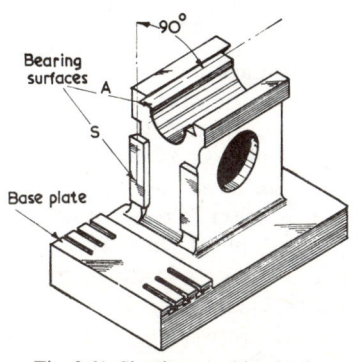

Fig. 3.61 Shaping machine body

surface shown as A, whilst the table support fits and slides on the surface indicated as S. It is evident that the bearing surfaces A and S on the body of the shaping machine must be in precise 90° alignment to each other. All that is now needed is to provide the ram with both forward and return motion, together with synchronised automatic feed or traverse of the table support. In other words, as the ram moves forward removing metal with a suitable single-point cutting tool, a simple mechanism prepares for a small linear movement of the table support; this movement to take place during the return stroke of the ram.

The assembled machine
Figure 3.62 shows an outline of the assembled shaping machine with

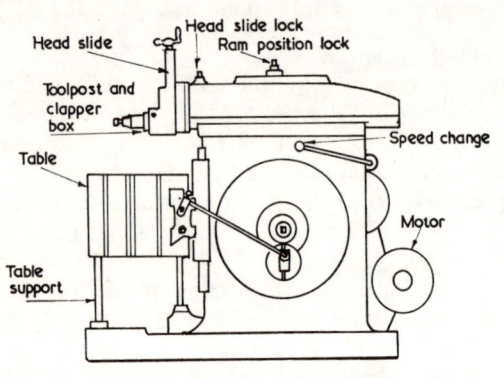

Fig. 3.62 The shaping machine

the main parts named. The following notes will not only help in identifying these parts, but also assist in recognising and carrying out the adjustments that may be required when shaping different types of work.

Length of stroke
This is the maximum distance moved by the ram and represents the length of the longest piece of work that can be machined. Shaping machines are available with length of strokes ranging from 350 to 800 mm; an average or medium sized general purpose machine has a stroke length of about 400 mm. Adjustment of the stroke adjusting screw enables the ram to be set to reciprocate at any particular stroke length to suit the job in hand. Figure 3.63A shows the principle involved. Rotation of the stroke adjusting screw turns the bevel gear shown in the diagram as A. This gear meshes with, and turns the bevel gear B, rotating the lead screw C. Rotary motion of the lead screw produces linear motion of the sliding member of the dovetail slide, and

Machine Tools

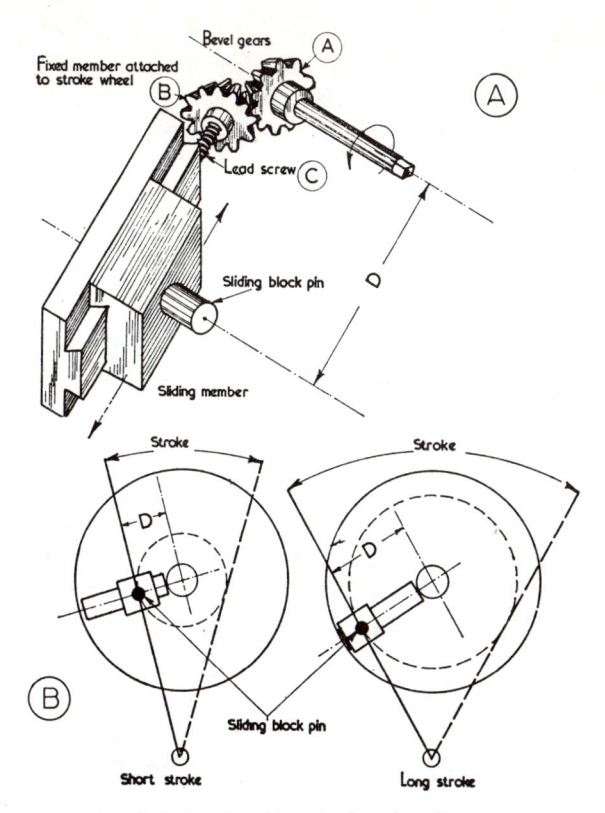

Fig. 3.63 Details of length of stroke adjustment

therefore this sliding member which carries a large diameter pin, can be brought inwards or outwards, according to the direction of the rotation of the stroke adjusting screw. The closer the pin is brought to the centre of the stroke wheel, which is simply shown in Fig. 3.63B, the smaller will be the stroke. When the pin is at its maximum distance from centre, the shaping machine ram will be set for its maximum length of stroke. Careful study of the simple representation of the mechanism shown in Fig. 3.63B will make clear the simplicity of the arrangement.

Movement of the ram

It is essential in the design and construction of machine tools, to ensure that as far as possible there are no moving parts on the outside of the machine which could be a source of danger to the operator. For this reason most of the mechanisms which activate the movements necessary to generate the surfaces required in engineering manufacture, are totally enclosed and out of harm's way within the body of the machine tool. This gives little opportunity to the technician to observe or study the basic principles that are used in machine tool construction, although an occasional visit to the maintenance department where machine tools are stripped down for repair or renovation would be well worth while. Figure 3.64 illustrates in a simple manner the

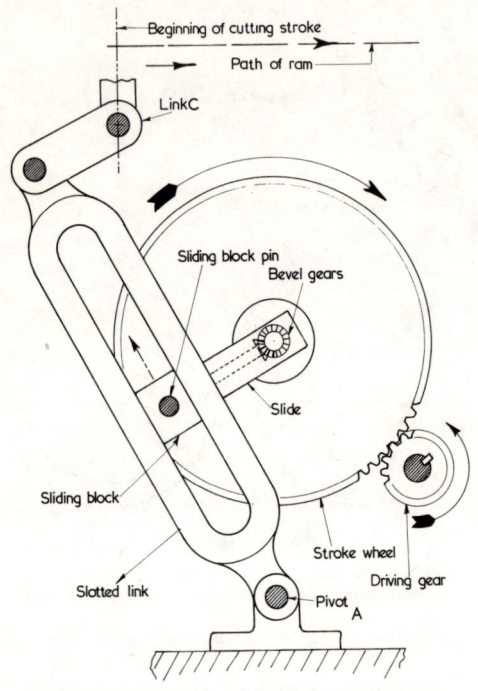

Fig. 3.64 Mechanism of the quick return action

basic movements needed to produce not only reciprocating motion of the ram, but also a *quick return* action. Reference to the diagram shows that the slotted link pivots at A, and the sliding block is constrained within the slot but free to slide along the slot. The stroke wheel is in a heavy spur gear which meshes with a smaller gear driven through a gear box by an electric motor. Rotation of the stroke wheel causes the sliding block to slide up and down, giving a reciprocating motion to the slotted link. The small connecting link at the top is needed because the point A at the top of the slotted link describes an arc, whereas it is necessary for the ram to move in a straight path.

The geometry of the quick return motion is shown in Fig. 3.65. Point A represents the top of the slotted link at the commencement of the cutting stroke. As the stroke wheel revolves in an anti-clockwise direction, point A describes an arc as it moves from left to right, and we

Machine Tools

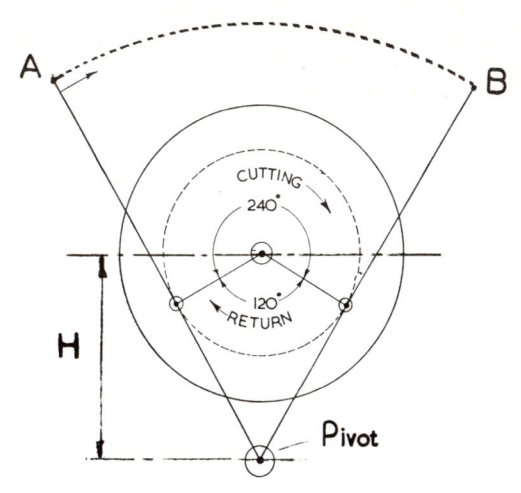

Fig. 3.65 Principle of the quick return action

see at once the need for a small connecting link. When the stroke wheel has rotated through 240°, point A will have moved to B as shown in the diagram, and this is the cutting stroke. Continuing rotation of the stroke wheel through 120°, brings the top of the slotted link back to A its starting position, and this is the return non-productive stroke. Because the stroke wheel rotates at constant speed, it is clear that the return stroke moves twice as fast as the cutting stroke, and we may express the ratio in mathematical terms as follows:

$$\frac{\text{Cutting time}}{\text{Return time}} = \frac{240°}{120°} = \frac{2}{1}$$

The value of the cutting/return ratio is determined by the distance shown as H in Fig. 3.65; the smaller this distance the greater will be the ratio between the cutting and return times, and the more efficient the shaping machine.

Positioning the ram

The positioning of the ram is a relatively simple matter, and its necessity can be seen by reference to Fig. 3.66. Clearly it is wasteful of both time and energy to use the set-up shown when shaping the top face of the component. We need not only to reduce the length of stroke, but also to position the ram so that the cutting tool starts and finishes its cut at about 10 mm from the job. Figure 3.67 shows the details of the means by which the position of the ram is adjusted. Slackening of the tightening nut shown as A, allows the ram to be pushed into any desired position, then locked by tightening the nut: a hard rubber or hide mallet is used for this purpose.

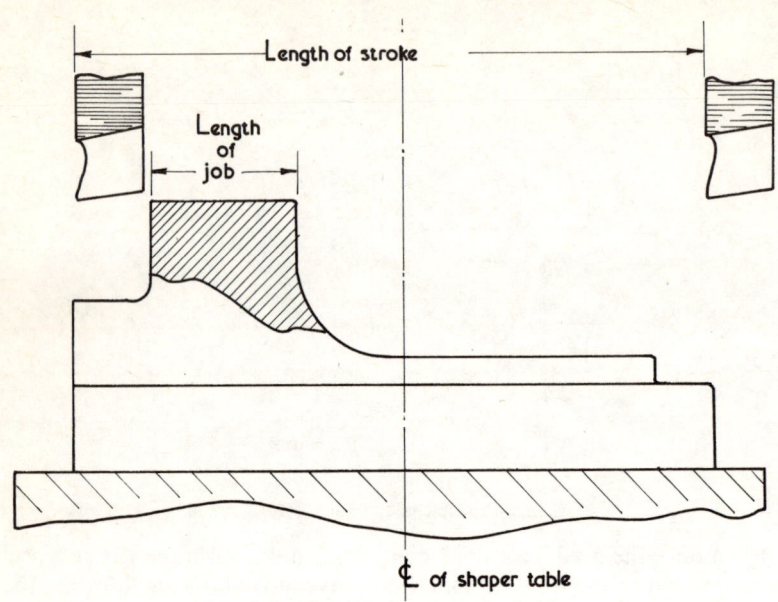

Fig. 3.66 Need for changing ram position

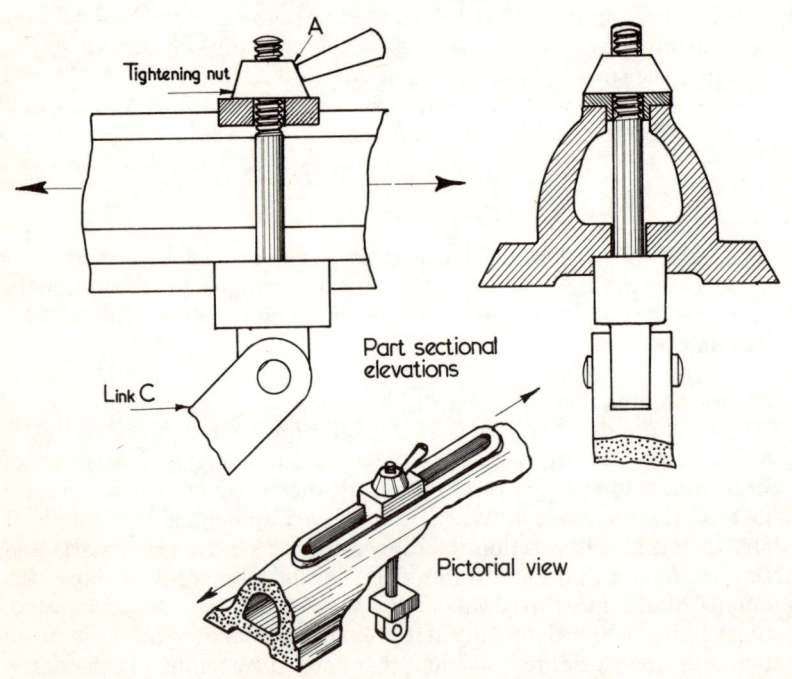

Fig. 3.67 Details of ram position adjustment

Ram speeds
In the same way that a drilling machine is provided with a range of spindle speeds to suit the different sizes of drill diameters, a shaping machine is equipped with a range of ram speeds. The speed of a ram refers to the number of strokes per minute, a stroke consisting of both cutting and return movements. A range of ram speeds suitable for a 400-mm stroke shaper is given below.

Table 3.2 Speeds available on a 400-mm shaper

Speed No.	1	2	3	4
Strokes per minute	18	30	57	109

It is unwise to use high speeds for large workpieces for the ram will travel at a fairly high speed, setting up undue strain and vibration. The correct procedure is shown in Fig. 3.68 where it may be seen that short strokes require fast speeds, whilst long strokes require low speeds.

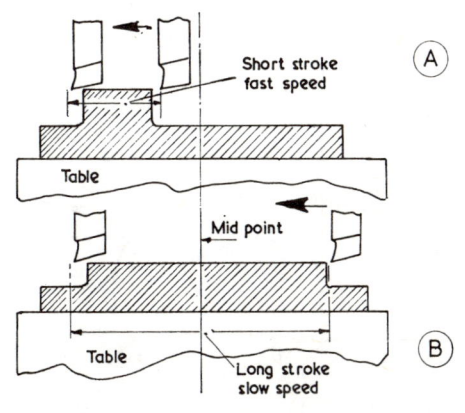

Fig. 3.68 Speed and length of stroke

Setting the tool
One of the secrets of successful machining is maximum rigidity of both cutting tool and workpiece. With regard to the single-point tools used in shaping machines, it is very important to ensure that the maximum section tool is used. It is customary to specify the maximum section to be used in a shaping machine of given capacity, for example the recommended tool section for a 400-mm stroke shaper is given as 20 × 30 mm. The reason for this apparent solidness is due to the fact that the tool is subject to an impact at the commencement of each cutting stroke, as may be seen by reference back to Fig. 3.68; the tool must be below the top surface of the work if any metal is to be removed. The

deeper the cut, the greater will be the impact, and as the object when roughing is maximum metal removal, it is clear that a strong solid large section tool is needed if it is not to deflect or bend when roughing-down using deep cuts.

The clapper box
The tool-holding device on a shaping machine is more commonly known as a clapper box, and is an interesting example of how engineers solve the problems that continually occur in the design and manufacture of machine tools. We have seen that the action of a shaping machine ram is to and fro, or to put it in engineering terms, the ram has reciprocal motion. This is unlike the drilling or turning of metal when the cutting action is continuous, and reference to Fig. 3.69 will show

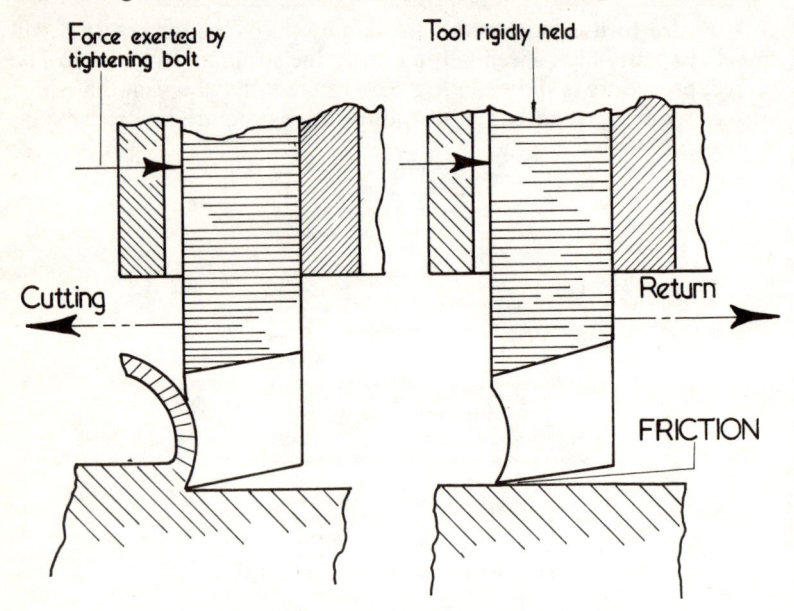

Fig. 3.69 Rigid holding of shaper tool

that there is severe friction and wear on the tool point during the return stroke of the ram. This will greatly shorten the useful life of the tool, necessitating more frequent tool-changing with consequent loss of production caused by the machine lying idle during the time the tool is re-ground. If however, the tool could be made to lift on the return stroke, all the above problems are solved, and the device that makes this possible is called a *clapper box*.

The principle of the clapper box is shown in Fig. 3.70, with the main parts named. It may be seen that the cutting tool is held in a toolpost

Machine Tools 177

which is a drive fit in the tool holder. This tool holder pivots on pin A, and during the cutting stroke the tool is rigidly held with the cutting force taken up on the face shown in the diagram as XY. On the return stroke the tool holder is free to rotate on pin A, causing the tool point to lift, and rest lightly on the work surface during the return stroke. It is evident from the diagram that tool changing is a rapid and simple affair; all that is required is slackening off of the screw shown as B, and removing the tool. Note that the side elevation shown in Fig. 3.70

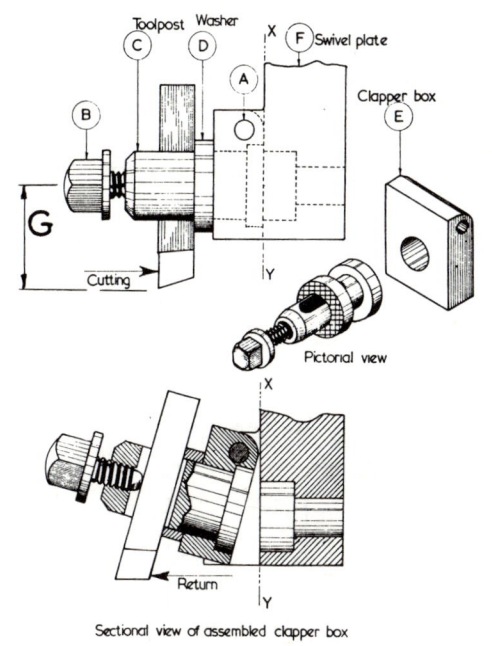

Fig. 3.70 Clapper box details

represents good machine technique, because the tool overhang shown as G in the diagram is at a minimum, so promoting maximum rigidity.

Tilting the clapper box
This is a further adjustment that can be made, and is carried out to reduce the frictional effects of tool rubbing on the return stroke when shaping vertical or inclined surfaces. It is not a simple principle that can be readily appreciated, and provides yet a further example of the complexities of a shaping machine. Careful reference to Fig. 3.71 may help to illustrate the principle involved. Let us assume that a vertical cut is being taken as shown in Fig. 3.71A. The axis of the *pivot* pin, around which the clapper box rotates slightly in order to lift the tool

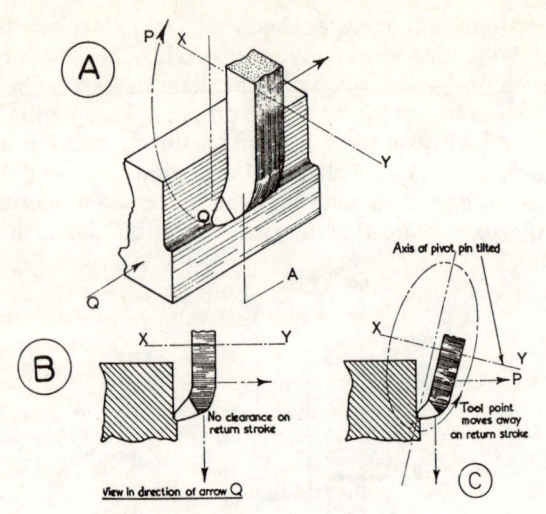

178 *Technician Workshop Processes and Materials*

Fig. 3.71 Tilting the clapper box

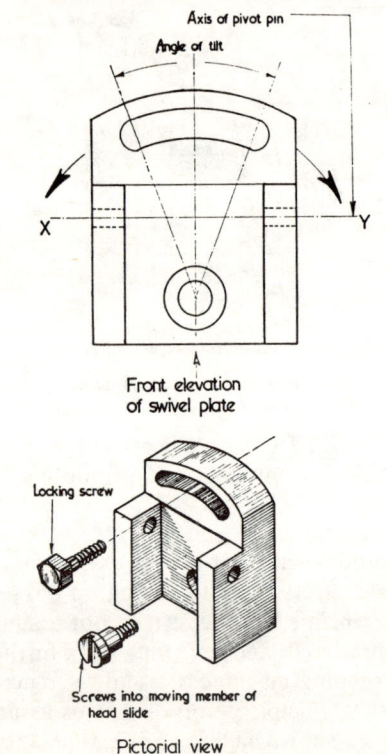

Fig. 3.72 Details of swivel plate

Machine Tools 179

for the return stroke is shown as XY. On the return stroke the tool point will follow the arc shown as OP in the diagram, and reference to the front view in Fig. 3.71B, shows that the tool point rubs on the return stroke because there is no movement *away* from the machined surface, although there is upward movement as the tool lifts. If now, the axis of the pivot pin is tilted as shown in Fig. 3.71C on the return stroke, lifting of the tool causes the point to follow an arc that lies on the plane OP, hence the tool point now has a side movement *away* from the machined surface, indicated by the arrow P.

Tilting of the axis pin is achieved by a small angular movement of the swivel plate, locked in position by a tightening screw located in a small radial slot. Figure 3.72 shows details of this swivel plate which locates on the moving member of the shaping machine head slide.

3.10 Shaping techniques

We have seen that the shaping machine possesses several ingenious mechanical devices, and is a good example of how engineers apply simple mechanical principles to achieve the essential geometry and movements that are inherent in all types of machine tools. Because of the ease with which the tool and work can be held, there is a tendency as previously stated, to regard the shaping machine as a somewhat inferior machine tool. There is no justification for this attitude, for the cost of a good quality shaping machine will differ little from that of a good quality centre lathe. If the technician is to use a shaping machine to best advantage, full use must be made of the geometric accuracy built into the machine, and the following shaping examples are intended to illustrate the correct approach to shaping technique.

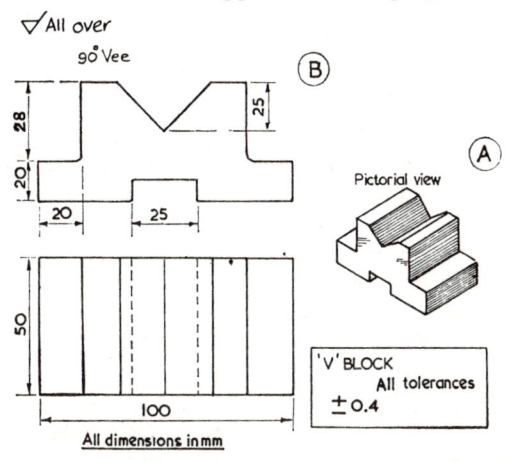

Fig. 3.73 Machining a Vee block

Examples of shaping

Figure 3.73A shows a pictorial view of a Vee block which is to be machined using a shaping machine. Although a pictorial view gives a clear picture of the finished job, it is extremely difficult to dimension, and for this reason, engineering components are always shown in what is known as *orthographic* projection. This is shown in Fig. 3.73B, and comprises of two views, a front elevation and a plan. Note the symbol on the top left-hand corner of the front elevation; this indicates that the Vee block is to be machined all over. All dimensions are in millimetres and the general machining tolerance is given as plus and minus four-tenths of a millimetre.

Operation 1

The size of the Vee block makes it quite suitable for holding in a machine vice, but we need to ensure that the cutting force is taken by the vice jaws as shown in Fig. 3.74B. Before machining however, we need to check that the vice jaws are truly at 90° to the path of the ram,

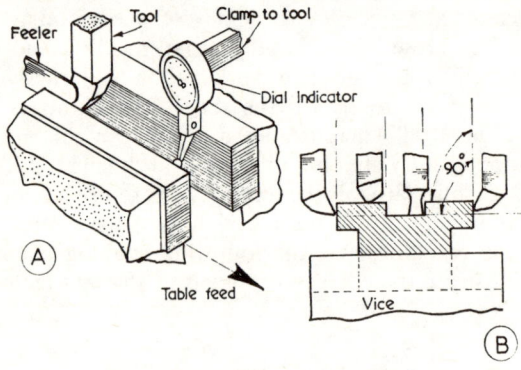

Fig. 3.74 First set-up for Vee block

and a feeler gauge or lever-type dial test indicator may be used as shown in Fig. 3.74A. The essential technique is to clamp the dial test indicator to the tool or toolpost, and adjust the ram until the pointer of the dial test indicator reads zero when in light contact with the vice jaw. The worktable is now fed by hand with a careful watch being kept on the pointer, which must maintain a zero reading while traversing the full length of the jaw. Any deviation from zero may be corrected by slackening off the vice clamping screws and gently tapping the vice body in the direction required.

It is a general rule when machining that the largest available area is selected for the first *datum* face, and at Fig. 3.74B, we see the set-up for the first operation. Although the datum face is first machined as shown in the diagram, note may be taken of the fact that both the recess and

Machine Tools

slides are also machined at the same setting, care being taken to remember that the clapper box needs tilting when shaping the two vertical sides. Both external and depth micrometers may be used during the whole of the first setting to ensure that the linear dimensions of width and depth are within the limits of size indicated on the drawing.

It is essential to appreciate that *four* different shaping tools are needed, as shown in Fig. 3.74B, but provided these tools are to hand, no problem ensues. This is due to the rapidity and simplicity of tool changing on a shaping machine, making possible a high degree of accuracy with regard to the parallelism and 90° relationship of the surface shaped in the one setting. It will be seen from the diagram that the end faces are machined with vertical feed. If the shaping machine table is fitted with automatic vertical traverse, the position of the head slide is of little importance, but if the faces are to be machined using vertical hand feed of the head slide, it is essential to ensure that this feed is truly 90° or vertical to the worktable. This is simply checked using a dial test indicator and an accurate try-square, as shown in Fig. 3.75.

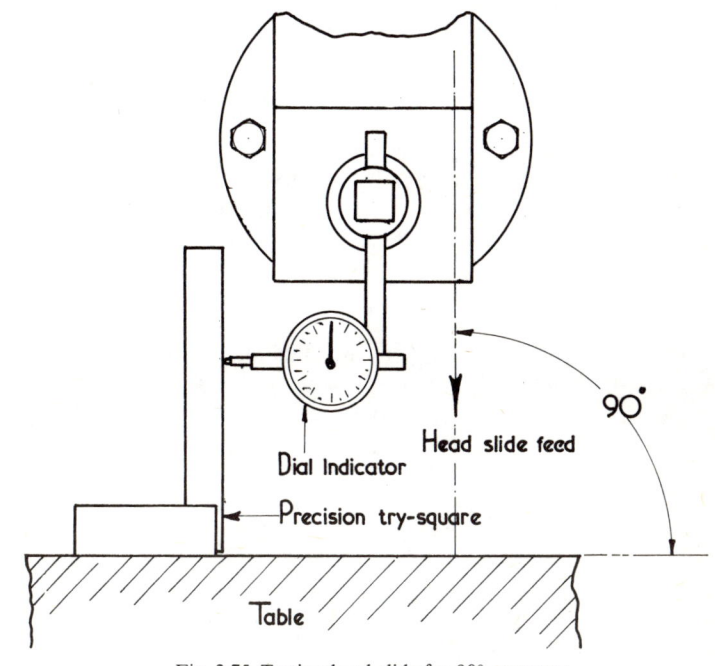

Fig. 3.75 Testing head slide for 90° accuracy

Operation 2
The second operation consists in shaping the front face of the Vee block as shown in Fig. 3.76A. This face requires to be 90° to the datum

face machined in the first operation and also 90° to the sides. These geometric relationships are shown in Fig. 3.76A, and at B we see the set-up for the second operation. The datum face is located against the fixed jaw of the machine vice, which is at 90° to the surface of the worktable, so that shaping the front surface will bring it at 90° to the datum face. The 90° relationship for the sides of the Vee block is obtained as shown as Fig. 3.76B, where an accurate try-square is used to square up the Vee block.

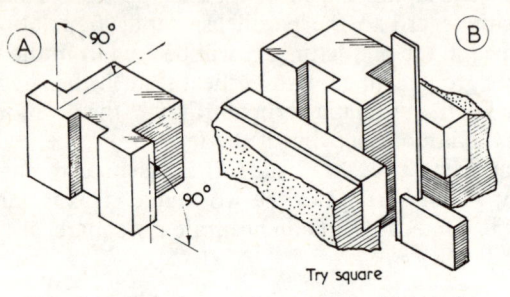

Fig. 3.76 Second set-up for machining Vee block

Operation 3

This is identical in all respects to the previous operation because we now need to shape the opposite face to that machined in Operation 2. Once again the datum face locates against the fixed jaw of the vice, with parallel bars used to locate the face machined in the second operation. We have in effect, turned the Vee block upside down for Operation 3.

Operation 4

This is the last set-up to complete the machining of the Vee block and comprises the machining of the top face, the Vee slot, and the side face F as illustrated in Fig. 3.77A. Note the use of parallel bars to ensure

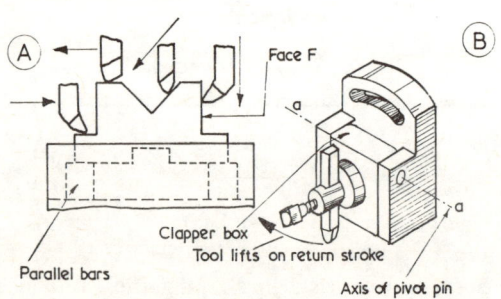

Fig. 3.77 Final set-up for machining Vee block

Machine Tools

that the datum face is parallel with the vice base; no other setting precautions are needed except that the Vee block rests firmly on the parallel bars. The fixed jaw of the machine vice has already been checked for alignment prior to Operation 1. We need to remember to tilt the clapper box for the vertical faces, in order that the tool point clears on the return stroke, and a front view of the actual tilt for the right-hand face is shown in Fig. 3.78B. When shaping the left-hand face the clapper box is tilted in the opposite direction.

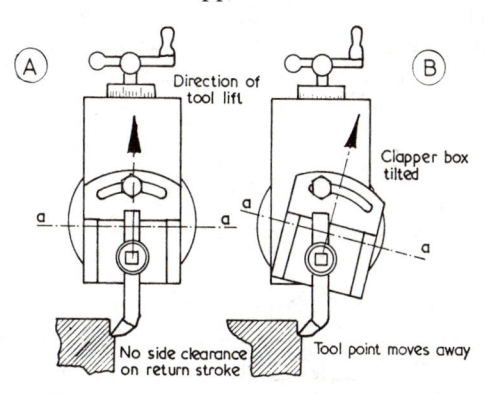

Fig. 3.78 Clapper box tilt

Shaping the Vee slot

The completion of the top face of the Vee block in one setting entails tool changing for the horizontal and vertical surfaces, together with tilting the clapper box. In order to shape the two sides of the Vee slot it is necessary to index the head slide at 45° each side of zero. This means that *two* settings of the head need to be made, and hand feed must be used. This is simply illustrated in Fig. 3.79; note also the clapper box

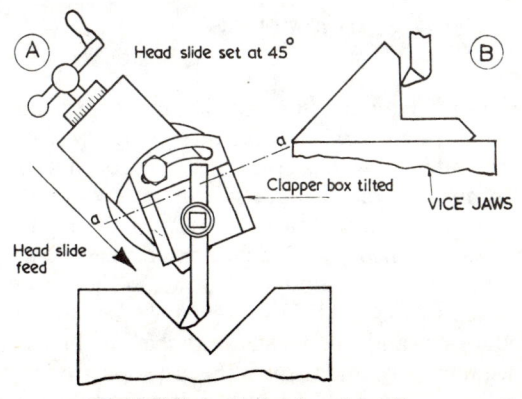

Fig. 3.79 Indexing head to shape Vee

tilt. An alternate method consists in re-setting the Vee block so that vertical feed of the head slide can be used, together with horizontal traverse of the table. This technique can be seen in Fig. 3.79B, and it is clear that the side face of the job must be set at precisely 45° to the worktable.

Best use of the machine vice
Machine vices may be rotated through 360° or indexed at any particular angle, and full use should be made of this device. A good example is the need to machine *flats* on round bars. This is most easily done on a shaping machine with the work held in a pair of Vee blocks. The technique is illustrated in Fig. 3.80, which represents a plan view of the

Plan view of set-up

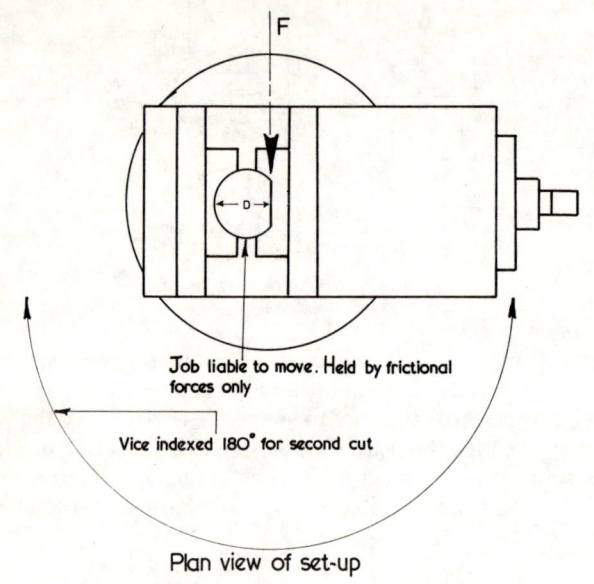

Fig. 3.80 Indexing shaper vice through 90°

shaping operation. The path of the shaping machine is indicated by the arrow F, and one side is first shaped, care being taken to ensure that the correct amount is removed. This is easily checked using an external micrometer to measure the distance shown as D. When this dimension is attained, the vice is indexed through 180° as shown in the diagram, and the other side shaped; once again a micrometer is used to ensure that the width is within the limits laid down in the drawing.

Reference back to Fig. 3.80 shows that the vice jaws are parallel to the path of the ram, and this means that the Vee blocks are held by frictional resistance only, and there is the grave possibility of the set-up moving under the influence of the cutting force shown as F in Fig. 3.80.

Machine Tools

This cannot be regarded as good machining practice, and a better method would be to use the vice jaws as solid stops or restraints. This basic, but vital principle is simply shown in Fig. 3.81, where it may be

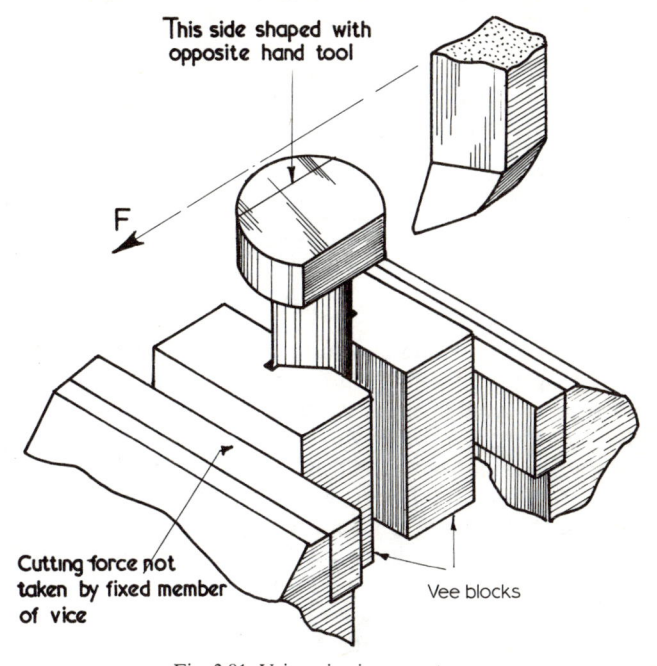

Fig. 3.81 Using vice jaws as stops

seen that the cutting force shown as F in the diagram is now taken by the vice jaw. Reference back to Fig. 3.80 shows that it is the moving member of the machine vice that takes the cutting force, and a better set-up would be to use the fixed jaw, but this is not possible because the tightening screw must face the operator.

Shaping profiled surfaces
Apart from shaping horizontal, vertical and inclined surfaces, the shaping machine may be used to machine a profiled component in fairly thick material. Figure 3.82A shows a tool-steel die used in a press tool. The thickness of this die is 25 mm, and skilful use of a shaping machine head slide permits reasonably rapid metal removal down to the scribed line. Note the use of a round-nose tool for the radiused corners, and it is clear that much use of the head slide is called for.

Safety at the shaping machine
Generally speaking the shaping machine is not considered as a potentially dangerous machine tool. Apart from the danger of chips leaving

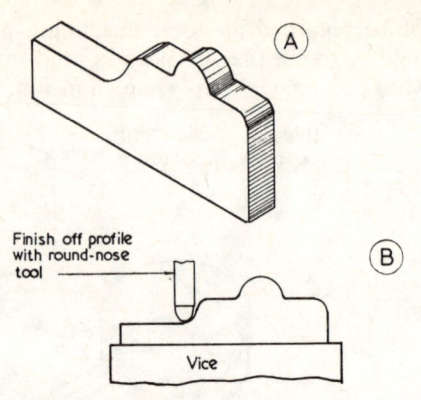

186   *Technician Workshop Processes and Materials*

Fig. 3.82 Shaping profiled surfaces

the workpiece when shaping at a fairly fast stroke speed, the hazards present when operating a shaping machine are readily identifiable. A small chip guard placed at the front of the machine will effectively stop the chips from flying across the workshop, while the obvious danger of the reciprocating ram ensures that hands are kept well away from the work area during the shaping operation. It is important to ensure that maximum rigidity of both tool and work holding is observed at all times, and Fig. 3.83 shows the correct approach. At A we see a good

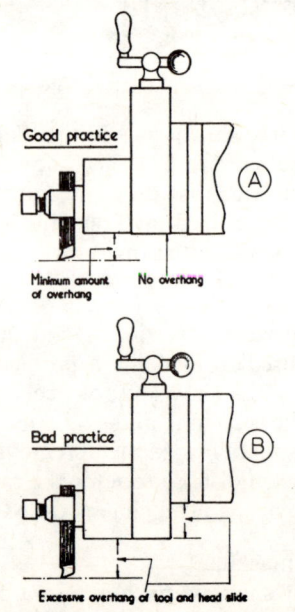

Fig. 3.83 Importance of maximum rigidity

set-up, while at B both tool and head slide have excessive overhang, leading to vibration, with consequent loss of accuracy and poor surface finish.

3.11 Cutting tool theory

A cutting tool is used to remove metal. If the tool is held by hand it is called a hand tool; examples of hand cutting tools are

  (i) cold chisels,
  (ii) files,
  (iii) hacksaw blades.

The wedge principle
Although the hand tools listed above are described as cutting tools, the action of metal removing is one of *shear* rather than cutting. A closer look at the cutting section of a flat cold chisel is given at Fig. 3.84A,

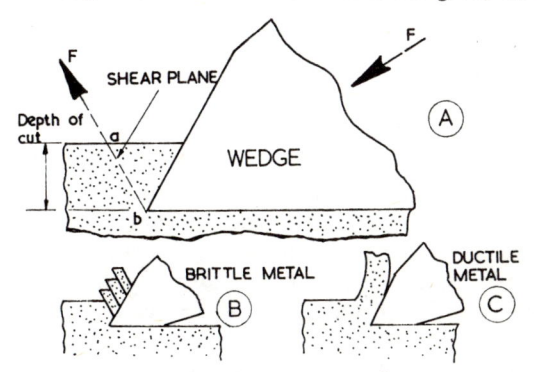

Fig. 3.84 The wedge principle

and represents a side view of the cutting edge in contact with the workpiece. The force of the hammer blow is shown as F, and it may be seen from the diagram that this force is transmitted from the chisel to the work along the plane *ab*, and the metal along this plane is subjected to a shear force. If the metal is brittle, as grey cast iron or brass, there is a tendency for the shear force to cause the metal to break or crumble, producing small separate chips as shown in Fig. 3·84A. If on the other hand, the metal is fairly ductile or soft, as mild steel or copper, the tendency will be to produce a continuous or unbroken chip.

Tool angles
If a cutting tool is to shear or cut metal efficiently it is important that the angles that make up the wedge shape be of the correct values

according to the nature of the machining operation and the type of metal being machined. Figure 3.85A shows a pictorial view of the wedge principle, with the angles clearly indicated and named.

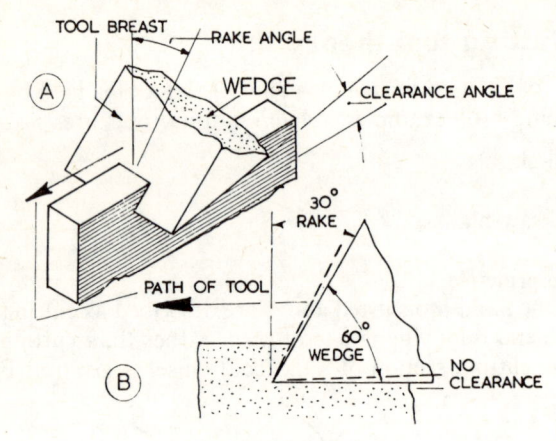

Fig. 3.85 Tool angles

Rake angle
The rake angle of a cutting tool is the angle or inclination of the tool breast to a line drawn perpendicular to the path or feed of the tool. This is more clearly shown in the side elevation shown in Fig. 3.85B. Note that the example given is that of the cutting point of a flat cold chisel, suitably ground for chiselling mild steel. The rake angle is 30°, the wedge angle 60°, and there is no clearance angle, as can be seen from the diagram. We may consider the rake angle as an *effective* rake angle, because any tilt of the chisel would reduce the rake angle, and the angle of tilt would be equal in value to the clearance angle produced by the tilt.

Clearance angle
The purpose of a clearance angle is to remove the friction arising from the rubbing action between tool and work. Clearly a slow moving tool such as a cold chisel will suffer little frictional resistance, and for this reason it is not strictly necessary to have a clearance angle. Better control over the depth of cut is obtained by keeping the heel of the chisel flat on the workpiece as shown in Fig. 3.85B. If, on the other hand, a cutting tool moves at fairly high speed, as is the case with twist drills and milling cutters, then the frictional resistance or rubbing action increases with the speed of movement, resulting in considerable energy being required to overcome this friction. Machining operations carried out on machine tools always involve high speeds with maxi-

Machine Tools 189

mum metal removal as the main aim, making clearance angles vitally necessary. It is a general rule that clearance angles are kept as small as possible, seldom exceeding 10°.

Rake and clearance angles on hand tools
The following examples are intended to show that the wedge principle forms the basis of all hand cutting, and that the rake angle has an important effect on the efficiency of the cutting action.

Hand files
All hand files are essentially multi-point cutting tools, that is to say their cutting faces are made up of a large number of small wedge- or tooth-like points. Figure 3.86A shows an enlarged section of a typical

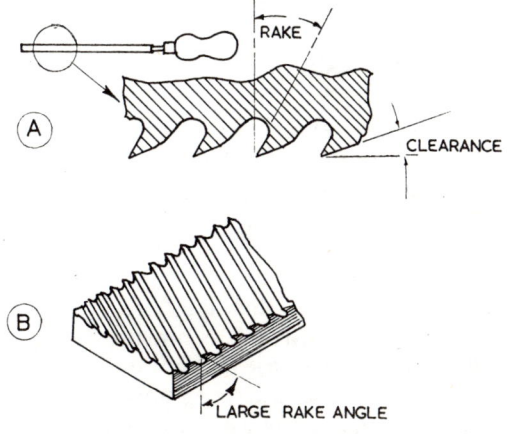

Fig. 3.86 Rake and clearance angles on files

file, note the rake and clearance angles for each tooth. Clearance angles are very necessary, for although a hand file is a slow moving cutting tool there are a very large number of teeth, and the frictional or rubbing action would be considerable, resulting in undue fatigue of the user. In general, the softer the metal, the bigger the rake angle; soft metals such as copper or aluminium need special files with large rake angles on the teeth. Such files are called *rasps* or *dreadnoughts*, and at Fig. 3.86B we see a part view of a rasp suitable for filing aluminium. Note the large rake angle and the sloping direction of the teeth across the file face.

Hacksaw blades
The cutting action of a hacksaw blade is identical to that of a hand file, and we may regard a hacksaw blade as a very thin-section file. Once

again clearance angles are very necessary because of the large number of teeth present on a blade, and Fig. 3.87A shows an enlarged view of a few hacksaw-blade teeth. It may be noted that each tooth has its own

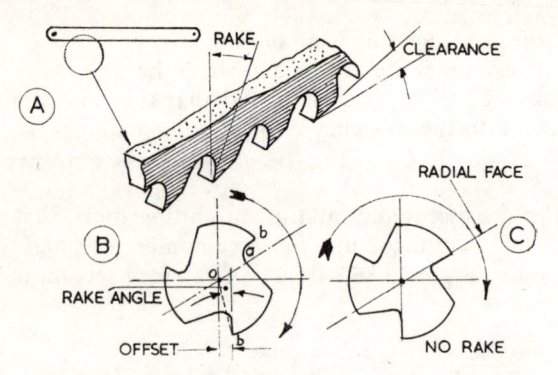

Fig. 3.87 Rake and clearance angles on hacksaw blades

rake and clearance angle. Hacksaw teeth are also slightly angled to the line of the blade, and this allows the blade to cut slightly larger than the width of the blade, giving clearance and an easier cutting action; this is known as the *set* of the teeth.

Taps
We have seen, in earlier work, that taps are used to cut internal threads. The tapping of a drilled hole is no easy matter, for taps are hardened and tempered, fairly brittle and not difficult to break. At the same time there are no clearance angles on taps, leading to a considerable amount of friction and resistance during the cutting action. The flutes of a tap provide cutting faces for the teeth and, provided these flutes are machined *off-set* as shown in Fig. 3.87B, each tooth is provided with rake angle. It is not a difficult matter to calculate the off-set in order to provide a tap with a required angle, and the following formula may be used.

$$\text{Amount of off-set} = R \sin A$$

where R = radius of the tap, and A is the required angle in degrees. For example if a tap of 15 mm diameter is to have a rake angle of 20°, then the off-set is found as follows:

$$\begin{aligned}
\text{Amount of off-set} &= R \sin A \\
&= \frac{15}{2} \times \sin 20° \\
&= 7 \cdot 5 \times 0 \cdot 3420 \\
&= 2 \cdot 565 \text{ mm}
\end{aligned}$$

Machine Tools

If, however, the flutes are milled with no off-set, that is to say the cutting faces are on a radial plane as shown in Fig. 3.87C; the tap would have *negative* or no rake angle. Such a tap would be most unsuitable for tapping threads in soft metals, but would be suitable for grey cast iron.

Button and loose dies
A button die is used in conjunction with a suitable holder or stock to cut an external thread on a round bar, as may be seen by reference back to Fig. 2.18. Once again the provision of rake angles on the cutting faces of the dies will greatly increase the efficiency of the thread-cutting ability of the die, and the principle adopted is simply illustrated in Fig. 3.88A. Note that the centres of the four holes are off-

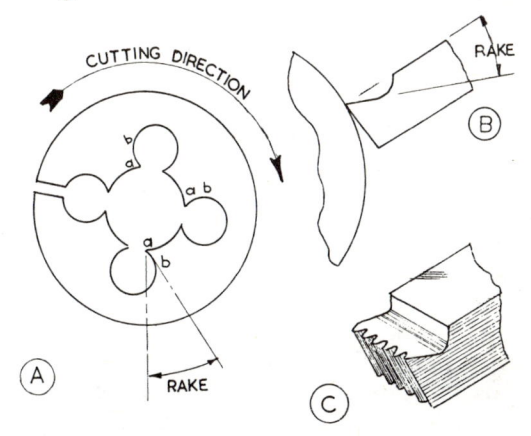

Fig. 3.88 Rake angles on thread cutting dies and taps

set from the centre-line of the die, and this has the effect of giving a rake angle shown as *ab* in the diagram, on the cutting faces of the dies. Clearly it is difficult, if not impossible to re-grind these faces, and this has led to the introduction of loose dies which greatly simplify sharpening. Fig. 3.88B shows how a loose die is applied to the workpiece, while at B we see a pictorial view of a typical loose die.

3.12 The single point cutting tool

Single-point cutting tools find a wide application on lathes, shaping machines, planing and slotting machines. It is important that the principle or theory of metal removal using single-point cutting tools be understood, and it is a good plan to consider the cutting tool as a wedge. This is a continuation of the technique adopted when considering the cutting action of a flat cold chisel when removing metal at the bench.

Figure 3.89A shows a single-point tool removing metal at 90° to the lathe centre-line. It will be seen that the cutting or shearing action is similar in all respects to that of a flat cold chisel, except that the force

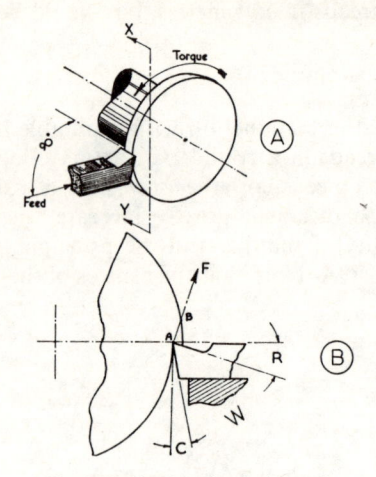

Fig. 3.89 Single-point lathe tool

causing a shear stress is the equal and opposite of the *force* component present in the *torque* or turning movement of the rotating workpiece. This is more clearly shown in Fig. 3.89B, where we see that the shear plane on the workpiece is represented by the line AB. The rake angle is shown as R, and the clearance as C; the wedge angle of the tool indicated as W. In order to make a closer comparison with the

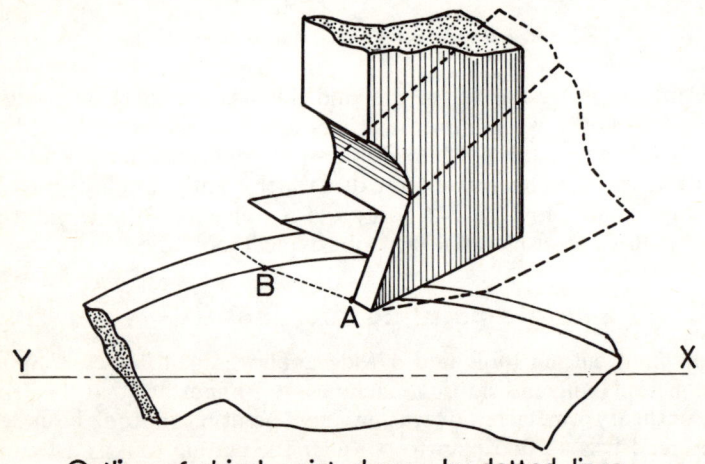

Outline of chisel point shown by dotted lines
Fig. 3.90 Comparison between lathe tool and cold chisel

Machine Tools

cutting action of a cold chisel, and to show clearly the wedge principle, Fig. 3.90 is a pictorial representation with the line XY on a horizontal plane. The dotted lines show the outline of a flat cold chisel, and it may be seen that the shearing action is identical. Note that the energy required to remove the chip will be proportional to the length of the shear plane AB, the longer this line, the greater will be the energy needed to shear the metal.

Importance of the Rake Angle

With aid of Fig. 3.91 we can appreciate the importance of the rake angle. The workpiece revolving about its centre has a turning movement or torque which is the product of force × radius.

$$\text{torque} = \text{force} \times \text{radius}$$

Clearly the revolving work exerts a tangential force on the tool, and provded the tool is rigidly supported, an opposite and equal reaction force acts on the work. This reaction force has a direction of about 90° to the cutting face or breast of the tool as shown in Fig. 3.91A, setting-up the shear plane AB. If the rake angle is now increased, as shown in Fig. 3.91B, the effect is to reduce the shear plane, and also the energy requirements to remove the metal. It is clear, that as the rake angle is increased the wedge angle is decreased, making the tool point considerably weaker and less able to stand up to the cutting stresses. We see now the need for ensuring that all clearance angles are kept to a

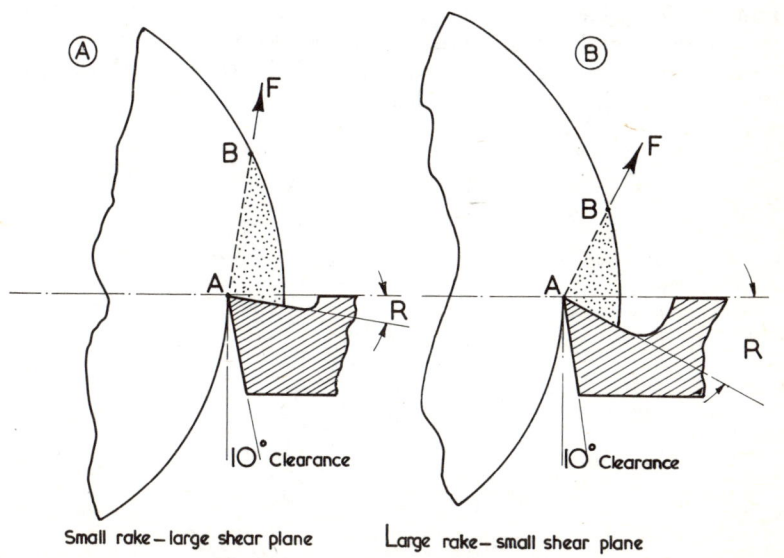

Small rake—large shear plane   Large rake—small shear plane

Fig. 3.91 Importance of the rake angle

minimum, for excessive clearance angles also reduce the wedge angle and hence the useful life of the cutting tool.

The actual rake angle ground on a cutting tool depends mainly on the metal to be machined, and will be a compromise between the maximum angle possible and the strength of the tool. The following table gives the rake angle and cutting speeds for the more common engineering materials, when using a high speed steel single-point cutting tool.

Table 3.3 Rake angles and cutting speeds.

Material	Rake angle	Metres per second	Metres per minute
Mild steel	20	0·3–0·45	18–27
Cast steel	12	0·2–0·3	12–18
Grey cast iron	0–6	0·3–0·45	18–27
Brass	0–6	0·8–1·25	48–75
Copper	35	1·25–2·0	75–120
Aluminium	35	1·25–2·0	75–120
Perspex	45	1·75–2·0	105–120
	Degrees	m/s	m/min

Radial cutting

In all the foregoing examples of the use of single-point cutting tools at the centre lathe, the radial cutting principle has been used. By this we mean that the cutting tool is presented radially to the workpiece, as shown in Fig. 3.92A, and it is clear that the force component of the

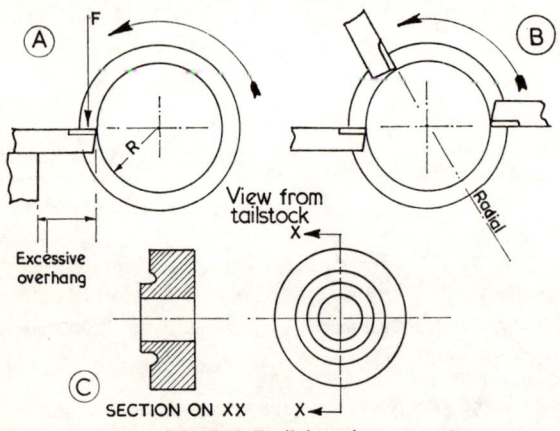

Fig. 3.92 Radial cutting

Machine Tools

torque acts downwards. There is, therefore, a tendency to deflect the cutting tool, and this leads to vibration and poor surface finish. The excessive overhang shown in the diagram represents poor machining technique, because it is vital that the cutting tool be given maximum support, although this is not always practicable; for example when parting-off at the centre lathe. It may be appreciated at this point, that several tools can be applied to a revolving workpiece, and provided each tool lies on a radial plane there is no variation in their cutting actions. This is simply illustrated in Fig. 3.92B, and the technique finds great use in *automatic lathes*, where several tools are applied to the workpiece in the manner shown in the diagram. In this way, automatic lathes are able to produce components at very high production figures; for example, the bronze pressure pad illustrated at Fig. 3.92C requires ten separate machining operations yet is produced in the remarkable time of seven seconds.

Tangential cutting

As the name suggests, the principle of tangential cutting consists of presenting a tool to the workpiece so that it lies along a tangential plane as shown in Fig. 3.93A. The cutting force is now taken through

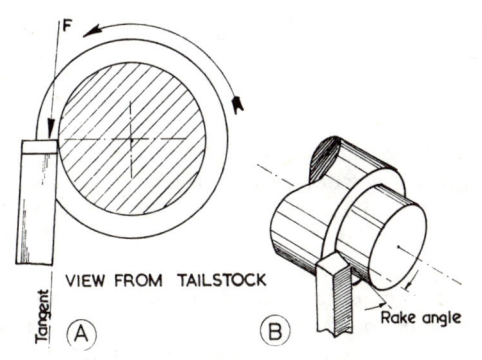

Fig. 3.93 Tangential cutting

the strongest section of the tool with no tendency for bending or deflection. Special tool-holders are needed for the application of this principle but high-metal-removal rates are possible, making the technique very suitable for production turning on automatic or capstan lathes. Figure 3.93B gives a pictorial view of a tangential tool, and it may be seen that whilst rake angle needs to be ground on the top face of the tool, clearance is obtained by presenting the tool at a slight angle or inclination.

Tool setting

In addition to ensuring maximum rigidity when tool holding, it is essential that the single-point cutting tool be properly set. Using an incorrectly set tool causes a change in the effective rake and clearance angles, and reference to Fig. 3.94A shows a pictorial view of a parting-off tool about to be fed at 90° to the centre-line of the work. At B we see an end view of the set-up, as viewed from the tailstock, and it is clear that provided the parting-off tool is on centre as shown, both rake and clearance angles are correct. Figure 3.94C shows the effect of

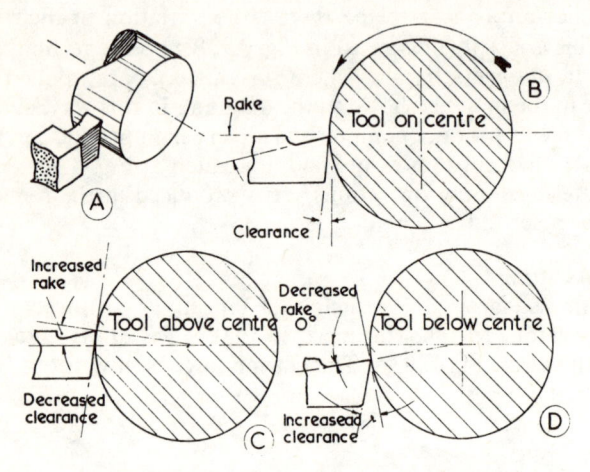

Fig. 3.94 Importance of correct tool setting

setting the tool *above* centre. The effective rake angle now increases, whilst the clearance angle decreases to a negative value. This results in the tool rubbing the work, producing a rough finish and causing loss of dimensional accuracy. At D we see the effect of setting the tool *below* centre, and the change in effective rake and clearance angles is now reversed. The rake angle has now decreased with an increase of the clearance angle, and once again the finish of the machined surface is likely to deteriorate. It is true, however, to state that the above examples refer to a badly set tool; slight variation from dead centre, say up to 0·25 mm, have little noticeable effect on the machining operation.

Orthogonal cutting

All lathe tools are variations of one standard single-point tool commonly known as a standard straight-edge knife tool. The cutting or shearing action of this tool is referred to as *orthogonal*; that is to say the rake angle lies along the same line as the tool feed or line of action. This can be seen in Fig. 3.95, where the front edge of the tool is 90° to the feed, thus making the rake angle parallel to the feed. The operation shown is known as a *sliding* cut, and the tool would be described as a right-hand straight knife-edge single-point lathe tool. A knife-edge tool

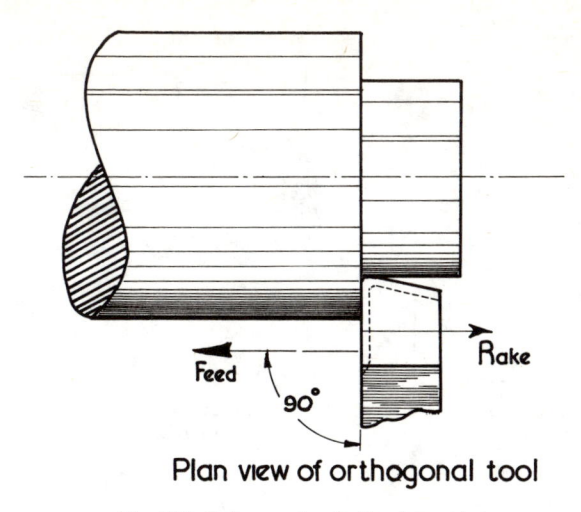

Fig. 3.95 Orthogonal or knife-edge tool

may be used for facing the end of a bar as shown in Fig. 3.96, and in this case the tool would be described as a left-hand straight knife-edge single-point lathe tool. All knife-edge tools are relatively easy to grind and equally easy to sharpen or re-grind, and are widely used for many machining operations.

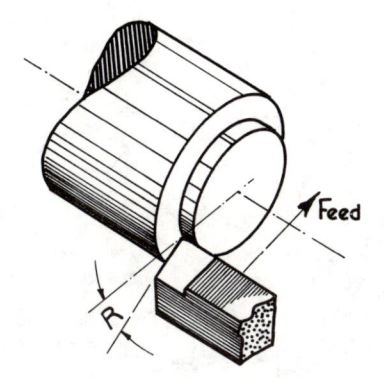

Fig. 3.96 Surfacing with a knife-edge tool

Oblique cutting
When roughing down work on a machine tool, the primary object is maximum metal removal, and *oblique* single-point tools are preferred. The advantage of the oblique principle of metal removal can be seen by reference to Fig. 3.97, which shows two bars receiving identical cuts. The depth of cut and rate of feed are identical in both cases, but the cutting force acts on a larger area in the case of the oblique cutting

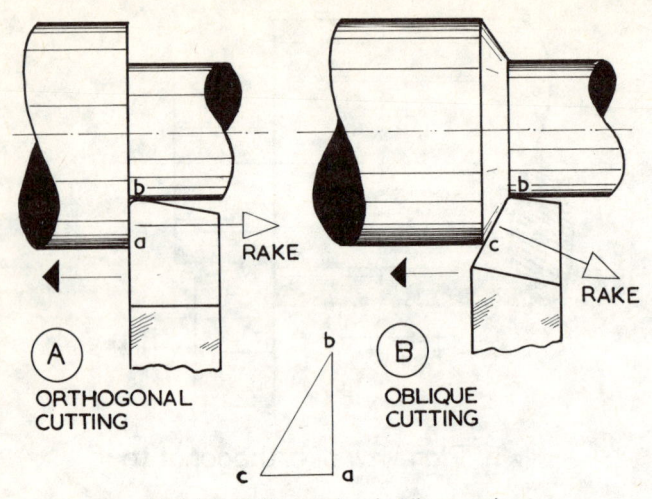

Fig. 3.97 Orthogonal and oblique cutting

tool, making possible a longer tool life. Alternatively the oblique tool shown in Fig. 3.97B, will remove more metal in the same life as the straight-edge or orthogonal tool.

Tool nomenclature
The correct names for the different angles of a standard lathe tool are laid down in the appropriate British Standard 1296: 1961. A typical

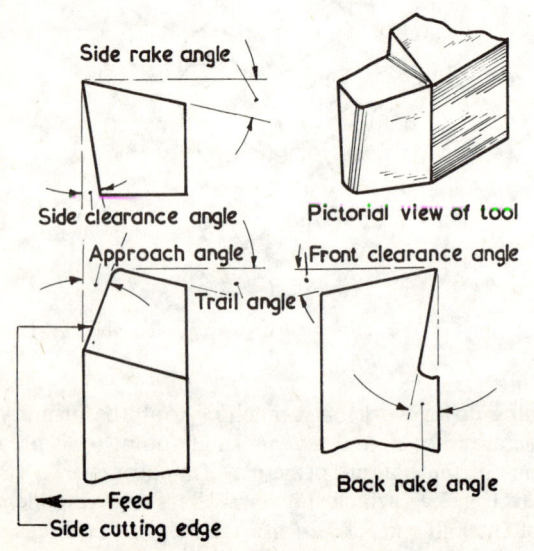

Fig. 3.98 Cutting tool nomenclature

lathe tool designated in the correct manner is shown in Fig. 3.98; note that this is a roughing tool because it has an approach angle, giving it an oblique cutting action.

Tool grinding and sharpening
The grinding and sharpening of single-point cutting tools are two separate and distinct operations. The purpose of grinding is to give the cutting point of the tool the correct shape and angles. The object of tool sharpening is to restore or re-furbish the cutting face or breast of the tool point. These two important operations are simply illustrated in Fig. 3.99. At A we see a high-speed steel tool-bit ready for grinding,

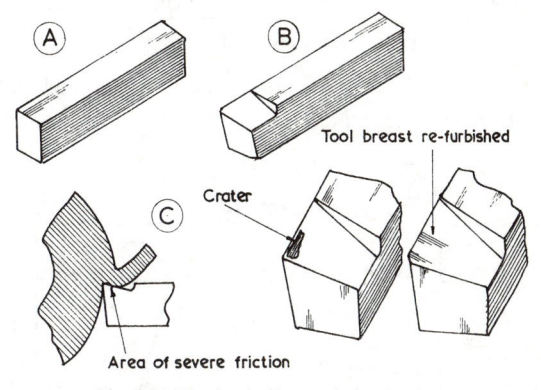

Fig. 3.99 Tool grinding and sharpening

and the finish ground tool point is seen at B. The view at C shows that as the sheared chip leaves the parent metal, it subjects the tool face to severe friction, resulting in wear and deterioration of the cutting edge as shown in Fig. 3.99C. The object of sharpening is to restore this face so that it possesses a smooth fine finish, and it is important that the minimum amount of metal be removed. It is seldom necessary to grind or sharpen the clearance angles.

Choice of single-point lathe tools

Straight knife-edge and straight rougher (right-hand)
These are shown in Fig. 3.100. The straight knife-edge used for average cuts, and the rougher used for heavy cuts.

Straight knife-edge and straight roughing facing
Shown in Fig. 3.101, these tools are similar in all respects to sliding tools except that they are left-hand. They may, therefore, be used for sliding cuts towards the tailstock but this would not be considered as

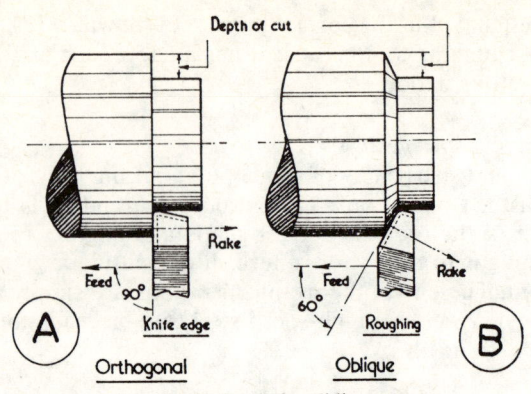

Fig. 3.100 Tools for sliding cuts

good turning practice, as all turning should be carried out with the tool moving towards the headstock.

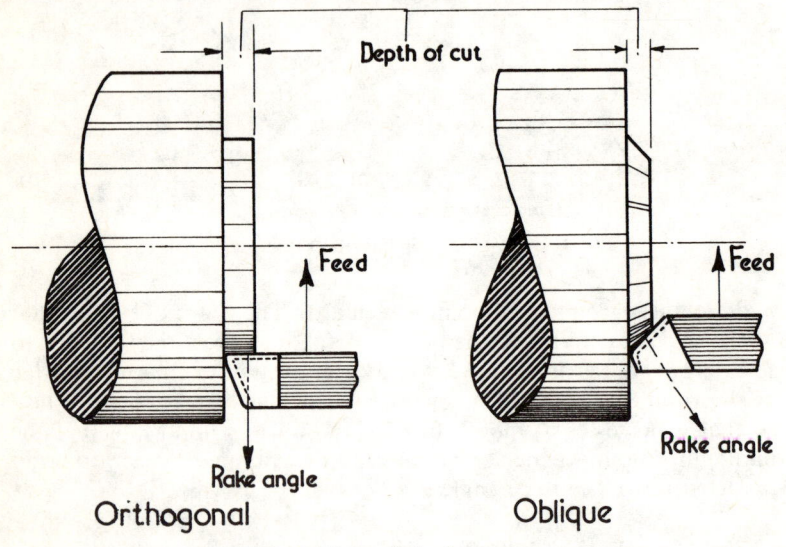

Fig. 3.101 Tools for facing cuts

Straight parting-off
As shown in Fig. 3.102 this tool is used to part-off work, with the tool feed at 90° to the lathe centre-line. Parting-off is not an easy operation, and it is essential that the centre lathe be in good condition, with the minimum of wear or play in the headstock bearings. Best results are obtained if the operation is carried out as close to the lathe headstock as possible.

Machine Tools   201

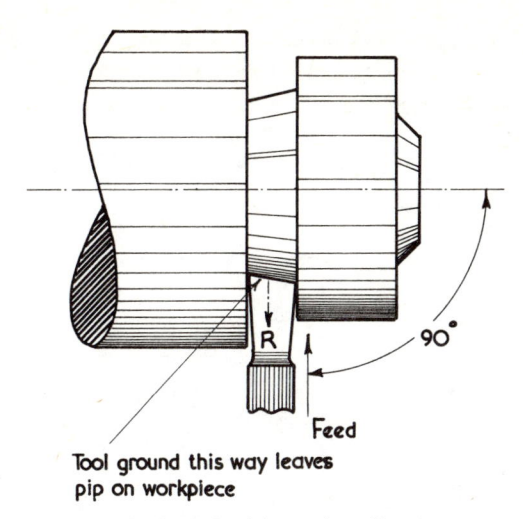

Tool ground this way leaves
pip on workpiece

Fig. 3.102 Straight parting-off tool

Straight round nose
This tool deserves a short discussion. We have seen that in both orthogonal and oblique cutting, the sloping face of the wedge determines the amount of shear plane, and thus the energy required to shear the metal. The *true rake* angle is the angle of greatest slope, and is parallel to the feed in the case of orthogonal cutting, and inclined at an approach angle in the case of oblique cutting. The use of a round-nose tool for sliding cuts, as shown in Fig. 3.103 represents bad practice,

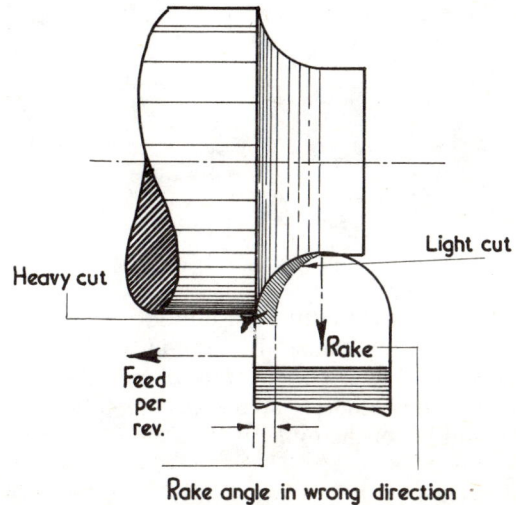

Fig. 3.103 Sliding cut with round-nose tool

because the rake angle is in the wrong direction, and if a fairly heavy cut is taken the finish will be poor, due to the fact that there is negative rake on the leading edge of the tool where most metal is removed. The correct use for a round-nose tool would be to form the spherical surface as shown in Fig. 3.104, the feed of the tool 90° to the lathe centre-

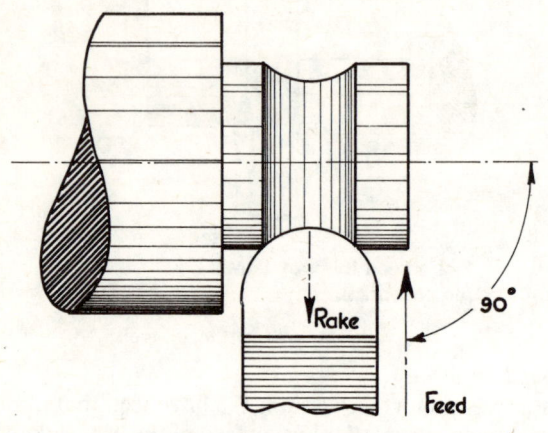

Fig. 3.104 Forming with round-nose tool

line and thus parallel to the rake angle, giving it an orthogonal cutting action. It is, however, quite permissible to grind a radius on the front of a single-point cutting tool in order to improve the finish on the workpiece. In Fig. 3.105A we see a straight knife-edge tool ground to a

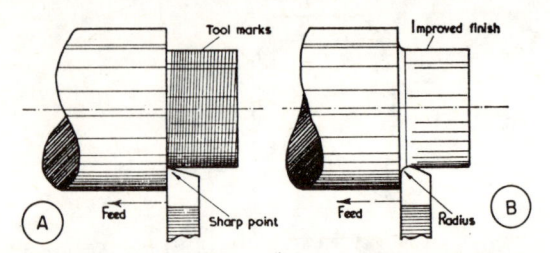

Fig. 3.105 Use of radius to improve surface finish

sharp point, and it is clear that the finish produced by this tool will be inferior to that produced by the tool shown in B. Although the cutting action will be slightly less efficient, this is more than balanced by the fact that the radius tends to strengthen the tool point and helps lengthen the useful life of the tool.

Straight finishing tool
A straight finishing tool is illustrated in Fig. 3.106, and it may be seen

Machine Tools 203

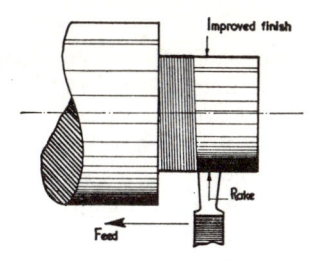

Fig. 3.106 Use of finishing tool

that the edge of the tool is in full contact with the work. Provided the lathe is in good condition, the depth of cut not exceeding 0·05 mm, with both tool and work rigidly supported, an excellent finish will result at medium spindle speeds and slow feeds. The amount of metal removal will be very small, the operation time-consuming and requires both a high quality centre lathe and a highly skilled turner; this means that very little practical use is made of lathe finishing tools.

Bent or cranked tools
It should be clearly understood that a bent or cranked tool is a straight tool that has had its shape changed to make the tool more serviceable, and to simplify grinding and sharpening. A simple illustration will serve to show the principle of bent or cranked tools, and reference to Fig. 3.107 illustrates two examples of the use of non-straight lathe tools. At A we see the turning operation known as facing outwards, and because the recess being machined is of reasonable depth, a cranked tool is needed to clear the side wall of the recess; the lathe tool used in this case is a right-hand cranked straight-edge tool. At B we see the turning operation known as facing inwards, and the tool used in this instance is a left-hand bent straight knife-edge single-point lathe tool.

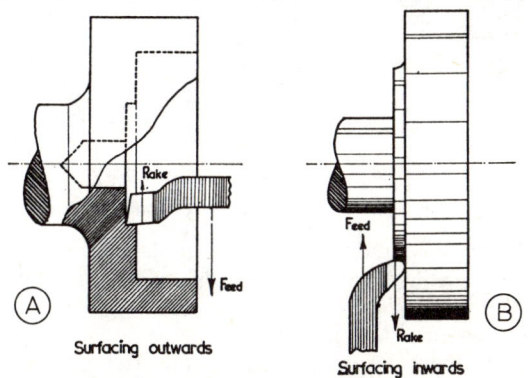

Fig. 3.107 Use of bent or cranked tools

Boring tools
Boring is the name given to the operation of generating an internal cylindrical surface or machining a hole. When boring at a centre lathe, the boring tool is held in the toolpost, and traversed along the bed with the workpiece rotating. Provided the path of the tool is parallel to the centre-line of the workpiece the bored hole will be a true cylinder and accurate in all respects. The principle is simply illustrated in Fig. 3.42A where we see a boring bar having a high-speed tool bit secured at one end. It is of importance to note that the cutting point of this tool bit is a small version of a left-hand straight knife-edge tool, and the cutting action is orthogonal. We see now that once the basic cutting action of a tool is understood, it can be applied to any form of machining, and the rake angle would be ground to suit the material in which the hole is bored. Tapered bores are also possible, and Fig. 3.42B shows a tapered bore machined by off-setting the compound slide at one-half the included angle of the taper. If the bore is of relatively small diameter, it may be necessary to grind a *secondary* clearance angle on the tool bit, and this technique is shown in Fig. 3.108A. A primary clearance angle of about 10° is first ground, fol-

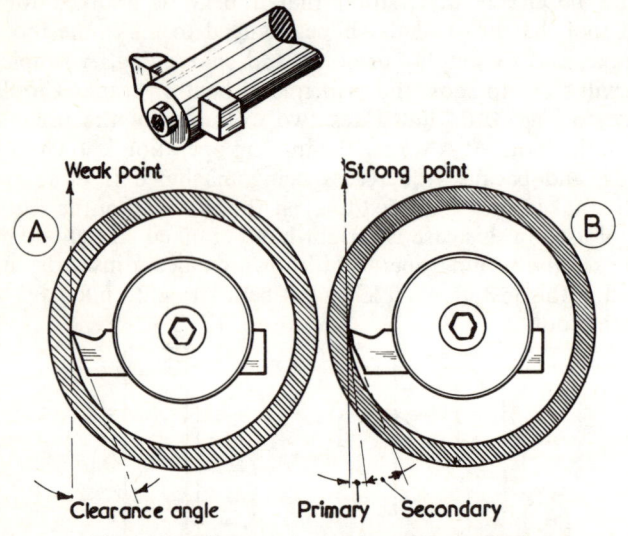

Fig. 3.108 Primary and secondary clearance angles

lowed by a larger secondary angle. This increases the strength of the tool point, and also prolongs the useful life of the tool.

The use of tool bits and tool holders
Tool bits, made from hardened and tempered high-speed steel are available in a wide range of sections and lengths, but are relatively

Machine Tools

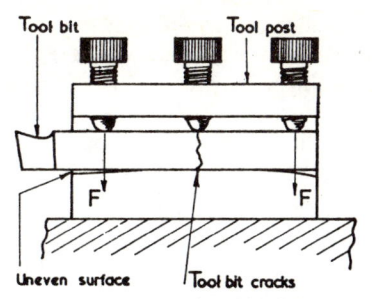

Fig. 3.109 Cause of tool bit fracture

expensive. They are also quite brittle and very likely to fracture if gripped in a toolpost as shown in Fig. 3.109. A popular method of holding tool-bits is shown in Fig. 3.110, but the set-up cannot be regarded as good machining practice, due to the excessive amount of overhang of the cutting tool when compared with a solid tool set-up. Provided only a small amount of material is to be removed, toolholders and tool-bits are satisfactory, but as we have pointed out repeatedly, a machine tool such as a centre lathe is designed to remove metal at maximum speed, and small cuts are justified only when the operation is a finishing one, with surface finish and dimensional accuracy the main aims.

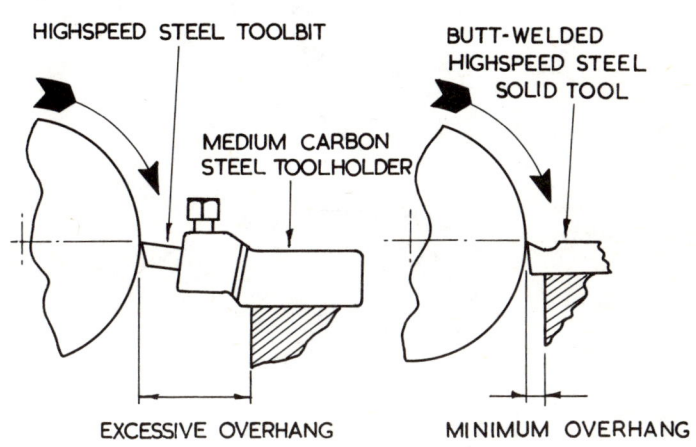

Fig. 3.110 Use of tool holders

3.13 The use of cutting fluids

Although the main purpose of a cutting fluid is to dissipate heat and provide some lubrication, several other functions are also performed, and they may be listed as follows:

(i) To conduct heat away from the tool,
   (ii) reduce friction between chip and tool face,
   (iii) improve surface finish,
   (iv) reduce energy requirements,
   (v) extend the useful life of the cutting tool,
   (vi) carry away swarf and chips.

It is clear that if metal is to be removed efficiently, a suitable cutting fluid is essential, and the type of cutting fluid used depends on the nature of the machining operation, and the metal being machined. We can divide cutting fluids into two main types

   (i) coolants,
   (ii) lubricants.

Coolants

A coolant is used when cutting speeds are high, and when depths of cut are considerable. Fast metal removal involves the generation of a large amount of heat, the main friction area lying just beyond the tool point as shown in Fig. 3.111A. The ease with which the sheared metal or

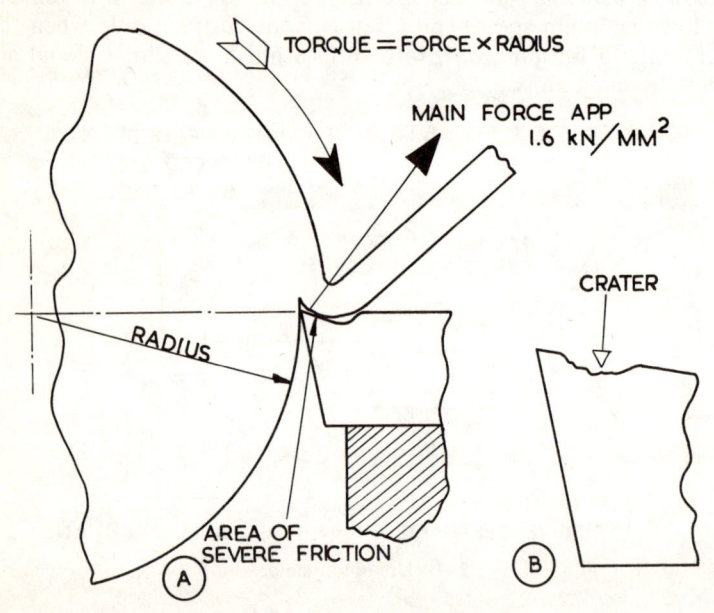

Fig. 3.111 Friction when cutting

chip moves across the tool face indicates that very powerful forces are at work, and in general, the shear force has a value of about 1·6 KN/mm². As we have seen in previous work, an equal and opposite force acts on the tool face, and if we consider also the velocity of the sheared chip as it moves across the tool face it is evident that an area of severe friction exists at the point indicated in the diagram. The effect of this friction is to cause a crater or hollow to appear on the tool face, the longer the tool is used the greater will be this *cratering* effect. Eventually the crater will extend to the tool extremity as shown in the diagram, leading to breakdown of the tool point with a consequent poor finish and loss of dimensional accuracy. The copious use of a suitable coolant will help carry away the heat generated by friction, and hence reduce, but not prevent the cratering effect.

Water-soluble oils
This is the cheapest type of coolant available, and is more commonly referred to in engineering workshops as *suds*. The cheapness of this coolant is due to the fact that the main constituent is water, with a small proportion of soluble oil added to the ratio of one part soluble oil to twenty parts water. Soluble oils are usually mixtures of mineral or fatty oils together with an emulsifier such as lime or soda soap. When added to water the familiar milky liquid results which will corrode neither the work nor the machine tool. Of greater importance perhaps, is the fact that it is *non-toxic*, that is to say it has no injurious effect on the machine-tool operator. Because of the water content, the specific heat is high, and provided an adequate flow of coolant is maintained on the cutting point, the greater part of the heat generated by the cutting action will be carried away by the coolant. Soluble oil is widely used for machine-tool operations such as turning, drilling, reaming, boring and milling, and this coolant needs to be suitably cleaned and filtered before it is returned to the workpiece from the storage tank.

Lubricants
Water-based coolants such as soluble oil do not possess very good lubricating properties, and would not be suitable for more complex machine-tool operations such as gear cutting and spline milling. Operations such as these involve maximum metal removal in minimum time, and because of the precision of the components produced, it is essential that the cutter wear be kept to a minimum. The use of a suitable transparent cutting oil will do much to preserve, not only the useful life of the very expensive cutting tools used, but also the quality of finish on the component.

Cutting oils
These oils consist of mixtures of mineral and fatty oils with small amounts of sulphur and chlorine added to improve the surface finish of the machined component. The transparency possessed by these oils

is an additional advantage, allowing the machine-tool setter or operator to observe the cutting action, and provided there is an adequate flow both heat and chips will be carried away from the tool point. The specific heat of cutting oils is lower than that of soluble oils, but this is more than compensated for by the excellent lubricating properties possessed by cutting oils, which effectively help to reduce the rate of cratering and consequently increase the life of the cutting tool.

Both coolants and cutting oils must have the ability to keep well, that is to say they must not tend to deteriorate over a period of time and become offensive in either appearance or smell. Neither must they cause corrosion or discolouration of either workpiece or the important parts of machine tools, such as bearing surfaces and guideways. The use of cutting oils is confined to those machine tools which specialise in the mass productions of machined components, and such machine tools include single- and multi-spindle automatic lathes, gear and spline cutting machines, and broaching machines. As previously stated, the primary purpose is the preservation of the cutting faces of the complex and expensive multi-point cutting tools, making possible long machining runs before tool changing is necessitated by the wear and cratering of teeth surfaces. On the other hand the use of water-based coolants such as soluble oil is better suited to the less productive types of machine tools, for example drilling machines and centre lathes.

Chapter 3 Assimilation Exercises

1. Make a neat sketch of an engineering component and indicate the geometrical surfaces, naming the machine tools used to produce these surfaces.
2. With the aid of a neat diagram indicate the movements necessary to generate a cylindrical surface at a centre lathe.
3. Show by means of neat diagrams, techniques of forming using a shaping machine.
4. Make a neat sketch of a centre lathe, indicating the main features and stating the materials used giving reasons.
5. Make a neat sectioned sketch of the bed of a centre lathe, showing clearly the means by which the saddle and tailstock are located and guided.
6. The lead screw of a lathe cross slide has a pitch of 5 mm, with 100 divisions on the indexing dial. Calculate the indexing to infeed the cutting tool a distance of 2·5 mm.
7. Sketch three different methods of work-holding at a centre lathe.
8. Sketch two different methods of tool-holding at a centre lathe.
9. Make a neat sketch of a component that would require the use of face plate.

Machine Tools 209

10 If the cutting speed for mild steel is given as 24 m/min, calculate the approximate rev/min when turning a 50-mm-diameter bar.
11 Explain the desirability of adopting the roughing and finishing technique when turning at a centre lathe.
12 Explain the advantages offered when completing a turned component in one setting at a centre lathe.
13 Sketch typical engineering components that would be machined using each of the following techniques:

   (i) turning between centres,
   (ii) use of a four-jaw chuck,
   (iii) use of a collet chuck.

14 Using Table 5, calculate the depth of an M22 thread.
15 Describe the set-up when turning each of the tapers given below:

   (i) a taper of 78° included angle,
   (ii) a taper of $\frac{5}{18}$.

16 Outline the main safety precautions to be taken when operating a centre lathe.
17 Make a neat sketch of a shaping machine, identifying the main features.
18 With a neat diagram show the principle of the quick-return motion of a shaping machine ram.
19 Explain the purpose of a clapper box on a shaping machine.
20 Make a neat sketch of a shaping operation that would necessitate tilting of the clapper box.
21 Explain how the two following shaping machine adjustments are carried out:

   (i) Altering the length of stroke,
   (ii) changing the position of the ram.

22 For what reasons are shaper tools of bigger section than lathe tools?
23 Make a neat sketch of a component that could be shaped in one setting.
24 With a neat diagram show how a single-point cutting tool shears metal.
25 By means of neat diagrams show how rake angle is obtained on the following hand tools:

   (i) button die,
   (ii) tap,
   (iii) dreadnought file.

26 Explain the importance of grinding the correct rake angle on a single-point cutting tool.

27　Explain why all clearance angles should be kept to a minimum.
28　By means of neat diagrams show the differences between radial cutting and tangential cutting.
29　By means of neat diagrams show the differences between orthogonal cutting and oblique cutting.
30　With a neat diagram illustrate a standard straight-edge knife tool suitable for sliding cuts at a centre lathe.
31　Show by means of diagrams how a roughing tool differs from a knife-edge tool.
32　Explain the advantages to be obtained when using a cutting fluid during a machining operation.
33　State the reasons why a cutting oil is sometimes chosen instead of a coolant.
34　Explain why all cutting fluids must be non-corrosive and non-toxic.

4
FASTENING AND JOINING

Objectives—The principles and applications of:
1 Threaded fasteners
2 Locking devices
3 Rivets
4 Soft soldering
5 Sweating
6 Fluxes
7 Hard soldering
8 Welding
9 Adhesives
10 Safety precautions

Engineers are concerned with making components, the finished product consisting of an assembly or a large number of separate items joined together in one way or another. A simple example of an engineering assembly is illustrated in Fig. 4.1, where we see a pedal and crank assembly for the humble bicycle, and it is evident that many techniques of metal joining are available to engineers. Motor cars, washing machines and motor cycles are all examples of finished products made up of a large number of separate parts. The principles adopted when joining parts together have changed little since engineers first constructed their primitive machines from wood, and Fig. 4.2 shows a comparison between the techniques of wood and metal joining. It is, of course, much easier to work and join wood than it is to machine and join metal, and wood is still very widely used for a large number of articles. Before having a closer look at the metal joining techniques in general use, it will be advantageous to separate them into two distinct groups:

  (i) temporary joints,
  (ii) permanent joints.

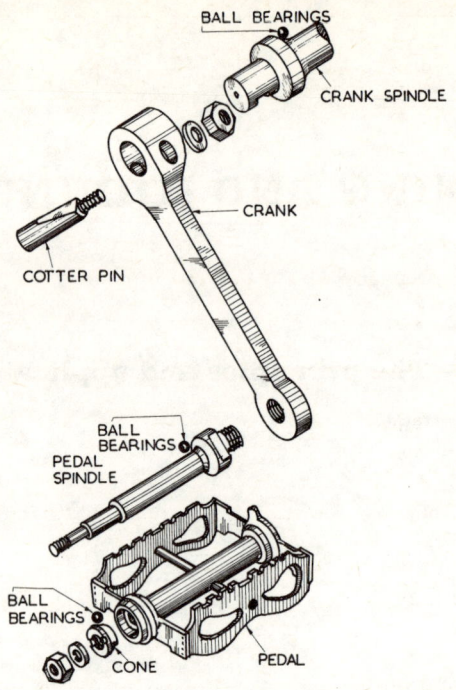

Fig. 4.1 Pedal and crank assembly

Temporary joints

We may define a temporary joint as one which will require dismantling or separating for purposes of renewal, repair or adjustment. Figure 4.3 shows a temporary joint familiar to most technicians, and it is clear that the joining techniques are provided by the use of threaded fasteners.

4.1 Threaded fasteners

Any device which makes use of the screw thread principle to achieve the tightening effect required to bring two or more parts together is called a threaded fastener, and the more common threaded fasteners are listed below:

    (i) nuts and bolts,
    (ii) screws,
    (iii) studs.

4.1.1 Nuts and bolts

Nuts and bolts are probably the most widely used of all threaded fastening devices. They are best used when both bolt head and nut are

Fastening and Joining

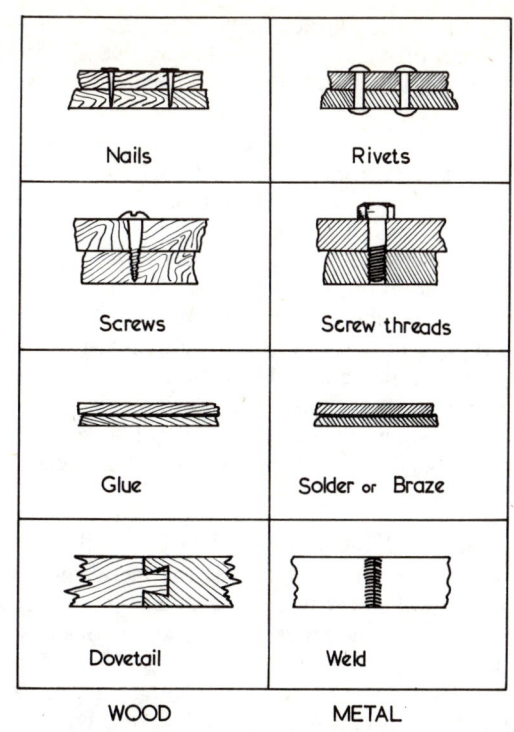

Fig. 4.2 Principles of wood and metal joining

readily accessible, allowing the use of spanners both to tighten the nut and also to prevent rotation of the bolt head. The use of nuts and bolts offers also simplification during the manufacture of the component as

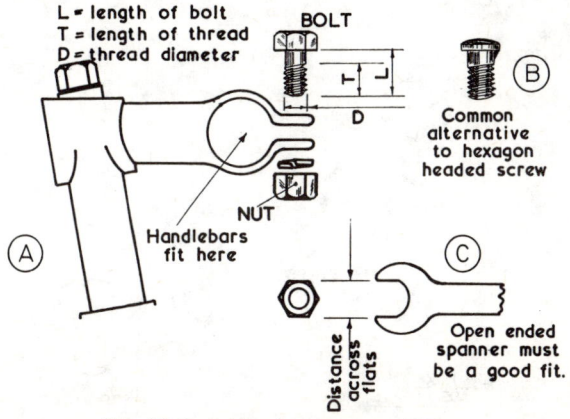

Fig. 4.3 Example of a temporary joint

clearance holes only are required, and no threads need be machined. In the event therefore, of a thread being damaged by overtightening (a common assembly fault), it is a simple and cheap matter to replace the faulty nut and bolt; whilst a stripped thread in a component may mean that the component is scrapped. Reference back to Fig. 4.3A shows a typical use for a nut and bolt, with both nut and bolt head easily accessible. At B we see an alternative bolt head, having a small dimple, which locates in a recess or indentation in the handlebar stem. This prevents rotation of the bolt and hence speeds up assembly by removing the need for an additional spanner to hold the bolt head. A simple spanner, shown at C, is more commonly referred to as an *open-ended* spanner. This type is quite suitable for tightening-up small nuts and bolts, say up to 8 mm diameter, but for diameters in excess of 8 mm it is better practice to use a socket-head type and thus reduce or remove the possibility of spanner slip, which is a prolific source of injury to operator and damage to the hexagonal nut or bolt head.

Types of nuts and bolts
Nuts and bolts are available in many metals and alloys such as brass, copper and aluminium, but mild steel is the metal most widely used. Mild steel nuts and bolts may be either *black* or *bright*; black bolts are produced by a hot forging process and have a black scaly surface, whilst bright bolts are either cold forged or machined. Black nuts and bolts are used for outside work, for example in structural work such as bridge building, whilst bright nuts and bolts are used for small assemblies requiring smaller diameters and are more expensive than their black counterparts.

Classification of nuts and bolts
A nut and bolt is classified as follows:

(i) Length,
(ii) diameter,

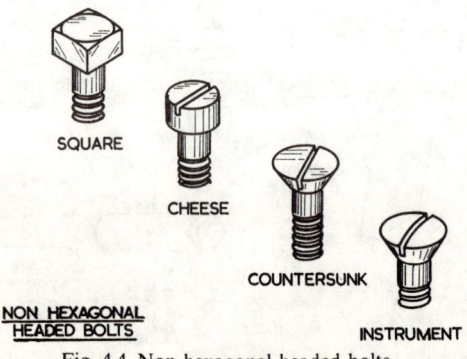

Fig. 4.4 Non hexagonal-headed bolts

Fastening and Joining 215

(iii) shape of head,
(iv) material,
(v) type of thread.

Several types of non-hexagonal shape heads are available and a small selection is shown in Fig. 4.4. A correct description of the nut and bolt shown in Fig. 4.3, and used to secure the handlebars to the bicycle stem would be as follows:

40 × 6 Metric coarse series hexagonal bright mild steel bolt.
The thread would be indicated as M6 on an engineering drawing.

4.1.2 The use of set-screws

Set-screws are used when it is either impossible or impractical to use nuts and bolts. A simple example is shown in Fig. 4.5, where we see a junction box used with a conduit pipe for the protection of electric wiring. The junction box, as the name suggests, represents a point

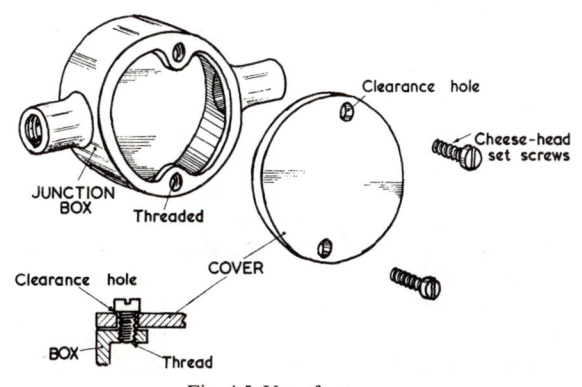

Fig. 4.5 Use of set-screws

where changes in the direction of the wiring take place, and its use saves the considerable time and labour that would otherwise be involved in bending a conduit pipe at 90° angles. Clearly we need a tight-fitting cover to the box, and this cover must, of course, be a temporary joint as we may need to check the wiring at a later date. It is at once clear that the use of nuts and bolts is an impossibility; set-screws must be used with threaded holes in the junction box as shown in the diagram. Each of these threaded holes, receives a set-screw which is threaded for the greater portion of its length. We see now that if excessive tightening pressure is used there is grave danger of stripping the thread in the junction box, and some care needs to be taken in the choice of spanners or screw drivers when tightening set-screws. Small set-screws need small screw drivers or small spanners and it is a great mistake to use an excessively large screw driver to tighten a relatively small screw. The range of set-screws available is very similar to the range of nuts and bolts; thus they may have hexagonal or non-hexagonal heads, while the screw threads will be chosen from the Metric range of threads, according to the use of application of the set-screw.

4.1.3 The use of studs

Studs are used when very heavy tightening forces are needed, for example when joining the cylinder head of a high compression motor car engine to its cylinder block. A typical stud is illustrated in Fig. 4.6A,

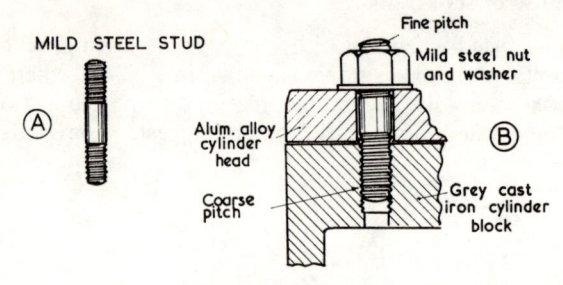

Fig. 4.6 Use of studs

and it can be seen that it is threaded at both ends. A sectional view showing the technique applicable to the use of studs is given in Fig. 4.6B. Note that the tightening effect is achieved by rotation of the mild steel hexagonal nut; this is done after the stud has been screwed securely home in the grey cast iron cylinder block. Because grey cast iron is a relatively brittle material, a coarse pitch thread is less likely to crumble or strip, and the stud-end that enters the grey cast iron cylinder block is given a coarse pitch thread for this purpose. On the other hand, a mild steel nut is well able to withstand the greater forces produced when the top portion of the stud is given a thread of finer pitch, and is also less likely to work loose under the effect of engine vibration.

4.2 Locking devices

The danger of nuts, set-screws and studs working loose due to vibration or other causes needs no emphasising, and many an accident has been caused by a nut working loose off its bolt. The greatest care needs to be taken in this matter of *locking* threaded fasteners, and the following notes are intended to indicate the basic techniques in use in engineering assembly. Two main principles are in use, *positive* locking and *frictional* locking.

Fastening and Joining 217

Positive locking
This principle is adopted when it is imperative that the fastening device remains firmly tightened under all conditions; that is to say it is impossible for it to work loose. For example, a heavy flywheel rotating at speed, *must* be joined to its shaft in such a way that it is never likely to work loose and, as threaded fasteners are certain to be used, these fasteners must be *positively* locked. The following devices or techniques form a vital part of what is known as final assembly.

Split pin and castle-nut
Provided the tightening torque is not critical, the split pin and castle-nut principle is a very effective method of ensuring a positive lock with no possibility of the nut working loose. Very slight adjustment of the nut is needed to ensure that the radial hole in the bolt is in line with a slot in the nut and the correct procedure is first to tighten the nut to the required torque and then ease it back until a slot lines up with the hole. The split pin is then inserted and bent over the nut as shown in Fig. 4.7A.

The use of a locking wire
A locking wire is a length of soft steel wire passed through a hole in each screw head before the wire ends are firmly twisted together. It is

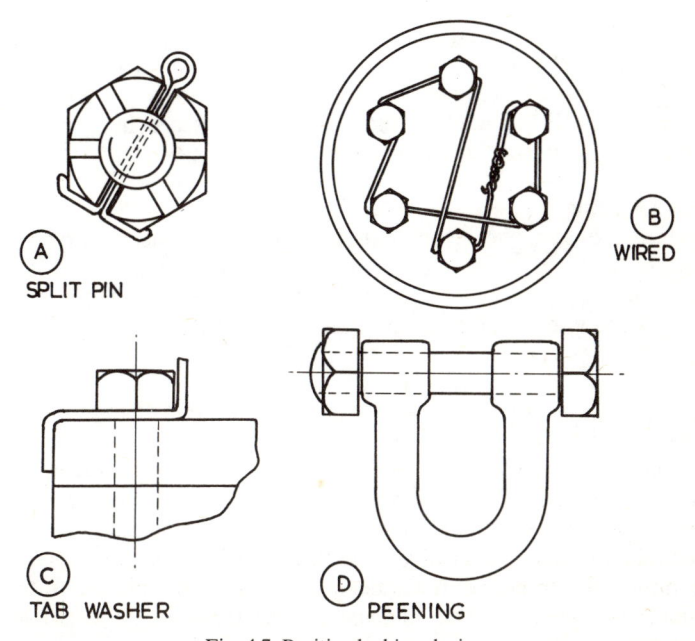

Fig. 4.7 Positive locking devices

always used when there are several tightening screws in close proximity, for example six hexagonal-head bright mild steel screws arranged on a pitch circle diameter, and used to tighten a flywheel onto its shaft as is simply illustrated in Fig. 4.7B. On assembly it is essential that the wire be kept reasonably tight as it passes through the hole drilled in the head of each screw, and is then firmly twisted in the manner shown in the diagram.

Tab washers
These are soft mild steel washers best used when there is a step or ledge adjacent to the screw head or nut. The principle is illustrated in Fig. 4.7C, and it is essential to make a rule of never re-using a tab washer which has had its tab straightened to unscrew the nut or screw. Re-bending a tab washer is almost certain to result in the washer cracking along the bend line with the serious danger of the nut or screw working loose.

Peening the thread
The three positive locking devices just described are designed for use with temporary or non-permanent joints. That is to say, in the case of the flywheel joined to the shaft, it is a relatively simple matter to remove the locking wire and then slacken-off the tightening screws in order to remove the flywheel. In the case where a threaded device is to be permanent, a standard method is to peen-over a small portion of thread as shown in Fig. 4.7D: this is a most positive method of ensuring that the thread remains in permanent position, and finds much application on lifting tackle where fatal injuries are likely to result when nuts or other threaded devices work loose.

Friction locking
By friction locking we mean that it is frictional resistance only that prevents the nut or screw from working loose. Although not positive in the true sense of the word, friction locking devices are highly efficient and are widely used for most mechanical assemblies.

Lock nuts
Reference back to Fig. 2.64B shows a simple use of a lock nut, though in this example both nuts are knurled to provide a reasonable finger grip. When hexagonal nuts are used, the lock nut is first screwed down tightly, followed by the standard nut which is screwed down tight onto the lock nut. With a spanner holding the standard nut, the lock nut is tightened back onto the standard nut, thus locking both nuts together by frictional force between the nut faces. The technique is shown in Fig. 4.8A and it may be noted that the locknut is about one half the thickness of the standard nut. This principle of using a lock nut is

Fastening and Joining 219

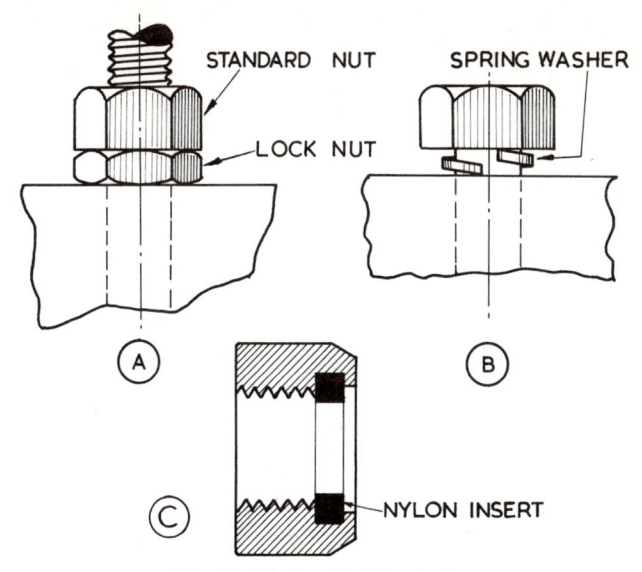

Fig. 4.8 Frictional locking devices

applied when the position of the standard nut is likely to vary along the length of its mating bolt, making it impossible to use drilled holes and split pins.

Spring washers
Spring washers are popular devices and may be used both with nuts and bolts and with set-screws. The washer, as may be seen in Fig. 4.8B is split and off-set, and is always placed next to the screw head or nut. The process of tightening compresses the spring washer which exerts an equal and opposite force on the nut, setting up frictional resistance between the mating threads. All spring washers must possess a good degree of elasticity, and are available as single coil or double coil. Another version of spring washers comprises a type of washer with a raised or serrated face, usually of fairly small diameter. On tightening, these small serrations are pressed flat, and in their attempt to regain their original shape they exert a force on the tightening nut.

Fibre or nylon insert nuts
This principle is simply illustrated in Fig. 4.8C, and is best suited for small diameter threads. As may be seen, a fibre or nylon band has been inserted within the nut, and when the nut is threaded on the bolt, the bolt is capable of cutting a thread on the insert. This thread will be an extremely tight fit on the bolt, making it unlikely that the nut will work loose when subject to vibration.

Permanent joints
4.3 Riveting processes

A rivet used to join together two metal components can be considered as equivalent to a nail used to join together two wooden components. However, unlike a nail, which is fairly easily driven into wood, holding or gripping with frictional force, a rivet needs a hole of suitable size and after insertion must have a head worked on similar in shape or proportion to the head on the other side. A simple example showing the use of rivets is illustrated in Fig. 4.9A; in this case four rivets are used

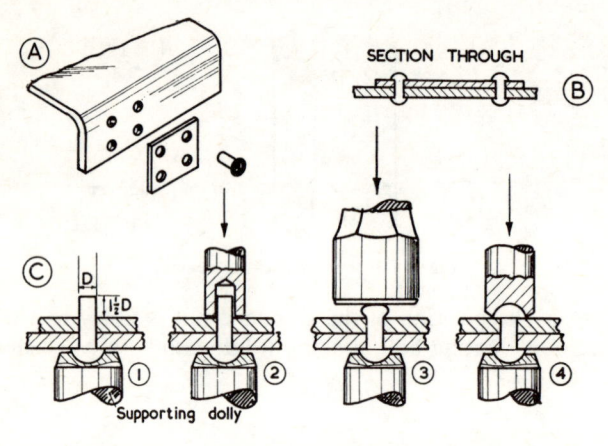

Fig. 4.9 Principles and application of riveting

to make a permanent joint of a metal identification tag to the metal chassis of a television receiver. A sectioned view of the joint is shown at B, while at C we see the essential stages in making the joint, assuming that the process is carried out at the bench using hand tools.

Stage 1   Here the rivet is inserted, note that the amount of rivet standing proud of the top surface of the metal is equal to one and a half times the rivet diameter. Hence the following formula is worth remembering:

$$\text{Amount of metal for riveting} = \frac{3D}{2}$$

where D = diameter of rivet.

Stage 2   This operation is known as drawing down, and a special tool is needed having hole in one end in which the rivet has a slack clearance fit. With the rivet supported, a few light hammer blows on the drawing down tool brings the two plates together.

Fastening and Joining 221

Stage 3 The flat end of a hammer is used in preference to the ball end, and a few blows only are required.

Stage 4 This is the final heading operation and a tool called a *snap* is required. The working end of this snap has a polished form identical to the head of the rivet, and some skill is needed to produce a well-shaped head.

Hot riveting
When joints have to be water tight or air tight as with boiler or ship's plates, it is usual to insert the rivet while it is red hot. This not only eases the task of forming a neat head, but also ensures a very powerful tightening force as the rivet cools down to normal temperature, bringing the two plates tightly together under the force of contraction.

Types of rivet
As with bolts and set-screws a fairly wide range of rivets is available, having different diameters, lengths and shapes of head, and manufactured from metals such as mild steel, copper, brass and aluminium alloy. A few typical rivets are illustrated in Fig. 4.10.

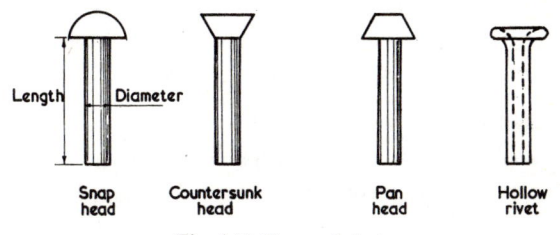

Fig. 4.10 Types of rivets

Metal to metal riveting
This is a cheap and efficient method of joining two metal parts, and reference back to Fig. 4.1 shows that the pedal assembly is an example of metal to metal riveting. This process is more clearly illustrated in Fig. 4.11A where we see an enlarged view of one end of the pedal assembly. Clearly it is a simple matter to insert the side piece into the end piece and rivet over, as shown in B. A strong assembly results and there is then no need for separate rivets.

4.4 The soldering process

The principle of soldering is similar to that of gluing two pieces of wood together as shown in the comparison made in Fig. 4.2. This means that a liquid film of metal is interposed between the two surfaces to be joined, giving on solidification a jointing compound be-

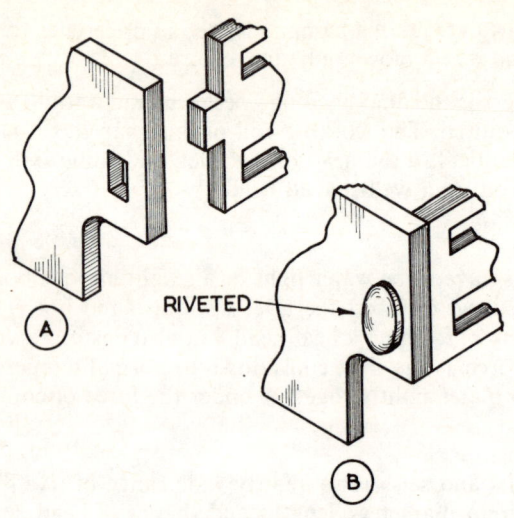

Fig. 4.11 Metal to metal riveting

tween the two metals. Soldering is divided into two main groups, *soft* soldering and *hard* soldering, and the type of soldering process used depends mainly on the strength required from the finished joint and the metals from which the joint is to be made.

4.4.1 Soft soldering

This process is suitable for joints which are not subject to excessive forces, as the strength of a soft soldered joint is not high unless special

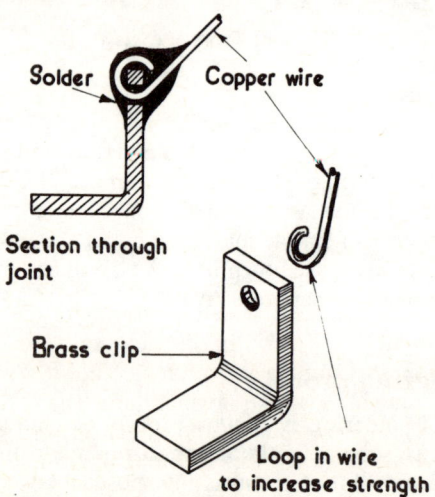

Fig. 4.12 Soft soldering a wire to a clip

Fastening and Joining 223

precautions are taken to strengthen the joint. The weakness of a soft soldered joint is due to the low tensile strengths of the metals used in the soft solder, namely tin and lead, and it is a general principle that every effort is made to strengthen soft soldered joints which have a low contact area. In the electrical industry a great number of joints are required in the manufacture of television and radio receivers, and perhaps the joining of a copper conducting wire to a small brass clip will serve to illustrate the technique of the use of a soldering iron. Figure 4.12 shows how the wire is first looped through the clip to provide a strong mechanical joint and then soft soldered to produce good electrical contact and permanency to the joint. The soldering of tin containers has been carried on for many years and is another good example of the use of soft solder to produce a strong joint aided by a mechanical principle. Once again the strength of the joint is obtained by interlocking the ends of the can as shown in Fig. 4.13; this was the

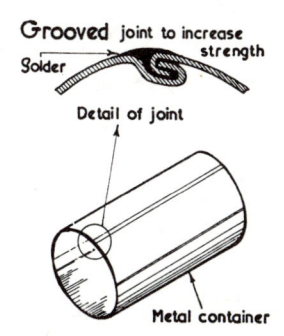

Fig. 4.13 Soft soldering a metal container

technique used by the Romans who made lengths of lead pipe by a similar method. Note that in the sectional view solder is drawn into the joint by capillary action.

Most of us are familiar with the serious accidents that may occur when the soldered nipple on the end of a bicycle brake cable breaks away. Provided we remember that the strength of the joint is not dependent on the solder it is not difficult to carry out a suitable repair. Figure 4.14 shows the principle involved. It will be seen that the end of the brass nipple is countersunk, with the strands of the cable splayed out in a uniform manner as shown in the sectional view. If now, we fill the gaps between these strands with molten solder, on solidification a solid wedge of solder and steel is formed, and the forces set up by application of the brake now act in the direction of arrows A as indicated in the diagram. The effect is to burst the nipple, and not pull the cable through. A joint made by just soldering the cable in the bore of the nipple would be most unsafe.

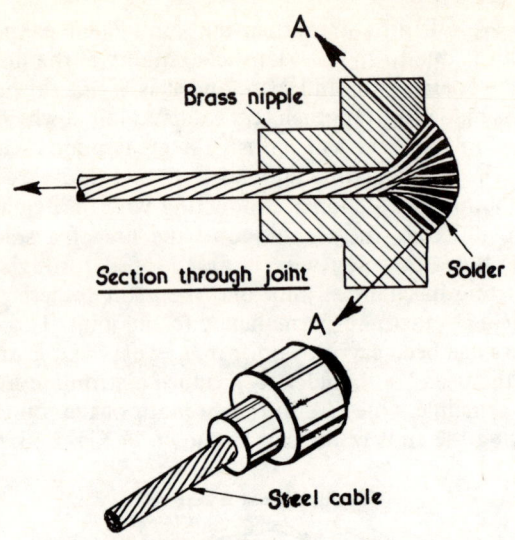

Fig. 4.14 Soldering a nipple to a steel cable

4.4.2 Use of the soldering iron

A soldering iron is always used when contact areas are small, for example when soft soldering the copper wire to the brass clip as shown in Fig. 4.12. It is evident that the amount of heat required will be only

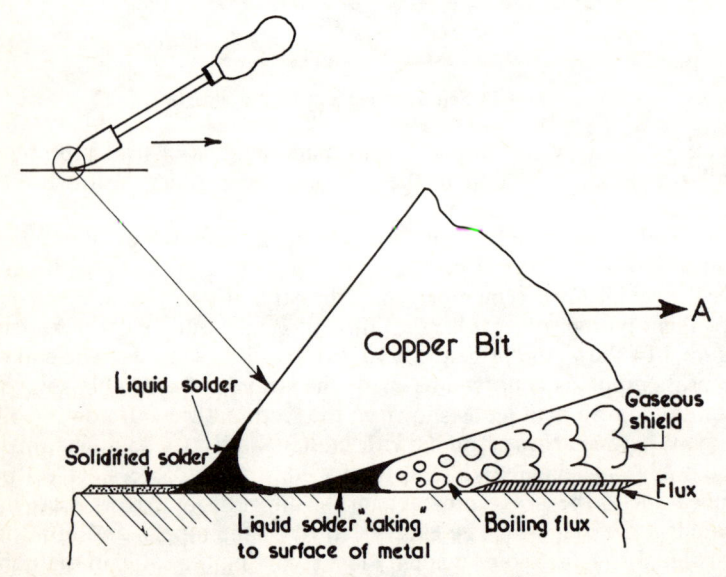

Fig. 4.15 Principle of the soldering bit

Fastening and Joining

that needed to melt the small amount of solder required, and the copper bit of a soldering iron is well able to supply this heat. The bit may be heated externally using a gas flame or coke fire, or can be heated electrically for production or very small work. The principles of soldering are important, and successful soldering is achieved only by strict adherence to the rules and techniques outlined below. Figure 4.15 illustrates the soldering action when using a soldering iron and the diagram magnifies the working end of the copper bit. Note the direction taken by the soldering iron as shown by arrow A. This direction is important because one of the functions of the copper bit is to transfer heat to the workpiece prior to the adherence of the soft solder. The effect of this heat is to cause boiling of the flux on the surface of the work, and this boiling action tends to remove the small amount of impurities or oxides that may be present. At the same time a small amount of gas is given off, forming a gaseous shield which helps to prevent the oxygen in the atmosphere combining with the hot surface of metal. Oxygen will combine easily with most metals, making successful soldering an impossibility, so the purpose of a flux is to remove impurities and prevent oxidisation of the surface of the hot metal.

Tinning the bit
In addition to transferring heat to the work, the bit acts as a reservoir for the molten solder, and for this reason a bit must be *tinned*, before use. This entails heating the bit carefully, then quickly giving a few strokes with a smooth file to remove the oxide films, dipping the hot bit in flux and then applying solder. The solder will immediately *take* to the bit, giving it a silvery appearance, hence the name 'tinning'. A well-tinned bit is capable of picking up and retaining solder from a piece of clean wood. Reference back to Fig. 4.15 shows that the following sequence of events takes place as the soldering iron is traversed from left to right.

   (i) The heat from the bit boils the flux and raises the temperature of the metal surface to that of the molten solder.
   (ii) The molten solder 'takes' to the clean metal surface left by the boiling flux, and protected by the gaseous shield.
   (iii) The solder solidifies behind the bit.

The soldering operation ceases when all the solder has been run off the bit.

The amount of solder deposited by the copper bit depends on the speed with which the bit traverses the joint. The faster the speed the less solder is deposited, and a skilled tinsmith is capable of working at a remarkably fast speed and yet producing high-class soldered joints. This results in a substantial saving of money, because solder is an alloy of tin and lead and tin at the present moment is one of the most expensive metals used by engineers. At the time of writing one metric tonne costs in excess of six thousand pounds. Provided the principles outlined above are observed, good soldering is easily achieved and very high production rates are possible using automatic equipment, where solder, flux and heat are applied by precision equipment. There is little doubt that the design, precision and amount of engineering skill involved in the machinery and equipment necessary for the soft soldering of so-called 'tins' on a mass production basis, would astound most customers of these products.

4.5 Sweating

We have seen that the use of a soldering bit is restricted to those applications where the area of contact is small and the amount of heat required is not excessive. Figure 4.16 shows the process adopted when

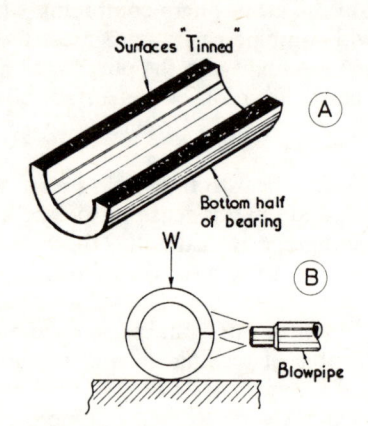

Fig. 4.16 Sweating two halves of a bronze bearing

two halves of a bronze bearing are to be temporarily joined for purposes of machining the bore. Clearly the heat given off by the bit of even a large soldering iron would be quite inadequate to both deposit a film of solder on each face and also melt these films to make the necessary joint. The joining of these two half bearings is therefore achieved by a process known as *sweating*, whereby relatively large areas of contact can be soldered.

We see from the diagram in Fig. 4.16A, that the surfaces of the half bearings need to be tinned, and this is achieved by cleaning the surfaces, applying flux and gently heating with the application of a small amount of solder which when molten can be spread over the surface with a dry cloth. With all faces tinned, the two halves are now placed in correct alignment with a heavy weight on the top bearing. Heat is

Fastening and Joining

now applied carefully and evenly around the joint using a gas–air blow lamp, and at the correct temperature the weight forces out the excess solder, whereupon the source of heat is removed and the solder allowed to solidify or freeze. After accurate boring the two halves are re-heated, the joint broken and the solder scraped off, with the bearing now ready for inspection and testing in its housing.

4.6 Fluxes for soft soldering

All soldering operations require the use of a suitable flux. The main purpose is to break up the oxide film that forms on the surface of clean or polished metal; this is achieved by the boiling action of the flux when subjected to heat. In general two kinds of fluxes are used when soft soldering:

(i) Passive flux
(ii) Active flux

Table 4.1 gives the types and applications of fluxes suitable for soft soldering.

Table 4.1   Fluxes and their uses

Group	*Name*	*Uses*
Passive (Protects only)	Tallow Resin	Wiped or plumber's joints. Electrical and small work in copper and brass.
Active (Cleans and Protects)	Zinc Chloride, Killed spirits, Baker's fluid.	Mild steel joints, tin plate. Never used for electrical work.

Passive fluxes
These fluxes have a resin or tallow base, and are usually in the form of a soft paste which is easily applied to the work. The surfaces of the work must be absolutely clean if the joint is to be successful. The great advantage of this type of flux is that there is no corrosive effect on the joint, and this means that passive fluxes are always used for large number of joints to be found in electrical fabrication and installation. For the rapid soft soldering of small joints using an electrically heated bit, a resin-cored soft solder in the form of a coil is used, and it is possible to arrange for automatic feed of this coil, this providing both resin, solder and heat to the job from *one* source, namely a suitable soldering gun.

Active fluxes
A popular active flux is known as 'killed spirits'. This liquid consists of hydrochloric acid in which some zinc has been dissolved. Glycerine may be added to reduce the corrosive action of this flux, but even so, all joints soldered with killed spirits must be thoroughly washed in hot water after completion of the soldering. This ensures that all traces of the flux are removed, so that there is no possibility of deterioration and break down of the joint at a later date caused by acid remaining in the joint. An active flux must never be used for electrical joints, and it is best used when soldering thin steel sheets more commonly known as *tinplate*, its cleaning action greatly assisting and simplifying the soldering action.

Types of soft solder
Although a large number of soft solders are available, all fall into two main types:

   (i) plumber's solder,
   (ii) tinman's solder.

Plumber's solder
The approximate composition of plumber's solder is two parts lead and one part tin, and its melting point is about 250°C. This solder takes an appreciable time to solidify from the liquid state; that is to say there is a short period of time when the solder is spongy or pasty. In this condition the solder is readily manipulated or worked into shape. A typical use for plumber's solder is simply illustrated in Fig. 4.17A

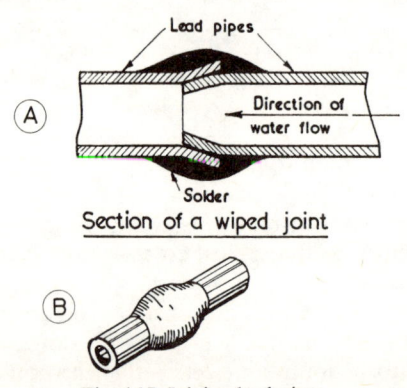

Fig. 4.17 Joining lead pipes

where we see a sectional view of the technique adopted to join two lead pipes; note the direction of the flow of water, and that one pipe fits into the other. This is another example of the use of a simple mechanical

Fastening and Joining 229

device, for if the pipes were simply joined end to end, the resulting joint would lack the rigidity and strength of the joint obtained by inserting one pipe end into the other. With careful manipulation of a blowlamp and also a stick of plumber's solder, the skilled technician is able to shape or wipe the solder whilst it is in its pasty or plastic state, producing what is known as a *wiped* joint.

Tinman's solder
This solder has a composition of about two parts tin and one part lead, and has a melting point of 186°C, which is lower than the melting point of plumber's solder. Tinman's solder solidifies very quickly, with no pasty or spongy state, making it very suitable for production soldering because the solder tends to solidify as soon as the soldering bit is removed, ensuring a firm joint.

4.7 Hard soldering

We have seen that soft soldering will make strong joints only when large areas are in contact or when mechanical aids are used to strengthen the joint. When solder contact only is possible, as in the copper jug illustrated in Fig. 4.18A, it is clear that a soft soldered joint cannot produce the required strength, and a stronger solder is needed. We may divide hard soldering into two main groups as follows:

 (i) Silver soldering,
 (ii) brazing.

In general, the brazing technique is restricted to the joining of steel to steel, whilst silver soldering is used to join copper, brass or silver.

Silver soldering
This technique is adopted when strong joints are required, for example the joining of the two copper handles to the copper jug shown in Fig. 4.18A. The solder used is known as silver solder, having a melting point of about 750°C and a typical composition as follows:

Copper	36%
Zinc	20%
Silver	44%

Silver solders have excellent penetrating powers, the increased fluidity of the solder being mainly due to the addition of the silver. In practice, the job is first cleaned and a flux made from borax powder and water applied, followed by a small strip of silver solder, preferably bent around the profile of the joint. Heat is now applied, a gas–air blow-

230   *Technician Workshop Processes and Materials*

lamp being quite suitable, and when the temperature of the joint reaches the melting point, the silver solder flows or runs between the interfaces of the joint. It is of course, essential that the melting point of any solder is below that of the metals to be joined. As the melting point of copper is 1080°C, and silver 960°C, it is clear that silver solder with its melting point of 750°C, is quite suitable for making hand soldered joints in the above metals.

Brazing
This is the term normally applied to the process of joining steel to steel. The solder used is usually referred to as *spelter* and is essentially a form of brass with a melting point of about 900°C, with a typical composition as follows:

$$\begin{array}{ll} \text{Copper} & 65\% \\ \text{Zinc} & 35\% \end{array}$$

Figure 4.18B shows a bicycle frame requiring the application of the brazing technique. The tubes and the lug are made from mild steel and

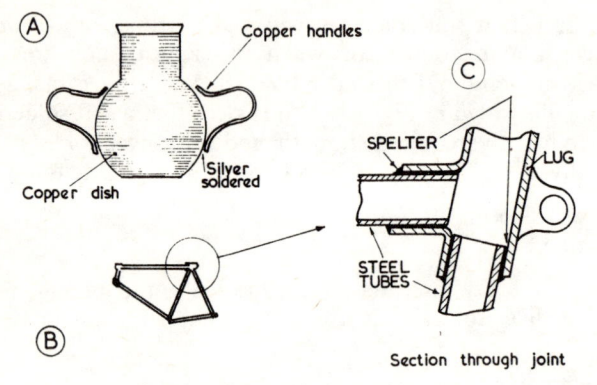

Fig. 4.18 Silver soldering and brazing

are therefore suitable for hard soldering. The essential technique involves cleaning the tube ends and the inside of the lug before applying the flux, which once again is a paste made from borax powder and water. A strip of spelter is cut and fitted around the joint and heated until the temperature is about 950°C, equivalent to bright cherry red. At this point the spelter melts and runs into the joint, producing on solidification a very strong and reliable assembly. It must be pointed out at this stage that it is essential to ensure that the whole joint is heated to a temperature slightly above that of the spelter, and this is achieved by ensuring that the flame or source of heat is *not* applied directly to the spelter. The penetration of spelter shown in the sectional view of the joint in Fig. 4.18C cannot be achieved unless the tempera-

Fastening and Joining 231

ture of the whole joint contact area is above that of the melting point of the spelter.

4.8 Basic welding principles

The great advantage offered by most welded joints is that no additional joining medium is required. Unlike soft and hard soldering which require the interposition of a suitable liquid solder between the surfaces to be joined, welding is achieved by *fusion* of the metal parts. This means that a relatively high temperature is a prime requisite for all welding processes, and two main sources of heat are employed, namely chemical and electrical.

Chemical sources of heat
The heat obtained from the combustion of coke is the result of a chemical change when the coke (carbon) burns with the oxygen of the atmosphere. Forge welding, that is to say the heating and joining of metal components by first inserting them in a suitable forge and then hammering the heated metals together, has been carried out by engineers for a great number of years. The metals best suited for forge welding are wrought iron and mild steel, and because the components must be taken to the forge, this makes forge welding a non-portable or static process Figure 4.19 shows some typical hand welded joints, and

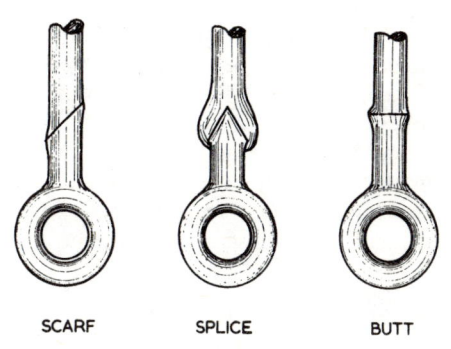

Fig. 4.19 Hand welded joints

it needs to be appreciated that a very high degree of skill is required for the proper execution of joints illustrated.

4.8.1 Oxy-acetylene welding

The development of the oxy-acetylene blowpipe, a device whereby oxygen and acetylene gas are burned to produce an exceedingly hot flame, gave to the welding process a much needed asset of mobility; in other words the source of heat could now be taken to the work, making possible a wide range of constructional and repair processes using the oxy-acetylene welding process. The essential equipment is simply shown in Fig. 4.20A, where it can be seen that two high-pressure storage vessels are needed, one for oxygen and the other for acetylene. Strong rubber pipes carry the gases to the blowpipe, and interchange-

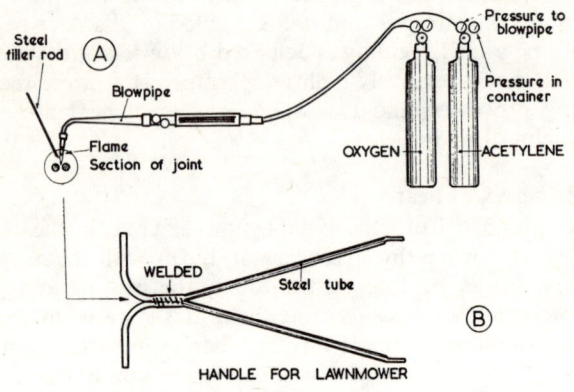

Fig. 4.20 Principle and application of oxy-acetylene welding

able nozzles enable a wide range of flame sizes to be used. Suitable regulators ensure that the gases are supplied to the blowpipe at the correct pressures since the proportion of oxygen and acetylene consumed in the chemical heating process has an important effect on the type of flame produced. Figure 4.20B shows a simple and typical application of oxy-acetylene welding. In this example, a filler rod is used to provide the additional metal required to ensure that a strong joint is produced between the two steel pipes which are welded together to make a handle for a lawnmower. Protective eye-shields *must* be worn by operators because of the intense glare at the heat source.

4.8.2 Electrical welding

Electricity or electrical power provides a useful and readily controlled source of heat easily applied to the welding process. This means that a greater degree of mobility is possible and there is a wide range of portable electrical devices available. The two main electrical welding techniques in use are:

(i) arc welding,
(ii) resistance welding,

and the following notes are intended to illustrate the basic principles involved.

Fastening and Joining 233

4.8.3 Arc welding

The principle of arc welding is illustrated in Fig. 4.21. We have in effect, an electrical circuit which is complete when the electrode makes contact with the work. This electrode is a mild steel rod coated with a layer of flux material which melts at the same time as the tip of the

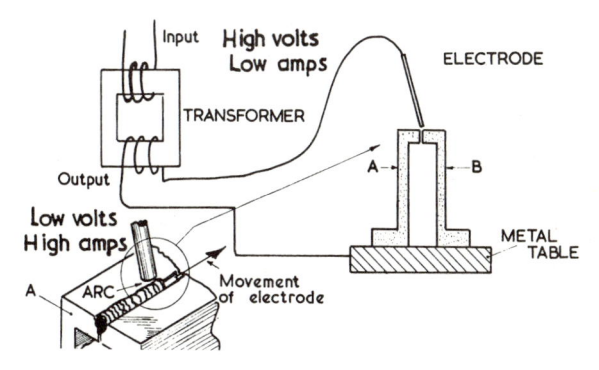

Fig. 4.21 Principle of arc welding

electrode due to the high temperature created by the arc across a small gap between the surface of the work and the electrode end. This is a simple method of welding fairly thick steel plates; the electrode as it melts acts as a filler material, whilst the boiling flux creates a gaseous shield preventing the oxygen in the atmosphere from contaminating the welded joint. As in oxy-acetylene welding it is important that the operator be shielded from the glare at the heat source, and this means that suitable protective eye-shields must be worn, with due protection of the cables and pipes carrying electrical power or gases.

4.8.4 Resistance welding

All metals vary in their ability to carry an electric current, with copper being a good conductor and steel a poor conductor. This means that if a high amperage current is passed through a steel bar, the bar is rapidly heated. This principle which makes use of the resistance of a metal to the passage of an electric current is widely used by engineers, particularly when steel components require permanent joints on a mass production basis. Although there are several types of resistance welding in use, the two examples given below will serve to demonstrate how production engineers continually seek to exploit simple basic principles.

Butt welding
When two parts are joined end to end the process is called butt welding. A good example of this technique may be seen in Fig. 4.22A, which

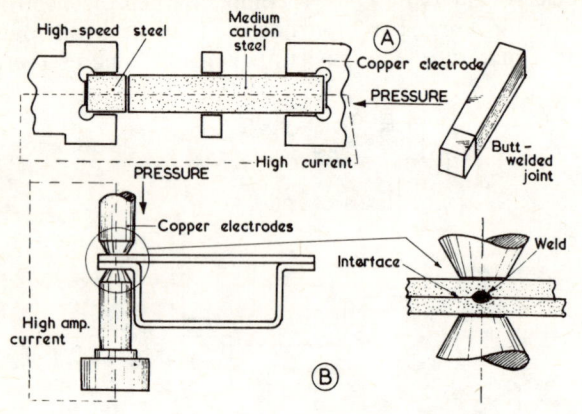

Fig. 4.22 Principle of resistance welding

shows a high speed tool butt-welded to a medium carbon steel shank. Because high-speed steel is an expensive metal, considerable economy is achieved by using it only where it is needed, namely at the cutting end, and the use of medium carbon steel as a strong tough shank makes possible a cheap, yet efficient single-point cutting tool. The means by which a butt weld is carried out is also simply illustrated in Fig. 4.22A. Both the high-speed steel bit and medium carbon steel shank are held in correct alignment and gripped by copper electrodes. They are brought together under pressure, and a high amperage current is passed through the joint, resulting in a rapid rise in temperature at the areas of contact. As soon as the temperature reaches the required value, making the joint semi-molten, further pressure is applied, resulting in a butt-welded joint. This technique is restricted to those joints having fairly low areas of contact, not exceeding about three square centimetres (3 cm^2), and it is essential that a good match or fit exists between the two faces to be joined.

Spot welding
A spot weld can be considered as equivalent to a rivet, and represents perhaps the fastest and easiest method of joining together two steel sheets of approximately the same thickness. The ease with which the joint is made is offset by the somewhat complex and expensive equipment required to make the spot weld and special-purpose spot-welding machines must be used. The principle involved is simply illustrated in Fig. 4.22B, where it may be seen that the work to be joined is placed between two copper-alloy electrodes. Pressure is brought to bear on

Fastening and Joining

the top electrode, keeping the joint in close contact, and a high amperage current passed through the joint. This results in a rapid rise of temperature at the joint interface, where the greatest resistance to the passage of an electrical current exists, causing the metal to melt as shown in the diagram. Electronic control ensures that the current is cut-off at the required time, and solidification of the molten pool of metal produces a strong and permanent joint, similar in effect to an internal rivet. To give an example of the efficiency of the spot-welding process, the underbody of a modern motor car, made from 0·7- and 0·9-mm-thick mild steel has about 1000 spot welds; yet the production time for these welds is less than *two* minutes. This very high rate of production is achieved with the aid of special jigs and fixtures, together with the application of banks of electrodes which move forward at predetermined intervals; in other words an automatic fully integrated spot welding assembly circuit.

4.9 The use of adhesives

The use of adhesives in the engineering manufacturing industry increases year by year. At one time the joining of separate parts using a liquid adhesive would be considered as *glueing*, a cheap and simple process of joining wooden components. Today, however, the use of adhesives goes a great deal further than simple glueing, and it is true to state that a properly bonded joint is equivalent and sometimes superior to a joint made by the mechanical fastening devices of riveting,

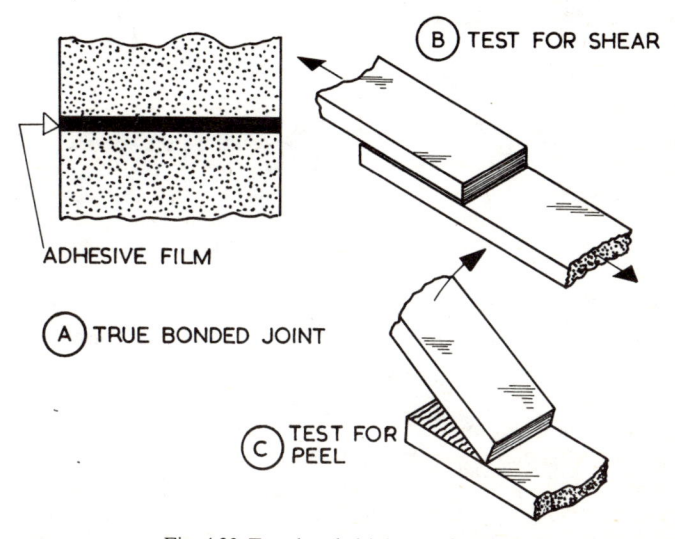

Fig. 4.23 True-bonded joints and testing

screwing and brazing. We need, however, to establish a few basic principles with regard to joints achieved with the aid of adhesives, and reference to Fig. 4.23A makes clear what is meant by a *true*-bonded joint. The sectional view shows that the two separate parts are joined by a thin film of adhesive. This principle is similar in many respects to soft soldering or hard soldering, except that the application of heat may not be necessary. Figure 4.23B and C show the conventional tests carried out to test the *shear* strength and also the *peel* strength. If the joint is properly made, it is usual for the material itself to fracture and not the joint. This ability of a joint to be stronger than the parent metal depends mainly upon the use of the proper techniques, and choice of the correct adhesive for the particular job in hand. Two basic types of true-bonded joints are in use:

(i) mechanical adhesion,
(ii) specific adhesion.

4.9.1 Mechanical adhesion

An adhesive joint achieved by mechanical principles makes use of a *keying* principle, that is to say an interlocking action exists between the adhesive surface and the component. A good example of mechanical adhesion may be seen in Fig. 4.24A, which shows a Ferodo brake lining true-bonded to its steel shoe. This is a popular method of joining brake linings to steel shoes, with the shear strength of the joint well in excess of service conditions, making it utterly reliable. The keying effect may be seen in the enlarged view shown in Fig. 4.24B, and it is

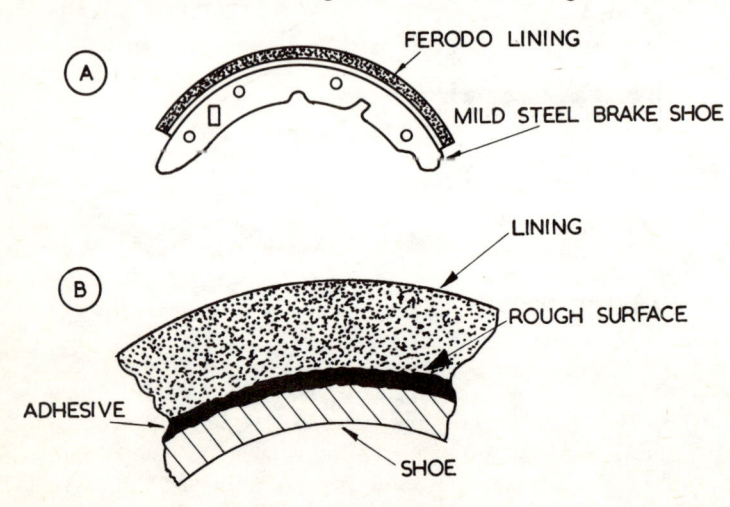

Fig. 4.24 Mechanical adhesion

Fastening and Joining 237

evident that both the lining and shoe possess a rough or uneven surface. In addition, the slight porosity of the brake-lining material assists in the penetration of the adhesive film, tending to increase the keying effect.

4.9.2 Specific adhesion

Figure 4.25A shows a phosphor bronze bush which is to be inserted in a steel bush plate. We may consider both the external surface of the

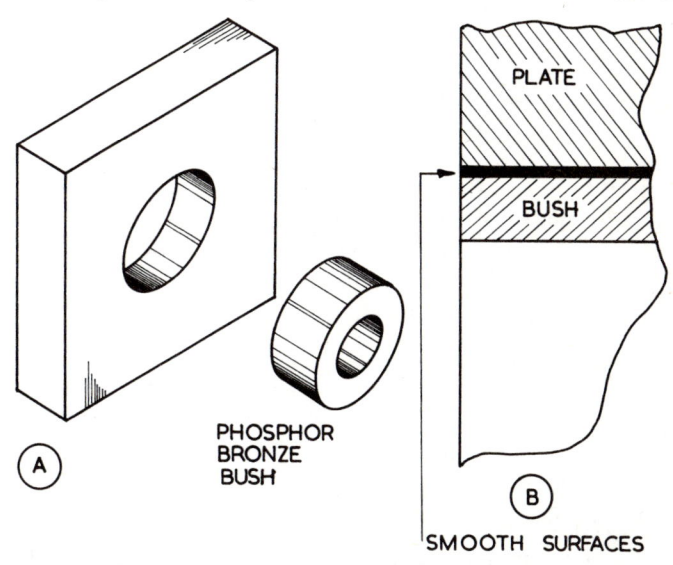

Fig. 4.25 Specific adhesion

bush and the internal surface of the hole as *smooth* surfaces, accurately machined to close limits of size. In a case like this, the strength of the joint is brought about by the molecular attraction, always present when two smooth surfaces are brought into close contact; for example when wringing two steel slip gauges together. At B we see an enlarged view of a section of the joint, where it is evident that very little interlocking or keying exists. Provided that the correct adhesive is used and the thickness of the adhesive film is within acceptable limits, the shear strength of this type of joint can be equivalent to the shear strength of the material.

Cemented joints
Care must be taken not to confuse cemented joints with true-bonded joints. Figure 4.26 makes a comparison between a cemented joint and a true-bonded joint, and it is clear that no adhesive film exists in a cemented joint. This is due to the fact that the material for the cement

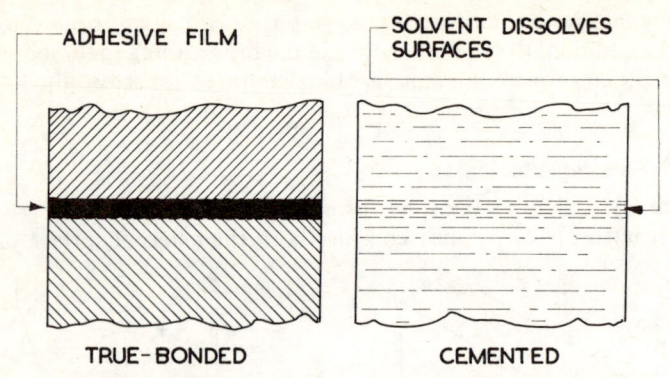

Fig. 4.26 Comparison between cemented and true-bonded joint

is identical to that of the material to be joined, a solvent being used to dissolve the material. Hence on application the intervening film of cement being of the same material as the two surfaces to be joined, produces a *welding* effect; that is to say the whole joint is composed of the same material or is homogeneous.

4.9.3 Types of adhesives

Early types of adhesives were derived from three main sources:

(i) Animal,
(ii) vegetable,
(iii) mineral.

Examples of animal-base adhesives include albumen and fish glue, but these are seldom used at the present time. Vegetable-base adhesives include starch, gums and resins; some of these are still widely used for paper jointing. Most mineral adhesives are bitumen based, and find little application for normal jointing purposes. The adhesive most widely used by engineers is *epoxy resin*, and the introduction of this modern glue has made possible tremendous advances in the manufacture of engineering components ranging from car brake linings to components for jet aircraft.

Use of epoxy resins

These are thermo-setting resins requiring the addition of a hardener of activator. This activator is added to the resin immediately before the joint is made, and when the activator has been thoroughly mixed with the resin it is essential that the mixture be used within a certain period of time. This period of time is known as the *pot-life* of the resin; as soon as this period of time is exceeded the resin hardens or cures and must be discarded. The great advantage offered by the use of epoxy

Fastening and Joining

resins is that they harden at room temperature, and no external source of heat is required. A bonded joint so made is immune from any solvent, including water, and as a result finds much application for marine use. Although possessing excellent shear strength, the joint is not suitable for applications where sudden blows are likely, or a certain degree of flexure is expected. In addition, the strength of the joint decreases at temperatures in the region of 100°C. However, a range of epoxy resins has been developed to meet servce requirements which include the need for flexibility or the capacity to withstand temperatures in excess of 100°C. These special epoxy resins are outlined below:

Table 4.2 Properties of some epoxy resins

Special properties	Resin type
Tolerates temperatures up to 250°C	Epoxy-Phenolic
Tolerates temperatures up to 350°C	Epoxy-Silicone
Increased resistance to impact	Epoxy-Nitrile
Increased resistance to oils and chemicals	Phenolic-Nitrile

The technique adopted when making a joint using epoxy resin is not unlike that used when making a 'sweated' joint using the soft soldering process. The surfaces to be joined should be clean and free from grease and other foreign materials. An optimum surface roughness is desirable, assisting in producing a maximum surface area for the adhesive film. Figure 4.27 illustrates the essential stages when true-bonding two metal components together; in the example shown a copper disc is being bonded to a mild steel disc. The process involved in each stage are:

Stage A
Both surfaces prepared by cleaning and roughing. Surfaces may be roughed by acid-pickling, shot blasting or by rough machining.

Stage B
Activator added to epoxy resin and thoroughly mixed. The amount must be sufficient to cover the jointing area only; any excess will harden once the pot life is exceeded, resulting in waste of a relatively expensive adhesive.

Stage C
Resin applied to surface of one component using a brush, serrated scraper or spatula.

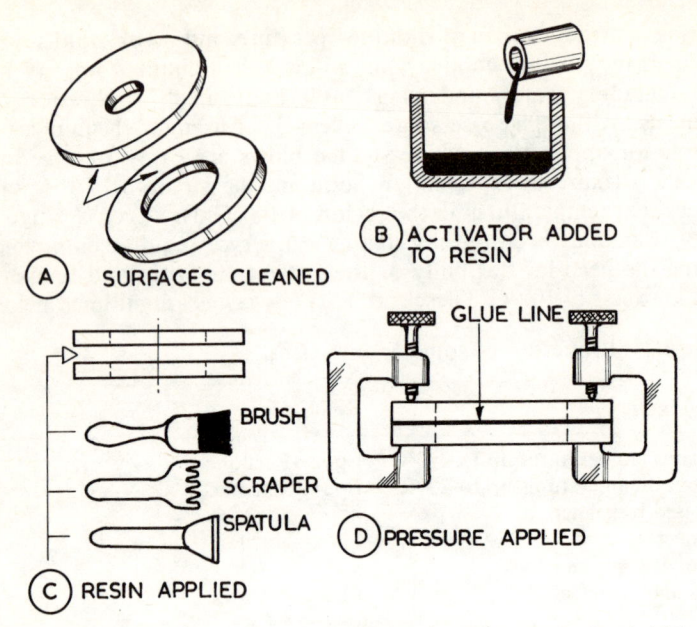

Fig. 4.27 Use of epoxy resin at room temperature to form a true-bonded joint

Stage D.
With the two components in correct alignment, external pressure is applied using clamps or other devices to keep the two components in close contact. The gap or distance between the two surfaces is often referred to as the *glue line*, and should have a thickness of somewhere between 0·02 mm and 0·2 mm. For best results the joint should remain clamped until the adhesive film of epoxy resin has hardened or cured. This may take up to twelve hours according to the type of resin compound used. A true-bonded joint produced in the manner just outlined is equivalent to a joint produced by either riveting or welding.

Use of pressure and heat
For large-scale production, such as may be found in the automotive industry, the use of both pressure and heat has much to offer in terms of simplified manufacture and reliability of the joint produced. It is true to state that most brake shoes and clutch linings, together with disc pads are true-bonded using adhesives, and failure or breakdown of these highly stressed and vital components are rare in the extreme and virtually unknown. Let us consider the bonding of a clutch lining to a mild steel disc or ring; an operation very similar to the joining of the copper and steel components just outlined and illustrated in Fig. 4.27. Clearly any method of joining a Ferodo disc to a steel ring

Fastening and Joining

without the need for drilling holes and the use of rivets is certain to result in considerable savings of both materials and machining time. Figure 4.28A shows the required assembly of both Ferodo disc and steel ring. Once again we may outline, in a simple manner, the basic stages required to make the joint, keeping in mind the need to ensure that the

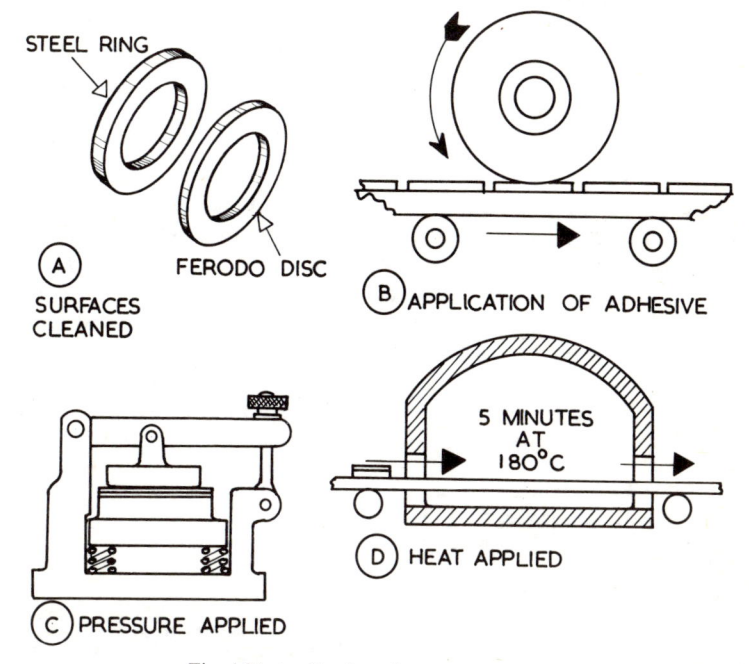

Fig. 4.28 Application of pressure and heat

technique is carried out on a mass production basis. That is to say, conveyor systems are used to process the components through the stages necessary to produce the bonding of the Ferodo discs to the steel rings.

Stage A
Preparation of the surfaces is again essential in particular the steel ring, which may require de-greasing and shot blasting.

Stage B
Application of the adhesive, in this case a *vinyl-phenolic*-type resin, is normally carried out by spraying, dipping or roller-coating, and Fig. 4.28B shows in a simple manner the roller-coating of the vinyl-phenolic adhesive on the joint surface of the Ferodo disc. These discs are passed beneath a roller suitably charged with adhesive, the whole

process taking place on a moving conveyor belt, ensuring speed of production together with the deposition of an even and constant layer of adhesive.

Stage C
The application of pressure is an important part of the bonding process, and it is usual for the clamping arrangements to include positive location of the Ferodo disc on the steel ring. The pressure, in the order of 0·4–1·5 N/mm^2, may be applied using spring-loaded clamps; the principle is simply illustrated in Fig. 4.28C.

Stage D
It is necessary to heat the joint whilst under pressure, and this is achieved simply by passing the clamped unit through a furnace at 180°C; about 5 to 10 minutes would be sufficient (Fig. 4.28D). As previously stated, the bond between the two components is extremely reliable, although achieved through mechanical adhesion, and shows once again how engineers are continually seeking and applying new principles and techniques to the art of engineering manufacture. Figure 4.29 shows a simple outline of the production line or layout required to bond the brake units on a flow-production basis.

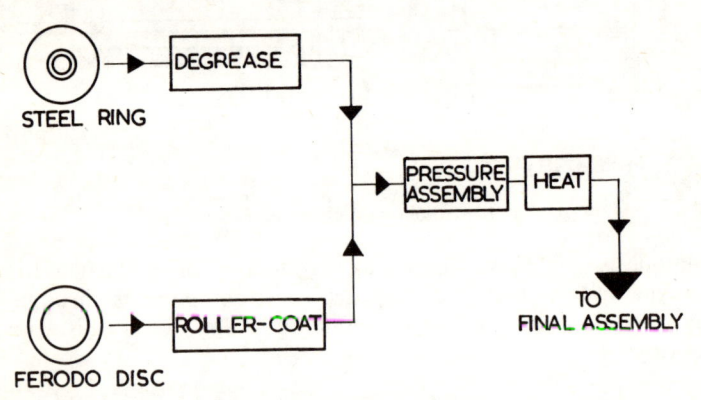

Fig. 4.29 Flowline bonding

4.9.4 Metal to metal bonding

In general, two types of metal to metal bonding are finding increasing use in engineering manufacture. In the first example the adhesion is in the form of a liquid or paste, and is applied by hand using a brush or comb. A typical example of this process is illustrated in Fig. 4.30, where we see the assembly of a mild steel flywheel to an alloy steel shaft. This technique eliminates the need to machine keyways in the

Fastening and Joining

shaft and flywheel, and also the time and cost involved in assembling the components; for the correct fitting of a key represents one of the most skilled and difficult of all fitting operations. Hot curing in an oven presents no problems, and it is certain that metal to metal bonding poses a great threat to more conventional methods such as shrink

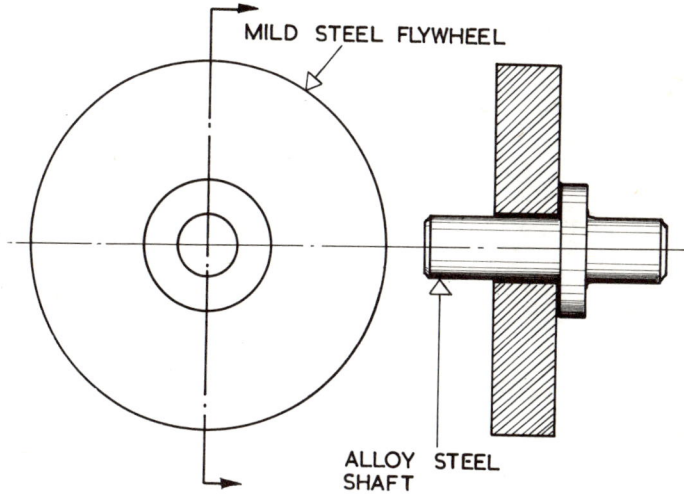

Fig. 4.30 Metal to metal bonding

fitting and keying. The second example of metal to metal bonding involves the use of the adhesive, not in paste or liquid form, but as beads, pellets or dry film. This method of making the adhesive available in precise shape and mass, ensures that the exact amount is applied to the joint, thus making possible considerable savings in both money and time. Simple press tools can be used to blank out complicated profiles from sheet adhesive film, with the resultant blanks easily placed between the surfaces to be joined. Both pressure and heat are needed to effect a proper bond, care being taken to ensure that the pressure is applied directly to the interface of the joints. Figure 4.31 illustrates the process of using adhesive film when joining two halves of an aluminium alloy housing.

4.10 Safety precautions when adhesive bonding

As we have seen, modern adhesives are most efficient in their ability to join together various materials, and they are equally efficient if, due to carelessness some adhesive comes into contact with the hands or skin. It is not uncommon to require plastic surgery in order to separate two fingers inadvertently bonded together because of lack of care in the use

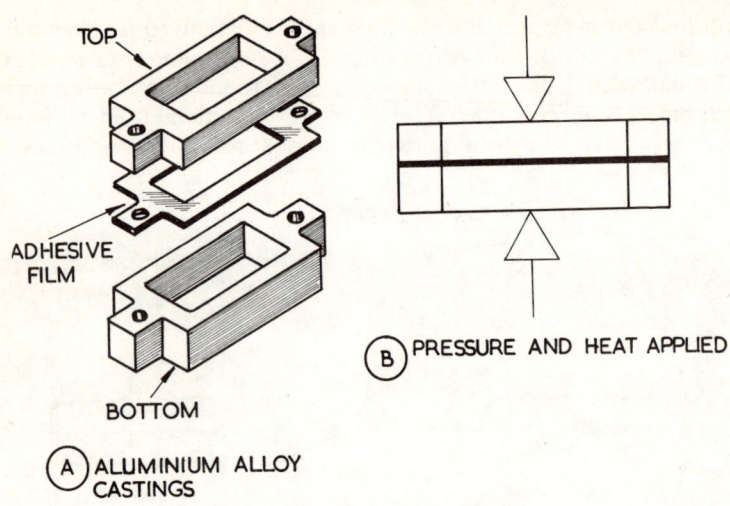

Fig. 4.31 Use of adhesive film

of adhesive. It is essential therefore, that the greatest precautions be taken both in mixing and application, and ensuring that the adhesive is not allowed to be in any place where it may fall into the hands of children or other persons not familiar with its potential dangers.

At the same time, if contact with the adhesive does not result in actual bonding, there is the grave risk that a skin reaction may be set up, leading to *dermatitis*, and it is as well to note that some persons are much more prone to skin reactions than others. The golden rule at all times is absolute cleanliness and good working methods, as described at the beginning of this book; for it is just as easy to acquire good working techniques as it is to acquire bad ones. Gloves are an essential part of the protection required, heavy duty aprons and adequate overalls are also needed; adhesives must be kept in proper containers and any spillage dealt with immediately. Any material which has been in contact with the adhesive must be properly and immediately disposed of, as must worn or discarded gloves and protective clothing. The increasing use of thin disposable protective wear, discarded at the end of each working shift helps to ensure that no contaminated clothing is left around the workpiece. Finally, hands must be washed regularly before, during and at the end of each working shift, and the application of a suitable barrier cream should be encouraged.

Fire and toxic risks

Many of the solvents required by modern synthetic adhesives are highly inflammable, and according to the degree of danger, storage reg-

Fastening and Joining 245

ulations are governed by Government regulations. In general it is essential that naked flames be prohibited from the working or storage area, and that workbenches and equipment are earthed to prevent any sparks produced by static electricity. Full ventilation is of vital importance; that is to say that all vapours and gases should be conveyed by ducts from the work and storage areas to outside the factory premises. Toxic risks arise mainly from the possibility of inhalation of the fumes or vapours associated with adhesive bonding, and in some processes face masks must be worn. It is commonly thought that a toxic vapour, that is to say one which can have an injurious effect on the health of an individual, is easily detected by its odour or smell, but this is far from the truth. In many cases, highly toxic vapours have no appreciable smell or odour, whilst on the other hand, operators of a particular process tend to become familiar with a particular vapour or odour over a period of time and can, as a result acquire a dangerous build-up of toxic material. At all times, good working habits need to be developed, with strict observance of the rules and regulations laid down for the protection and welfare of the operators. Anything different or unusual about the working environment must be reported without delay.

Chapter 4   Assimilation Exercises

1 Make a neat sketch of an engineering assembly joined using nuts and bolts.
2 Describe the circumstances which necessitate the use of set-screws to join two components.
3 Explain why some mild steel studs have different threaded ends.
4 Sketch or describe a simple engineering assembly where it is essential to ensure that locking devices are used to prevent threaded fasteners from working loose.
5 Explain the basic difference between positive locking and friction locking.
6 Sketch a typical engineering assembly that would require the use of a split pin and castle nut.
7 Describe the type of assembly that would best be secured using a locking wire.
8 Describe the circumstances where tab washers would be used to prevent set-screws from working loose.
9 Outline the basic principles involved in the following friction locking devices:

   (i) lock nuts,
   (ii) spring washers,
   (iii) nylon inserts.

10 Outline the correct procedure to adopt when making a riveted joint.
11 Make a neat sketch of two metal components joined by soft soldering.
12 Explain the purpose of a flux, and give two types of fluxes used when soft soldering, stating the best application for each type of flux.
13 Under what circumstances would hard soldering be chosen in preference to soft soldering?
14 Sketch two devices that would strengthen soft soldered joints.
15 Make a neat sketch of an engineering assembly that would be brazed together.
16 State the type of work for which silver soldering is best suited.
17 Outline the advantages offered by the use of oxy-acetylene welding over conventional forge welding.
18 With a neat diagram show the basic principle of arc welding.
19 Sketch typical engineering examples of each of the following resistance welding techniques:

  (i) Butt welding,
  (ii) spot welding.

20 Explain the difference between mechanical adhesion and specific adhesion when using adhesives for joining purposes.
21 With a neat diagram illustrate a simple cemented joint for two plastic materials.
22 Sketch an engineering example of a metal to metal joint achieved with adhesive.
23 Outline any safety precautions that need to be observed when using plastic adhesives.
24 Write down three dangers that may be present when working with toxic materials.

5
ENGINEERING MATERIALS

Objectives—The principles and applications of:
1. Engineering metals and alloys in common use
2. The physical properties of engineering materials
3. The mechanical properties of engineering materials
4. The uses of engineering materials
5. The heat treatment of carbon steels
6. The use of non-ferrous metals and alloys
7. The use of plastics materials
8. The protection of metal surfaces

5.1 Engineering metals and alloys

The development of all civilisations is inevitably linked with the materials from which the tools of warfare and agriculture have been fashioned. The first engineers made great use of suitable stones and flints, from which crude knives and axes were laboriously shaped, and many thousands of years were to pass before the discovery of the art of metal smelting led to an era of bronze manufacture, later to be followed by extensive use of iron and steel. The range of materials in use nowadays is very great indeed; we write on wooden desks using plastic pens having tungsten writing points; we ride on metal bicycles or drive metal cars. Metal aircraft fly overhead and metal ships cross the oceans; metal wires conduct electricity over, and metal pipes carry liquids and gas under, many hundreds of miles of countryside. The aircraft are light in weight but strong, the ships are able to withstand the most violent storms, and the electrical energy, liquids and gases are safely transmitted both over and under the ground, with no danger to the population. Clearly different materials serve a wide range of uses, and making the correct choice of a material for a specific purpose is one of the most important duties of an engineer. Figure 5.1 shows in diagrammatic form the main types of materials used at the present time. Note that there are two main groups, metallic and non-metallic, with further

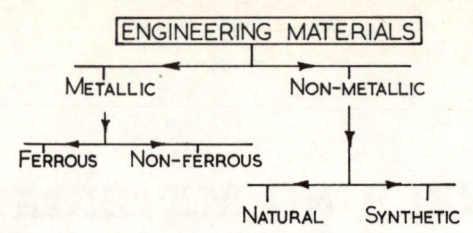

Fig. 5.1 Engineering materials

subdivision of the metallic materials into ferrous and non-ferrous, and of the non-metallic into natural and synthetic.

5.1.1 Ferrous metals

Figure 5.2 shows the more common ferrous and non-ferrous metals used in engineering manufacture. We may describe ferrous metals as those which contain iron; they are so named because the Latin word for iron is *ferrum*, from which is derived the chemical name, ferrite, and the chemical symbol Fe. Most metals are obtained by the process called *smelting*, during which the metallic ore is heated in a suitable furnace where it is separated from all impurities, to be followed by a refining or purifying process. Let us take a closer look at the ferrous metals in Fig. 5.2.

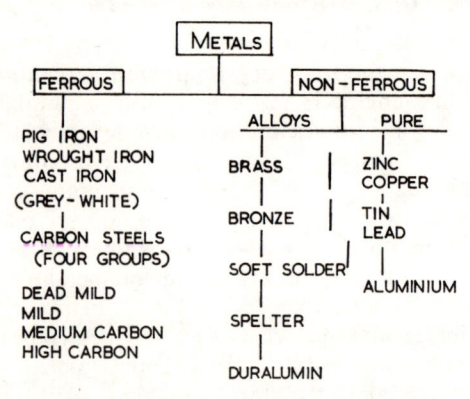

Fig. 5.2 Ferrous and non-ferrous metals

Pig iron
This is the product of the blast furnace and forms the basis of all ferrous metals; it is used in the manufacture of cast iron, carbon and alloy steels. Pig iron is never used as a structural material. It is weak and brittle, and is easily broken with a sudden blow; this is due to the

presence of approximately 4% carbon absorbed from the coke used as fuel in the smelting process.

Wrought iron
If the carbon is removed from pig iron we are left with a metal called wrought iron, which may be considered as pure iron with a ferrite content of 99·95%, the remaining 0·5% consisting of slag inclusions picked up during the smelting process carried out in a puddling furnace. Wrought iron is a soft yet tough metal, very easily bent or twisted while in the cold state, yet possessing excellent resistance to fracture when subjected to sudden shocks or blows. Its resistance to corrosion is very good, owing to the formation of a film or scale of black iron oxide which tends to prevent further oxidisation or scaling.

Grey cast iron
Pig iron reheated and refined in a small furnace called a *cupola* produces grey cast iron. Provided the liquid cast iron is allowed to cool slowly, the resulting metal is called *grey* cast iron, having a carbon content of approximately 3·5%, most of it in the free form known as graphite. The presence of this 3·5% carbon confers the property of excellent fluidity to the molten cast iron, making it possible to pour the cast iron into suitable moulds to produce engineering components known as castings. Remember that wrought iron contains no carbon, and when heated becomes spongy or pasty and cannot be poured into moulds to produce castings. The carbon present in grey cast iron, is in the form of flakes, and represents a source of serious embrittlement, making grey cast iron relatively weak in both tension and shear. Grey cast iron is so named because the fractured surface of the metal always presents a dull or grey appearance.

White cast iron
If liquid cast iron is rapidly cooled or chilled from the molten state an extremely hard cast iron results which is quite unmachineable using standard cutting tools such as files and twist drills. A sudden blow would shatter a thin piece of white cast iron into many fragments, all of which would display the characteristic white fracture from which the metal derives its name. Engineers have little use for white cast iron except when certain parts of a casting require to be dead hard; this is achieved by the insertion of steel or iron sections (called chills) into a sand mould. The sand, a poor conductor of heat allows the molten iron to cool slowly, whilst the steel 'chills', possessing a higher thermal conductivity cools that part of the molten iron in contact more quickly, thus producing white cast iron. The principle is simply illustrated in Fig. 5.3.

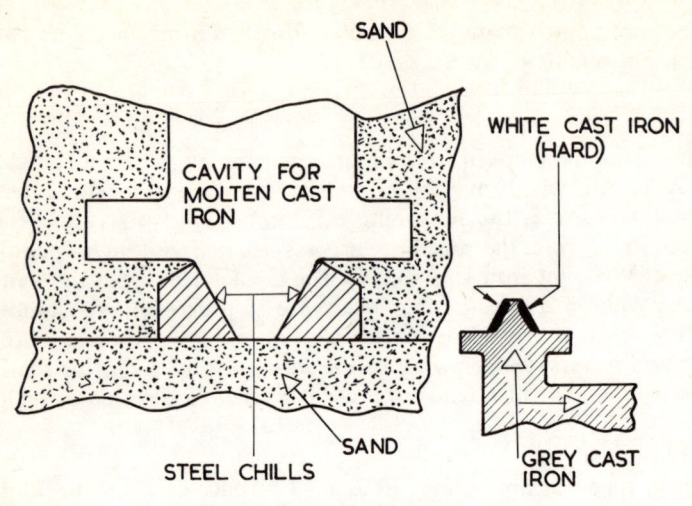

Fig. 5.3 Grey and white cast iron

Carbon steels
Carbon steels are made in steel-making furnaces, where close control is maintained on the amount of carbon present in the steel. Provided the amount of carbon does not exceed 1·5%, it combines chemically with the iron producing a strong structure with no free carbon or graphite present. The actual percentage of carbon present determines the type of carbon steel produced, and the four main groups are as follows:

  (i) dead mild steel,
  (ii) mild steel,
  (iii) medium carbon steel,
  (iv) high carbon steel.

Alloy steels
Alloy steels are used when carbon steels are unsuitable for the kind of conditions to which the component will be subjected. For example, if a very strong steel is required, a nickel–chrome steel would be chosen in preference to a plain carbon steel, because nickel–chrome steels are considerably stronger and better able to stand up to severe stressing. If a component is required to resist corrosion, in other words to be stainless, an alloy steel is used. Cutting tools, both single and multi-point are made from alloy steels and are capable of removing metal at very high speeds. The addition of alloying elements such as nickel, chromium, tungsten, molybdenum and vanadium change a carbon steel into an alloy steel.

5.1.2 Non-ferrous metals

In general, non-ferrous metals are much weaker than ferrous metals, and are seldom used for components that are subjected to tensile or shear forces. They have, however, a great advantage over most ferrous metals, namely their ability to resist corrosion; in other words they do not rust or oxidise easily. Reference to Fig. 5.2 shows that non-ferrous metals are divided into two groups:

   (i) pure metals,
   (ii) non-ferrous alloys.

An alloy, as we have seen, is composed of two or more metals, for example an alloy steel such as a nickel–chrome–vanadium steel.

Pure non-ferrous metals

Copper
Easily recognised by its attractive red colour, copper derives its name from the island of Cyprus where high purity copper was originally mined. A soft ductile metal, copper is an excellent conductor of both heat and electricity.

Tin
Silvery in colour, tin is a soft metal with an excellent resistance to corrosion. It is however, most expensive and is seldom used in its pure state except as a protective coating for mild steel sheet, more commonly known as *tinplate*.

Lead
Also silvery in colour, and another good resister of corrosion, lead is much used because of the ease with which it can be cold-worked. It is easily bent, twisted or hammered into shape with little risk of fracture or cracking.

Zinc
Another silvery metal, easily bent and with good resistance to corrosion. Also used as a protective coating for mild steel sheets.

Aluminium
About one third the weight of steel, aluminium is a soft silvery metal easily rolled into very thin sheet or foil. Most protective coverings for cigarettes and chocolates are made from aluminium foil.

5.1.3 Non-ferrous alloys

All the pure metals listed so far posses very little strength; they are used mainly because of the ease with which they can be cold- or hot-worked, or because of their superior resistance to corrosion. If however, certain other metals are added, some of the alloys produced are quite strong and this is due to the chemical changes that take place between the parent metal and the alloying metal.

Bronze

Having a dark red-brown colour, bronze is an alloy of copper and tin. Copper is the parent metal, and the amount of tin added seldom exceeds 15%. Bronze was highly prized in early civilisations, finding extensive use in swords, spear-tips and other war-like weapons. Today its use in the engineering field is restricted, except as a decorative metal, although with small amounts of phosphorus, a very useful bearing metal is produced. Bronze has a good resistance to corrosion.

Brass

The addition of zinc to copper produces the useful alloy known as brass. Several types of brass are in use, with colours ranging from rich gold to bright yellow; the amount of zinc added to the parent metal copper determines the type of brass produced. All brasses have a good resistance to corrosion.

Duralumin

This is a strong light metal with a strength far superior to that of the parent metal aluminium. A small amount of copper is present not sufficient to change the colour of the duralumin which closely resembles aluminium.

5.1.4 Non-metallic materials

Reference to Fig. 5.4 shows that non-metallic materials may be divided into two main groups:

(i) natural,
(ii) synthetic.

```
              NON-METALLIC
              /          \
         NATURAL       SYNTHETIC
            |              |
          STONE          GLASS
            |              |
          WOOD          PLASTICS
            |
          SAND
```

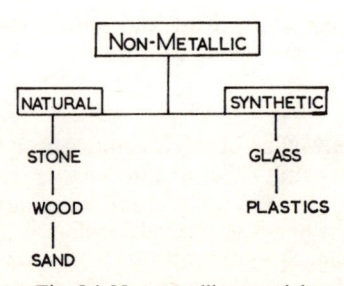

Fig. 5.4 Non-metallic materials

Engineering Materials

Natural materials
We may define natural materials as those supplied by Nature or natural means; wood, sand and stone provide good examples of natural materials that have been, and still are, in extensive use for all civil engineering projects. Wood is still widely used in engineering manufacture. It is strong, light and easily worked. Engineers make much use of suitable woods in the manufacture of patterns for gravity sand casting in cast iron and steel, whilst sand is needed for the moulds which are constructed to take the molten metal and producing a casting on solidification.

Synthetic materials
Synthetic materials are those made by man, they do not exist in the natural state. All synthetic materials have a great advantage over natural materials in that they can be made to meet specific needs. A wide range of plastics is now available, from which engineering components, radio parts, toys and fabrics are manufactured.

Cost of metals
It is of interest to have a comparison of the cost of the more common engineering metals, and Table 5.1 gives the price per metric tonne at the time of writing:

Table 5.1 Metal costs in 1977

Metal	Cost per tonne
Tin	£6300
Copper	860
Aluminium	630
Lead	420
Zinc	400
Mild steel	230
Cast iron	120

Clearly the cost of a material plays an important part when choosing a metal for a specific job, for the cost must influence the selling price of the finished product. We can be sure that engineers are ceaseless in their search for economies that lead to saving in both metal costs and machining times. It must be remembered that a metric tonne is the equivalent of 1000 kilogrammes, and as one kilogramme equals 2·22046 pounds, one metric tonne is equal to 2204·6 pounds and is therefore, 35·4 pounds short of an Imperial ton which is 2240 pounds. Figure 5.5 illustrates in graphical form the present cost of metals.

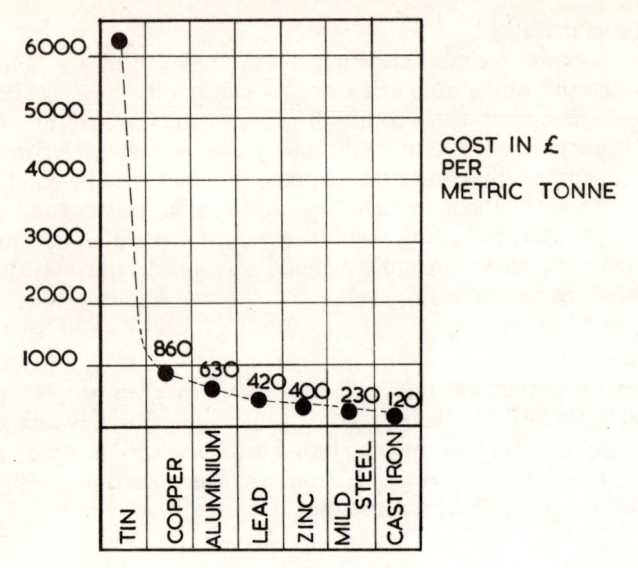

Fig. 5.5 Cost of metals

5.2 Physical properties of metals

The main physical properties of metals and materials of interest to engineers may be listed as follows:

(i) melting point,
(ii) density,
(iii) electrical conductivity,
(iv) heat conductivity.

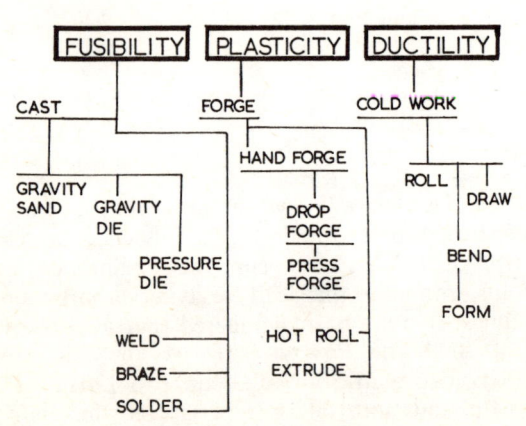

Fig. 5.6 Manufacturing techniques

Melting points

The importance of melting points may be seen by reference to Fig. 5.6, where we see the three main primary processes utilised by engineers when considering the methods available for the manufacture of components. The fusibility of a metal is one of its physical properties, and may be defined as its ability to change into a liquid or molten state when heated to a certain temperature. All metals that become liquid when heated to their melting points are capable of being poured into moulds; this technique is known as *casting*, a manufacturing process which has been practised by engineers for many thousands of years. Metals with low melting points are easily cast; for example zinc and aluminium alloys, whilst a metal with a high melting point such as mild steel is much more difficult to cast making mild steel castings an expensive product. Figure 5.7 shows the melting points of the more

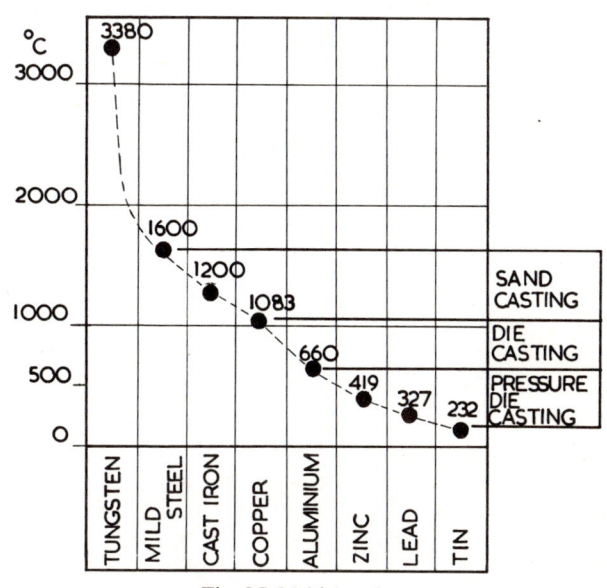

Fig. 5.7 Melting points

common engineering metals together with the methods or processes adopted to produce castings.

Density

The density of a metal may be defined as its *mass per unit volume*, that is to say the number of kilogrammes per cubic metre, or grammes per cubic centimetre. The *relative* density of a metal is the *ratio* of the density of a metal with the density of pure water at a temperature of 4°C, and because one gramme of pure water has a volume of one cubic centimetre at 4°C, then the density of one cubic metre of water is 1000 kilogrammes. This is simply illustrated in Fig. 5.8A, whilst at B we see a table of densities of the more common metals. Figure 5.8C gives a simple example of the use of the table, and it is required to calculate

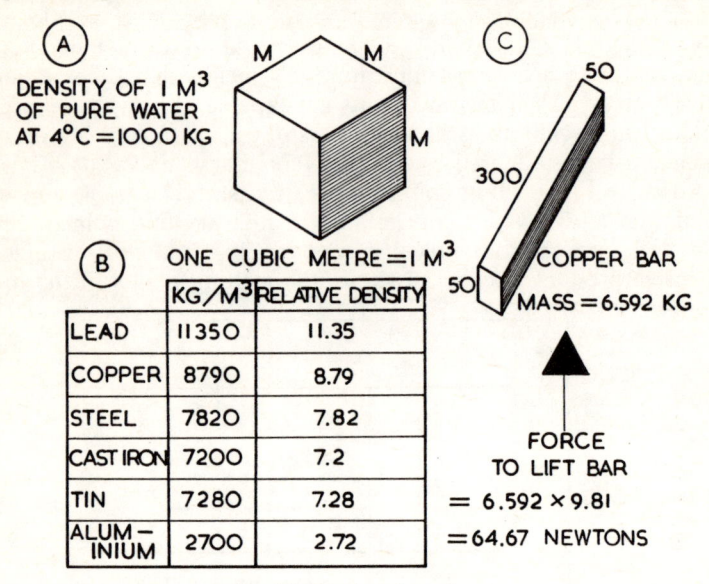

Fig. 5.8 Density of metals

the *mass* of the copper bar shown. The calculation is carried out as follows.

$$\text{MASS} = \text{Volume} \times \text{density}$$
$$\text{Area} \times \text{length} = \text{volume}$$
$$50 \times 50 \times 300 = 750\,000 \text{ cubic millimetres (mm}^3)$$
$$= 750 \text{ cubic cm (cm}^3)$$

From the table we see that the density of copper is 8·79 grammes per cubic centimetre, therefore the mass of 750 cm^3

$$= 750 \times 8\cdot79 = 6592\cdot5 \text{ grammes}$$
$$= 6\cdot592 \text{ kilogrammes.}$$

Hence a bar of copper, 50 mm × 50 mm × 300 mm length has a mass of 6·592 kilogrammes. As seen in the diagram, the *force* required to lift this mass of copper is calculated as follows:

$$\text{Force in Newtons} = \text{Mass} \times 9\cdot81$$
$$= 6\cdot592 \times 9\cdot81$$
$$= 64\cdot67 \text{ Newtons}$$

Electrical conductivity

All metals tend to resist the passage of an electric current, and those metals that offer little resistance are said to be good *conductors*. On the other hand, a metal that offers a high resistance to an electrical current is said to be a good *insulator*. If we consider that electrical energy is the prime source of power for machine tools, it is clear that efficient conductors are needed to conduct the electrical energy to the machine tool, but at the same time it is essential that the operator is protected from the dangers inherent in the use of electricity, and this means that efficient insulators are needed. Figure 5.9A shows the need for both

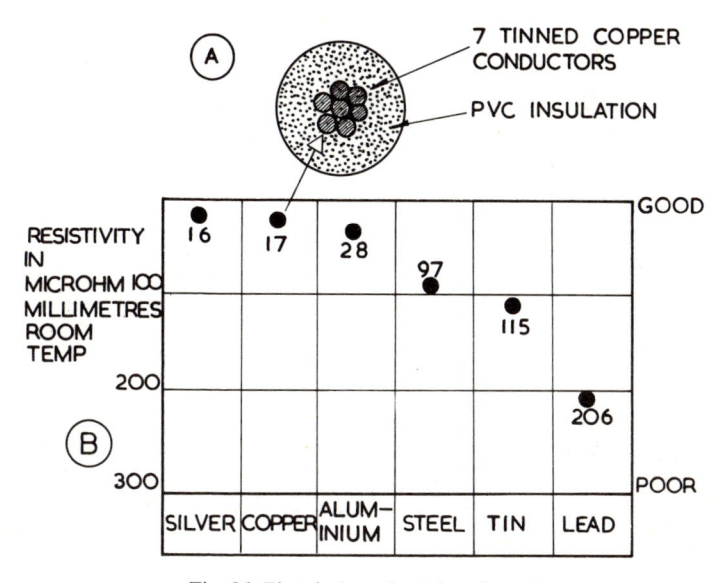

Fig. 5.9 Electrical conductivity of metals

good conductors and good insulators, while at B we see a simple graph indicating the electrical conductivity of the more common engineering metals and materials. It needs to be remembered that the electrical conductivity of a metal is affected by the purity and temperature of the metal.

Thermal conductivity

This is the ability of a metal to transfer or conduct heat from areas of higher to areas of lower temperature. Most metals are very good conductors of heat, whilst liquids and gases are relatively poor conductors; air for example, is a poor conductor of heat. Figure 5.10 indicates in a simple manner the thermal conductivity of some engineering materials.

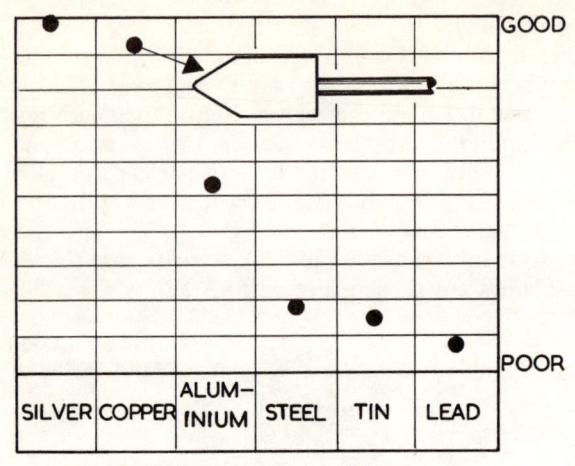

Fig. 5.10 Thermal conductivity of metals

5.3 Mechanical properties

We may define mechanical properties as those which concern designers and engineers when considering a material for a specific duty. For example, a hook or chain which is to be used for lifting heavy components *must* be able to withstand the tensile stresses imposed on it. Figure 5.11A illustrates the process of ingot-pouring at a steel works. The ladle, containing about 81 tonnes of molten steel at a temperature of 1600°C is supported by a chain and lifting hook, both taking the total force exerted by the ladle of molten steel. The safety and lives of the men whose work it is to fill the ingot moulds depend on the tensile strength of the hook and chain. This is only one instance when the

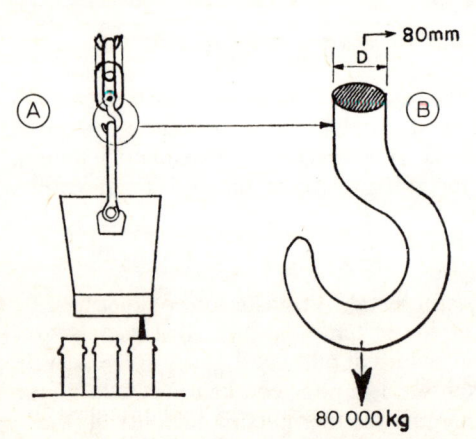

Fig. 5.11 Tensile forces on chain and hook

Engineering Materials 259

lives of men depend on the ability of metal to stand up to heavy loads, and there are many occupations in which the worker or operator depends on the strength of metal to ensure his safety. Simple examples include flying and testing aircraft, crossing a steel bridge and ascending or descending in a lift.

Ultimate tensile strength
When an engineering component is subject to tensile loads, the main property required is the ultimate tensile strength (UTS) of the metal. This is the maximum load or force the metal can withstand without fracture, and the units used are Newtons per square millimetre (N/mm^2) or Newtons per square metre (N/m^2). The greater the ultimate tensile strength, the stronger is the metal, and the more suitable it is for the purposes of lifting or pulling heavy loads. In general, the ferrous metals possess the highest tensile strengths, as can be seen by reference to Fig. 5.12, which compares the ultimate tensile strengths of

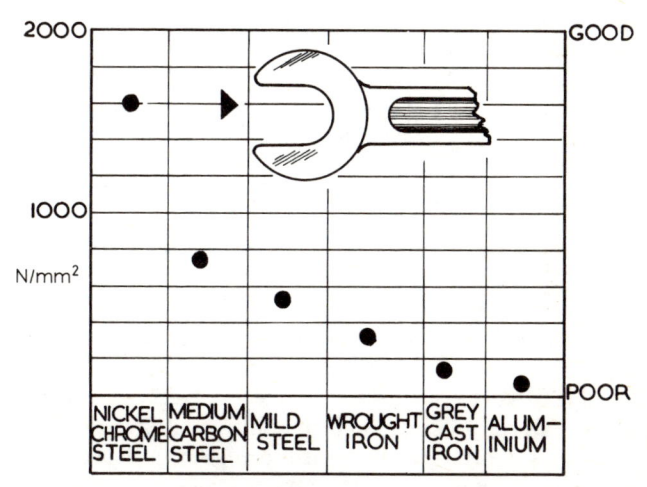

Fig. 5.12 Ultimate tensile strengths

the more common engineering materials. Calculations on tensile stress are not difficult, and one simple example will serve to show the value of the table given in Fig. 5.12. Let us assume that we wish to calculate the tensile stress on the shaded area of the lifting hook shown in Fig. 5.11B. The formula to use is:

$$\text{Stress} = \frac{\text{Force acting}}{\text{Area taking the force}}$$

Once again it is necessary to multiply by 9·81 the mass to be lifted, in order to find the force required,

$$\text{Force in Newtons} = \text{Mass in kg} \times 9\cdot 81$$
$$= 80\,000 \times 9\cdot 81$$
$$= 784\,800 \text{ Newtons}$$

$$\text{Stress on hook} = \frac{\text{Force}}{\text{Area}}$$
$$= \frac{784\,800}{\pi R^2}$$
$$= \frac{784\,800}{\pi \times 40^2}$$
$$= \frac{784\,800}{5026\cdot 5}$$
$$= 156\cdot 13 \text{ N/mm}^2$$

We see from the diagram that the material specified for the lifting hook is wrought iron, and reference to the table in Fig. 5.12 tells us that this material has an ultimate tensile strength of 320 N/mm². This is about *twice* the actual working stress of 156·13 N/mm², giving a *factor of safety* of about 2 to 1. It is extremely doubtful whether such a low factor of safety would be permitted, and the remedy would be to increase the diameter of the bar from which the hook is forged, or use a stronger material. The consequences arising from the sudden failure of the lifting hook when raising a ladle containing a mass of 81 metric tonnes of molten steel would be of a most serious nature.

Elastic limit
Most metals possess a reasonable degree of elasticity, that is to say they have the ability to stretch when subject to a force, yet regain their original shape or size on removal of the force. This is a very useful and essential mechanical property, since inability to stretch or bend would result in a metal of little structural use to engineers. The elasticity of a metal makes possible its use as a threaded fastener, for example a nut

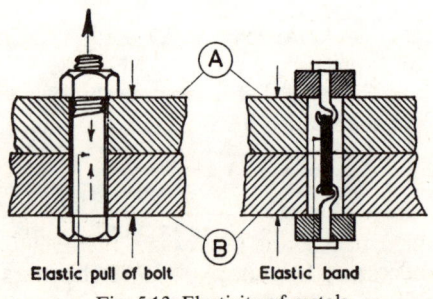

Fig. 5.13 Elasticity of metals

Engineering Materials 261

and bolt. This is simply illustrated in Fig. 5.13 (left) where we see two plates A and B held tightly together because the bolt, although stretched by the tightening action of the nut, is attempting to return to its original length. This condition is identical in all respects to the device also shown in Fig. 5.13 (right), where two plates are held together by an elastic band in tension. We see now the importance of elasticity in a metal, or its ability to return to its original shape under the influence of an externally applied load. The flexing of a good quality engineer's steel rule will provide ample testimony to the elastic property of high quality steel. All engineering assemblies possess a degree of elasticity. Steel bridges, bicycle wheels, the wings of a jet air liner, all have the ability to flex or bend when subject to external loads or forces, yet return to their normal shapes when the load is removed.

Ductility

Ductility may be defined as the ability of a metal to be bent, twisted or stretched whilst in the cold state. This is another important mechanical property for it allows the rapid and economical manufacture of a great number of components. For example the body of a motor car is made up of several sections that have been formed from mild steel sheets using large press tools, and it is the ductility of mild steel that allows the shaping of a car roof to be achieved in a matter of seconds. The ductility of a metal is usually expressed as the *percentage elongation* or *percentage reduction in area*. Figure 5.14 gives a close-up view of a

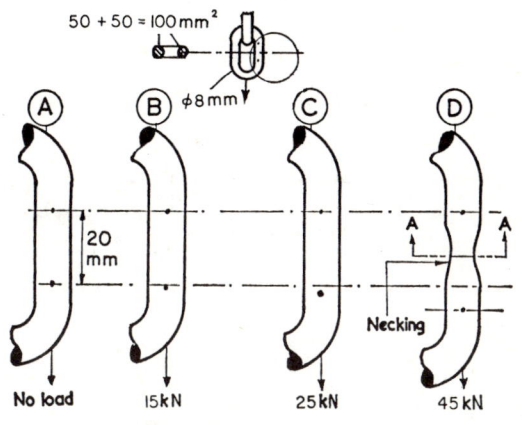

Fig. 5.14 Ductility of metals

chain link. Let us assume that this link is made from 8-mm-diameter mild steel rod, bent into shape by forging, then welded. Note the two small centre-dots about 20 mm apart. At B we see the effect on the link when a load of 1 500 kg or 1·5 tonnes is applied; the distance between the centre-dots has increased, indicating that the link has stretched. At C we increase the load to 2·5 tonnes and once again the distance between the centre-dot increases. If, on removal of the load, the centre-dots return to their original positions, it is clear that the link has not been loaded beyond its *elastic limit*, for it has regained its original shape or size. At D we see the effect of applying a load of 4·5 tonnes. Not only has the distance between the centre-dots greatly increased but the link has suffered a reduction in area also, and both the stretching and area reduction are permanent. The link is now permanently deformed and will not regain its original shape or size. This effect is known as *ductile* extension or *necking*, and is a most dangerous condition in any chain link, for it is evident that the reduced diameter produced by the 'necking' process will impart a serious weakness to the link and render the chain totally unsafe for use. The calculation of the percentage elongation involves the use of the formula

$$\% \text{ elongation} = \frac{\text{Amount of extension}}{\text{Original distance}} \times 100$$

If we assume that the distance between the two centre-dots on a chain link, just before fracture, measured 25 mm, then the amount of extension is the final distance minus the original distance

$$25 - 20 = 5 \text{ mm, and}$$

$$\% \text{ elongation} = \frac{\text{Amount of extension}}{\text{Original distance}} \times 100$$

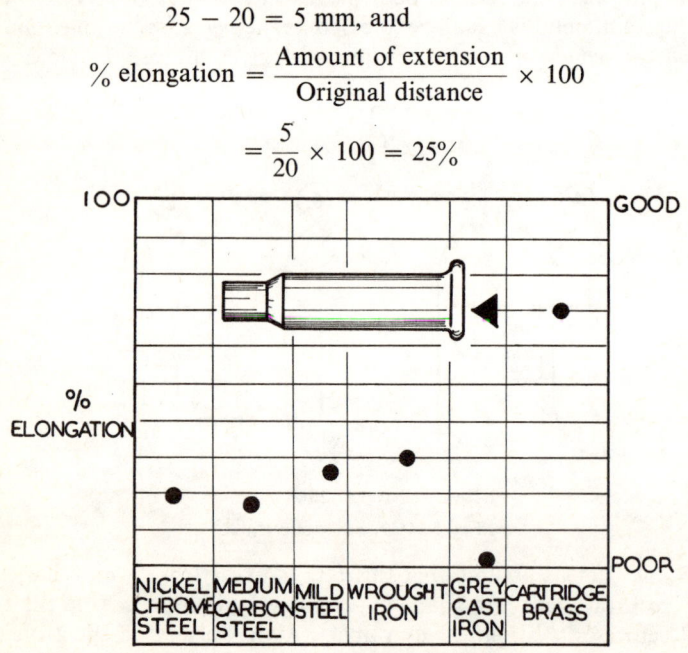

$$= \frac{5}{20} \times 100 = 25\%$$

Fig. 5.15 Percentage elongation of metals

Engineering Materials

Figure 5.15 gives the % elongation of the more common engineering materials.

5.3.1 Hardness

The hardness of a metal may be defined as its ability to withstand abrasion or indentation. This is an essential property of all cutting tools such as files, drills and milling cutters. There are several standard tests that may be carried out in order to determine the hardness of a metal and these are listed below.

>Brinell Test,
>Vickers pyramid hardness test,
>Rockwell hardness test,
>Shore sclerescope hardness test.

All the above tests have one common objective, and that is to obtain a measurement of the resistance of the metal under test to indentation by a harder body.

Brinell test

Devised by a Swedish engineer, Dr. August Johan Brinell, this test involves measuring the area of indentation produced when pressing a hardened steel ball under a constant load into the surface of the metal under test, with the actual hardness number obtained by direct calcu-

D = Dia of ball
d = Dia of impression
P = Applied load

$$\text{Brinell Number} = \frac{\text{Load}}{\text{Area}}$$

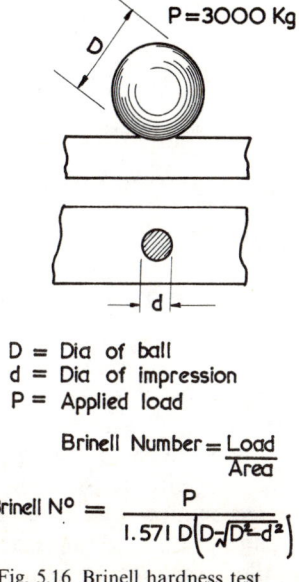

$$\text{Brinell N}^\circ = \frac{P}{1.571\,D\left(D - \sqrt{D^2 - d^2}\right)}$$

Fig. 5.16 Brinell hardness test

lation or tables. Some practice and skill is required for the correct use of the Brinell principle, and the machine used is often referred to as a fixed load hardometer. Different materials will also require different fixed loads in conjunction with different ball diameters, and it may now be appreciated that the Brinell hardness testing machine is best suited for the metallurgical laboratory or standards room, where it will be handled and set by skilled technicians. It is vital that the operator's handbook supplied with the machine, be consulted and studied at all times, but the basic principle is shown in Fig. 5.16.

Vicker's pyramid hardness test
Similar in principle to the Brinell hardness test except that a diamond indenter is used in place of a hardened steel ball. Best suited for hard components such as high-speed steel milling cutters, twist drills or lathe tools, the Vicker's machine is best located in a standards room or laboratory, and once again the operator's handbook is essential to its proper operation. Figure 5.17 shows the basic principle.

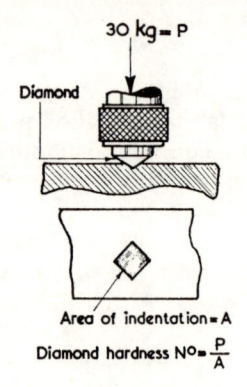

Fig. 5.17 Diamond hardness test

Rockwell hardness test
So far, both Brinell and Pyramid tests relate the hardness of the surface under test in terms of the load applied and the area of indentation, in other words,

$$\text{Hardness Number} = \frac{\text{Load}}{\text{Area of indentation}}$$

The calculations involved make these tests quite unsuitable for production work, that is to say periodic hardness testing of components that are being heat treated on a production basis. What is needed here is a simple direct-reading hardness testing machine, capable of being used by relatively unskilled personnel and showing the hardness num-

Engineering Materials

ber on a large easily read dial, without the need for calculations of any sort. This direct reading ability is made possible because the principle now adopted is related to the *depth* of penetration of a diamond indenter, or a hardened steel ball for soft materials. The principle of the Rockwell hardness test is simply illustrated in Fig. 5.18. Two scales are

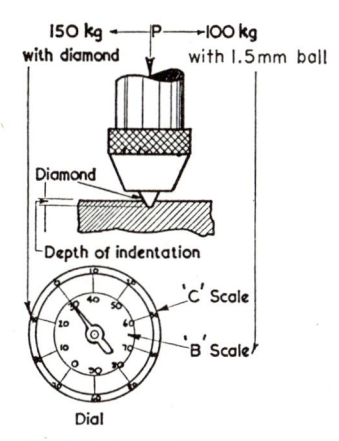

Fig. 5.18 Rockwell hardness test

present on the dial, the B scale for the hardened steel ball and the C scale for the diamond indenter, and the calibration is such that the harder the metal, the higher is the reading or the Rockwell number.

Shore sclerescope hardness test

If a table-tennis ball is allowed to fall on a carpet it does not bounce or rebound; if, however, the same ball is allowed to fall on a smooth concrete floor, a considerable bounce results. In other words, the harder the material on which the ball falls, the higher will be the resulting bounce, therefore the height of rebound may be taken as a measure of hardness. This is the principle of the Shore sclerescope hardness tester which consists of a portable glass tube containing a hardened pointed weight of 2·5 grammes mass, which can be made to fall through a distance of 250 mm, and the resulting rebound read off by comparison against a fixed scale attached to the glass tube. The mobility of this hardness tester means that it can be set up on a large casting or forging, and the hardness read off without much trouble.

Toughness

The toughness of a metal may be described as its ability to stand up to sudden shocks or blows without fracture. The head of a hammer or sledge is a good example of a component that is constantly subjected to sudden shocks and blows, and it is important that engineers have a relative measure of toughness. A simple workshop example of toughness testing is illustrated in Fig. 5.19, where we see a mild steel bar placed in a bench vice and struck a fairly severe blow with a hammer. Note that we have filed a small notch at A, and this will provide a source of weakness, the bar breaking or fracturing at this point if it is

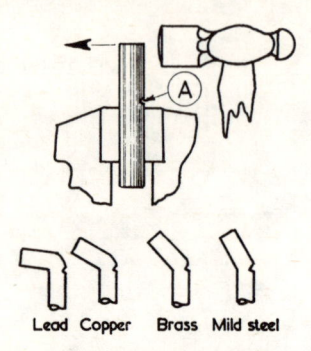

Fig. 5.19 Practical toughness test

unable to withstand the impact of the hammer blow. Assuming we are able to maintain constancy of the hammer blows on different materials, it is evident that the toughness would be indicated by the amount they have bent, the greater the bend, the less tough the material, whilst a brittle material is likely to be completely fractured. A very similar principle of toughness testing is used in the Izod impact test.

Izod impact test
In this test, a suitably notched specimen machined to specified sizes is subjected to a violent blow imparted by a heavy pendulum, as shown

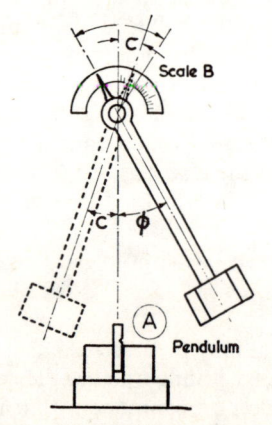

Fig. 5.20 Izod impact test

in Fig. 5.20. Release of a stop causes the pendulum to fall under the influence of gravity, and if the angle of the pendulum is φ before release, it will swing to the dotted position as shown in the diagram, with the angle from the vertical almost equivalent to φ. If however, a test piece is placed in the path of the pendulum, energy will be required to bend or break the test piece, and this must result in a smaller angle of upward swing of the pendulum. The energy lost by the pendulum is the energy absorbed by the specimen, and can be read off on the scale B. The reduced angle of upward swing is shown as C, and the smaller this angle, the tougher is the specimen; the scale calibrated in *joules*. Table 5.2 gives the main mechanical properties of some engineering materials.

Table 5.2 Mechanical properties

Metal	Elastic limit	UTS	% elongation	Izod value
Nickel–Chrome steel	784	1600	20	81
Medium carbon steel	450	720	18	54
Mild steel	320	480	25	41
Grey cast iron	Nil	176	Nil	Nil
Wrought iron	160	320	30	93
Units	N/mm^2	N/mm^2	%	Joules

5.4 Uses of engineering materials

Grey cast iron
Figure 5.21 shows two typical components easily recognised by an engineering technician. At A we see the bed of a centre lathe and at B a medium sized surface plate. Before choosing a metal suitable for the manufacture of these components, the following questions need to be asked.

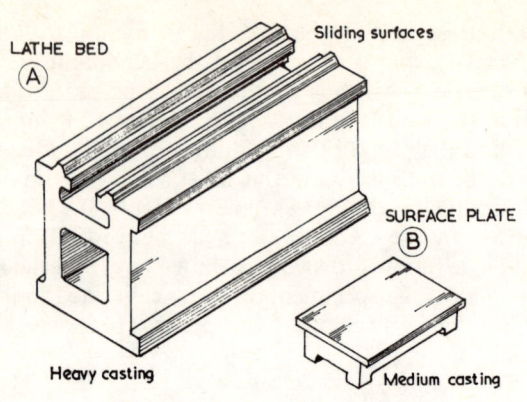

Fig. 5.21 Castings in grey cast iron

(i) Are the components subject to tensile forces?
(ii) Is a large mass of metal required?
(iii) What primary process is to be used in their manufacture?
(iv) What are the main purposes of the components?

We may outline the answers to the above questions in a simple manner as follows:

(i) There are no tensile forces exerted on the components.
(ii) Yes, large masses of metal are needed.
(iii) Both components are to be gravity sand cast.
(iv) To provide support and sliding surfaces for other components.

A suitable metal, therefore, needs to have the following properties,

1. low tensile strength,
2. must be a cheap metal,
3. must be easily cast,
4. must have low coefficient of friction surfaces, and good compressive strength.

It is evident that grey cast iron is ideal for the manufacture of both the centre lathe bed and the surface plate. Its low cost, low melting point, fluidity in the molten state and the presence of carbon in the free form of graphite all combine to make grey cast the correct choice of metal for the two components shown.

Types of grey cast iron

Table 5.3 gives the approximate composition of grey cast irons suitable for different types of castings. Note the variation in the silicon and manganese content to suit the type of casting required. As we have seen earlier, rapid cooling of the molten metal tends to produce white cast iron, and because small castings tend to cool more quickly than large ones, increase of the silicon with decrease of the manganese content will prevent the formation of white cast iron.

Table 5.3 Types of grey cast iron

	Percentage by type of casting		
Constituent	Heavy	Medium	Light
Carbon	3·5	3·5	3·5
Silicon	1·0	1·5	2·0
Manganese	1·0	0·5	0·5
Phosphorus	0·3	0·5	1·0
Sulphur	0·05	0·05	0·05

Effect of alloying constituents

Silicon
Silicon promotes the formation of free carbon in the form of graphite flakes, producing a grey cast iron with good sliding surfaces due to the self-lubricating effect of the graphite.

Manganese
Manganese has the opposite effect to silicon tending to prevent the formation of graphite by promoting iron carbide, a very hard and brittle substance, more commonly known as white cast iron.

Phosphorus
The presence of phosphorus greatly increases the fluidity of molten cast iron necessary when light, intricate castings are required. It tends however, to weaken the cast iron, and for this reason the phosphorus content is kept as low as possible.

Sulphur
This is an undesirable impurity picked up from the coke used in the smelting process carried out in a blast furnace. The sulphur content must be kept as low as possible, and seldom exceeds 0·1%. Any sulphur in excess of 0·1% will have a serious weakening effect on the cast iron, making it most unreliable in use.

Carbon

We see from Table 5.3 that grey cast iron contains about 3·5% carbon. Provided the silicon content is high and the manganese content low, most of this carbon will be present as graphite, a weak and brittle material, representing a severe source of weakness, and greatly reducing the strength of grey cast iron when subjected to tensile or shear forces. We see from Table 5.2 that the UTS of grey cast iron is 176 N/mm^2, whilst its elasticity and ductility are nil. This means that if struck a sudden blow, a bar of grey cast iron is easily broken, giving little or no warning before fracture. The compressive strength of grey cast iron however, is quite good, and Fig. 5.22 will help to explain why

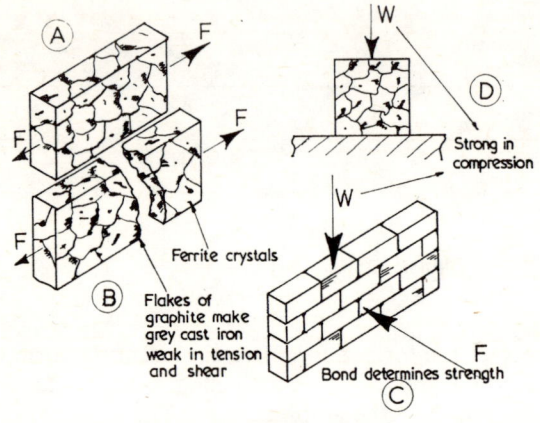

Fig. 5.22 Effect of graphite

the compressive strength is relatively unaffected by the presence of the graphite. At A we see a much simplified picture of the structure of grey cast iron; note the graphite flakes distributed throughout the ferrite crystals. Clearly if the cast iron is subjected to the tensile forces shown as F in the diagram, the graphite flakes have a serious weakening effect causing the metal to fracture as shown at B. The brick wall shown at C will serve as a simple analogy. If a force (shown as F) is applied at 90° to the line of the wall, the strength of the wall or its ability to resist fracture is determined not by the strength of the bricks, but by the strength of the cement or bonding material holding the bricks together. A weak bond, therefore, results in a wall totally unable to withstand a force in the direction of arrow F in the diagram. On the other hand, if a force is applied vertically or downwards, the ability of the wall to resist fracture is now determined by the compressive strength of the bricks; the presence of a weak bond having little effect. We see now that the presence of graphite in cast iron, whilst seriously reducing the tensile and shear strength, has little effect on the compressive strength,

Engineering Materials

which is about five times the UTS. At the same time the presence of the carbon allows cast iron to be poured easily into moulds. In addition, it gives the property of *anti-resonance*, that is to say the ability of grey cast iron to absorb vibrations, a useful property for beds and component parts of machine tools.

5.4.1 The use of wrought iron

The engineering uses of wrought iron are limited to those applications calling for an easily forged and welded metal having a good resistance to corrosion and impact, and possessing reasonable tensile strength. There is no carbon in wrought iron, therefore it cannot be brought to a liquid state, but remains plastic or pasty and most suitable for welding and forging. Figure 5.23 illustrates some typical components made

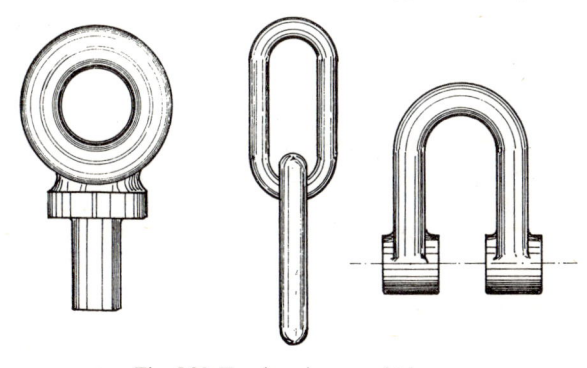

Fig. 5.23 Forgings in wrought iron

from wrought iron, all produced by the forging process. These components will have an excellent resistance to sudden shocks, as may be seen by the high Izod value given in Table 5.2, and the fact that good quality wrought iron tends to take on a film of black iron oxide which inhibits further oxidisation makes it eminently suitable for all outside applications. Unfortunately, good quality wrought iron is a relatively expensive metal, and its use is restricted to those components where the highest quality is needed, for example lifting hooks and tackle upon which the lives and safety of operators depend. The composition of wrought iron is 99·95% ferrite and 0·05% slag inclusions.

5.4.2 The use of carbon steels

It has been found that if the carbon content is controlled at between 0·1% and 1·5% a range of strong carbon steels results; the carbon now combining chemically with the ferrite or iron to produce a strong constituent known as *Pearlite*. The amount of carbon determines the type of carbon steel, and Table 5.4 shows the main types of carbon steel in use.

Table 5.4 Carbon steels

Type of steel	Percentage of carbon	UTS	% elongation	Brinell hardness number
Dead mild	0.1–0.15	320	30	80
Mild	0.15–0.35	480	25	120
Medium carbon	0.35–0.65	720	14	160
High carbon	0.65–1.15	920	8	220
Units	%	N/mm^2	%	HB

Dead mild steel (app. 0.1% Carbon)
A soft, ductile steel, very easily cold-worked by bending, twisting and rolling, also forges and welds easily. It is often used as a cheap substitute for wrought iron, because it is about one third the cost, but it will rust red and flake, necessitating painting or other means of protection. Dead mild steel is much used for tinplate, boiler plates, rivets, and cheap pipe and wire.

Mild steel (app. 0.25% Carbon)
This is the engineering metal of today. Having good tensile strength, good cold-working properties, fairly easily forged and welded, mild steel is in popular demand for a wide range of products. It is available

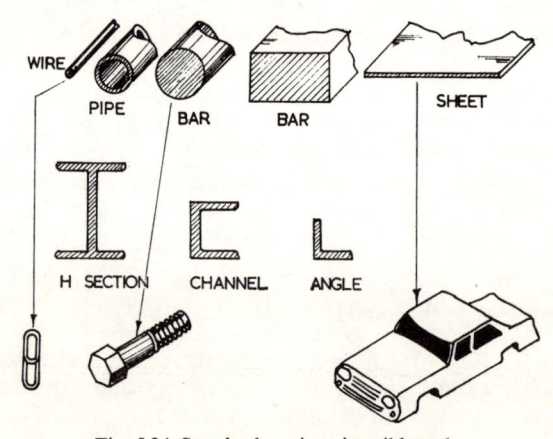

Fig. 5.24 Standard sections in mild steel

Engineering Materials 273

in a range of shapes and sections, as illustrated in Fig. 5.24, which shows also the type of products produced from the sections illustrated. The great disadvantage of mild steel is that it is subject to corrosion when exposed to the atmosphere; it is estimated that in Great Britain alone, the corrosion of mild steel involves an expenditure in excess of £50 000 000. We shall see in later work, some of the processes adopted by engineers to protect mild steel from the rusting or oxidising effects of the oxygen present in the atmosphere. Typical components produced from mild steel are nuts and bolts, motor car bodies, and sections for bridges and other constructional work. Mild steel may also be cast, but it does not possess the fluidity of cast iron, and it is not easy to produce intricate castings in mild steel. Due to the high melting point of about 1600°C, castings in mild steel are certain to be more expensive than similar castings in grey cast iron, which is not only about one third the price, but melts at a lower temperature of about 1150°C.

Medium carbon steel (0·45% Carbon)
Reference back to Table 5.4 shows that medium carbon steel is stronger, less ductile and harder than mild steel, and is therefore suitable for components that are likely to be subject to bending stresses. Figure 5.25A shows a toolholder used on a centre lathe, and it is important that no bending or flexing takes place under the influence of the cutting force during the machining operation. Medium carbon steel is a better choice than mild steel for this kind of condition; and a similar condition is shown at B, where we see the main force acting on an ordinary steel rail. The mass of the locomotive acts downwards tending to bend the rail, and in addition there is the abrasion or wear caused by the driving action and rolling of the locomotive wheel.

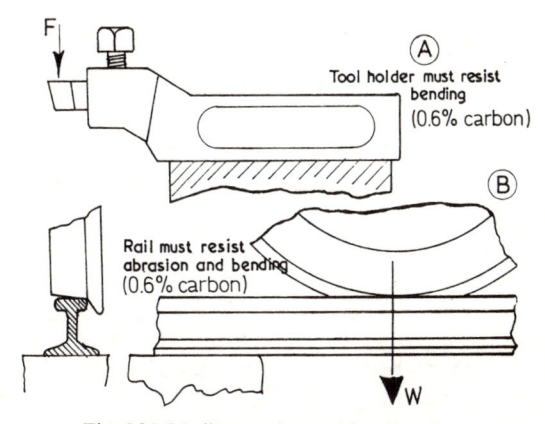

Fig. 5.25 Medium carbon steel components

Medium carbon steel is the material used, having a carbon content of about 0·6%, the rails are hot-rolled whilst in a red-hot condition.

High carbon steels (0·9% Carbon)
These steels are often referred to as straight, cast or tool steels. If the carbon percentage exceeds 0·9% a brittle constituent called *cementite* is present in the structure as a grain boundary, making the high carbon steel unable to withstand a sudden blow or impact. It must be appreciated at this point, that high carbon steel is *never* used as a structural material, that is to say, a metal which will be subject to tensile or shear forces, for example a nut and bolt, or parts of the steering arrangements of a motor car. The great advantage possessed by high carbon steel, and the sole reason for its use, is its excellent *hardening* property; the higher the carbon content, the harder is the steel after the hardening process. Table 5.5 gives the types and applications of high carbon steels, together with the conditions for which the hardened steel is suitable.

Table 5.5 Types and applications of high carbon steels

Percentage Carbon	Conditions	Tools
0·7–0·8	Blows and sudden shocks	Hammer heads, axes, springs
0·8–0·9	Medium blows with cutting ability	Chisels, punches, shears
0·9–1·1	No blows but good cutting ability	Drills, taps
1·1–1·4	Keen edge and some pressure	Scribers, scrapers

High carbon steels in the 'as-received' or unhardened condition do not possess much ductility, as can be seen by reference back to Table 5.4 which shows that the percentage elongation is 8, making it exceedingly difficult to cold-work them. Both welding and forging become more difficult, as does any hand work such as hacksawing or filing. As previously stated, the use of high carbon steel is confined to the manufacture of cutting tools, in particular those tools that are to be used at the bench, and more commonly known as hand tools. Figure 5.26 shows a typical selection of hand tools made from high carbon steel; note that

Engineering Materials 275

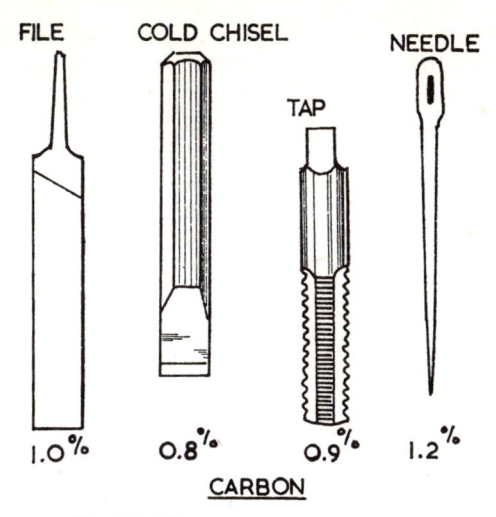

Fig. 5.26 High carbon steel components

the percentage carbon varies according to the type of work or cutting required from the tool.

5.4.3 Alloy steels

Figure 5.27 shows two engineering components that are subjected to considerable stressing when in use. At A we see a ball race and at B an open-ended spanner. For the whole of its working life, the steel balls

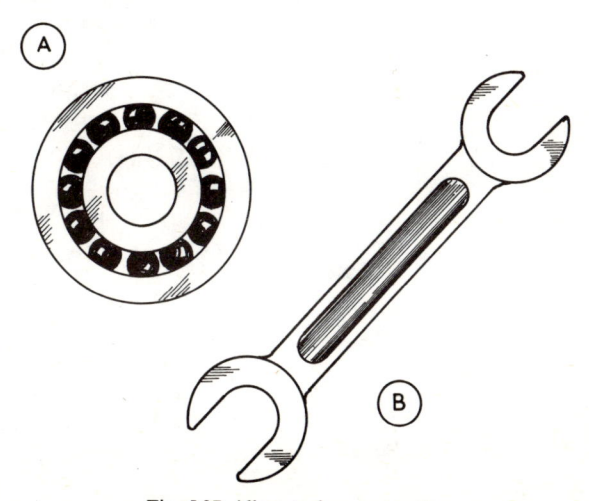

Fig. 5.27 Alloy steel components

will roll around between the inner and outer races of this bearing, whilst every time the open-ended spanner is used, it will be subject to forces tending to open the jaws. If both components are to give reliable and long service they must be made from a strong steel with superior mechanical properties, and the choice is certain to be a nickel–chrome

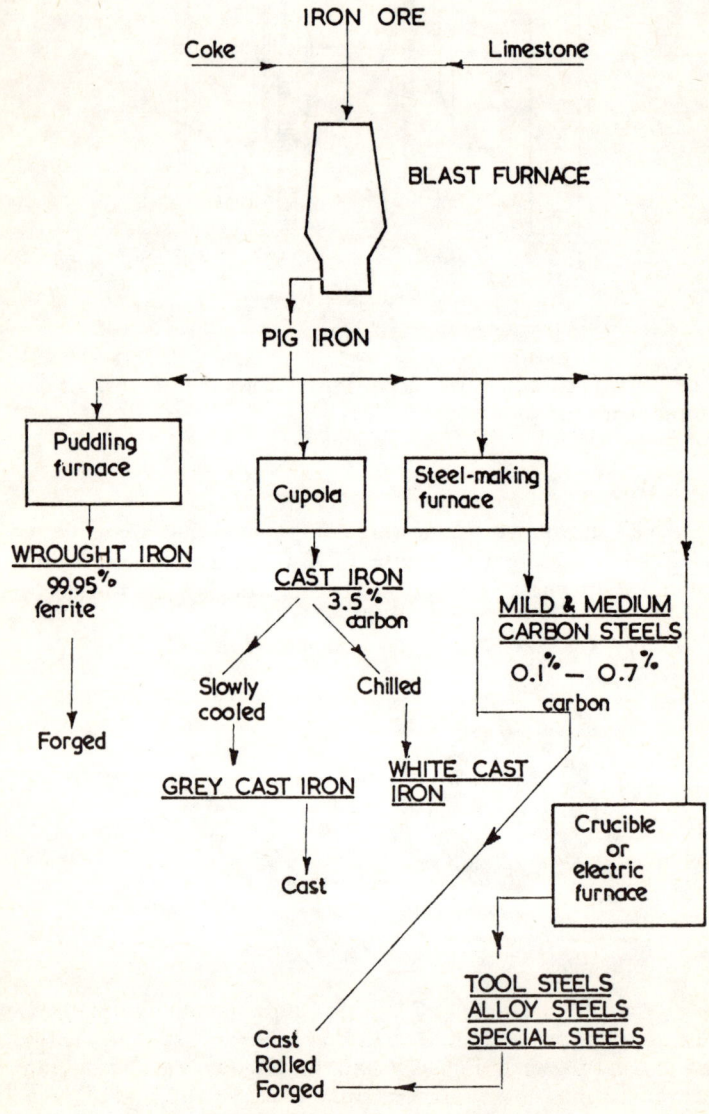

Fig. 5.28 Origin of ferrous metals

Engineering Materials 277

alloy steel. It is important to remember that alloy steels belong to the same family as the carbon steels and the cast irons, and Fig. 5.28 traces the origin of the ferrous metals used in engineering manufacture.

5.5 Heat treatment of carbon steels

The object of all heat treatment processes is to improve the properties of the metal so that it can be more easily manufactured, or better able to tolerate the working conditions likely to be encountered in use. We may consider the hot-forging of steel as a heat treatment process, because at the forging temperature for mild steel, about 1000°C, the steel has increased plasticity and is much easier to forge. Figure 5.29 shows

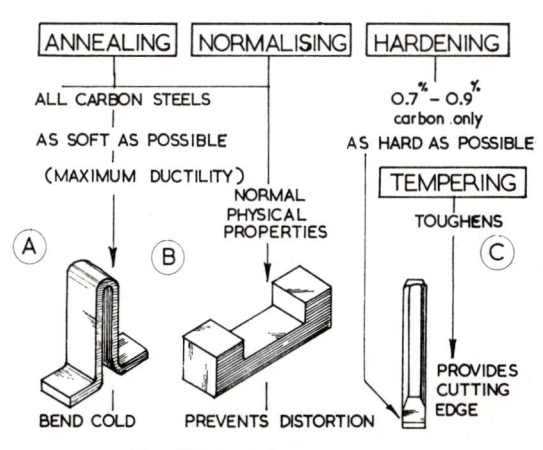

Fig. 5.29 Basic heat treatments

in diagrammatic form the heat treatments carried out on plain carbon steels together with the advantages gained. We see that the following heat treatments are indicated in the diagram:

(i) annealing,
(ii) normalising,
(iii) hardening,
(iv) tempering.

5.5.1 Annealing

The object of annealing steel is to promote maximum ductility, or in other words to make the steel as soft as possible. For example the mild steel bracket shown in Fig. 5.29, and indicated as A in the diagram, has been subject to considerable bending whilst in the cold state, with a serious risk of cracking along the bend lines. If the steel strip from which the bracket is made is bright mild, then it has been *cold*-rolled at the steelworks, and will be work-hardened as a result, with a distinct grain effect in the direction of rolling.

This may be more clearly understood by reference to Fig. 5.23 which shows a strip of bright mild steel which has been bent both with and

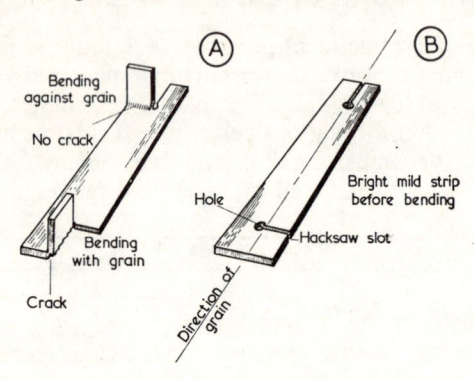

Fig. 5.30 Grain in bright mild steel

against the grain. Any attempt to bend the bright mild steel with the grain is certain to result in cracking along the bend line, and for this reason it is essential that all bright mild steel is annealed before bending.

The annealing operation is relatively simple; all that is necessary is to heat the steel to about 850°C (bright cherry red) and allow to cool as slowly as possible. If the heating is carried out in a furnace the best plan is to allow the heated steel to cool with the furnace; that is to say as soon as the steel has reached the required temperature, the furnace is turned off with the steel left in it. All furnaces have excellent insulating properties; their design is such that the minimum amount of heat is allowed to escape, and the furnace takes a long time to cool, as will any component left inside it.

If the steel is heated with a blowpipe or at the forge, it should be buried in dry ashes or sand after heating to bright cherry red. It needs to be remembered that it is not necessary to anneal *black* mild steel before bending. Black mild steel is produced at the steelworks by *hot*-rolling, which means that the steel is white hot as it passes between the rolls. No work-hardening can take place while the steel is at this temperature, therefore there is no tendency for black mild steel to crack when bent in the cold state.

5.5.2 Normalising

The object of normalising is to restore to steel its normal properties.

Engineering Materials

For example, as we have stated, cold-rolling stresses the surface layers of the steel sheet, distorting and elongating the crystal structure with the result that the strength and hardness increases. Any attempt to carry out extensive machining of bright mild steel is certain to result in severe distortion of the component being machined. Consider the component shown in Fig. 5.31A. This is a 100 mm sine bar made from

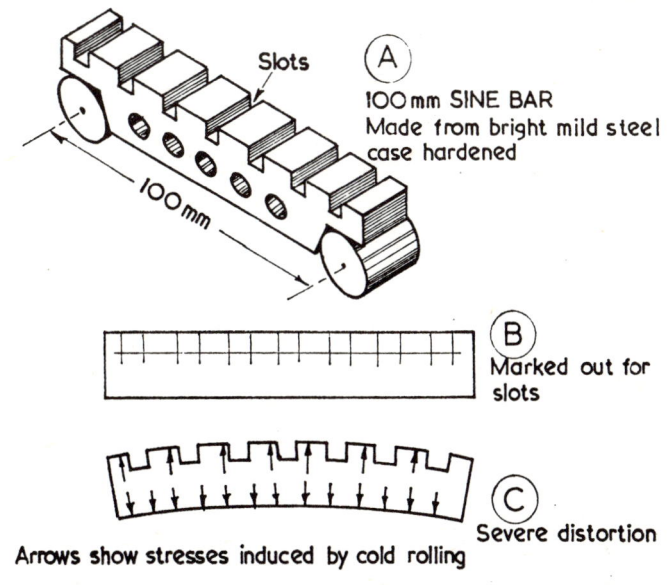

Fig. 5.31 Normalising to prevent distortion

bright mild steel and case-hardened, and is a measuring instrument used for the accurate determination of angular faces using a trigonometrical principle. We see that the top face has several slots, intended to reduce the surface area, and there are also several holes drilled through the body to reduce the mass of the sine bar.

Any effort and time spent in grinding the bar to size and squareness prior to marking-out and machining will be brought to nothing, because as soon as the bar is milled and drilled, severe distortion takes place due to a redistribution of stresses as shown in Fig. 5.3.1C, with considerable bending of the bar. If, however, the bar is first normalised, that is to say heated to bright cherry red and then allowed to cool in still air (by which we mean placing the bar on one side to cool naturally), all the stresses set up by cold-rolling are removed. Normalising removes all the changes in properties caused when cold-rolling by restoring the deformed crystals to their normal shape, and the normalising process is identical to that of annealing except for the rate of cooling.

5.5.3 Hardening

This is the most important of all the heat treatment processes. Steels which have a carbon content of about 0·8% and upwards have excellent hardening properties, and reference back to Fig. 5.26 shows some typical high carbon steel components that require hardened faces. Provided simple rules are followed, hardening high carbon steel is a relatively simple process, and one example will serve to illustrate the technique to be adopted.

At Fig. 5.32A we see a typical flat cold chisel made from hexagonal

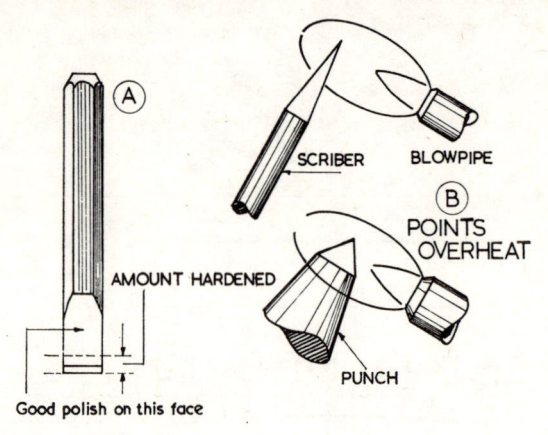

Fig. 5.32 Hardening high carbon steel

high-carbon steel of about 0·8% carbon; this is in accordance with the applications of high carbon steels given in Table 5.4. Note that the amount of chisel actually hardened is quite small; no more than about 12 mm from the cutting edge, and the heating technique is simply shown in Fig. 5.33. Although the hardening procedure is simple, namely heat the portion to be hardened to cherry red and quench in cold clean water, there are one or two mistakes to avoid.

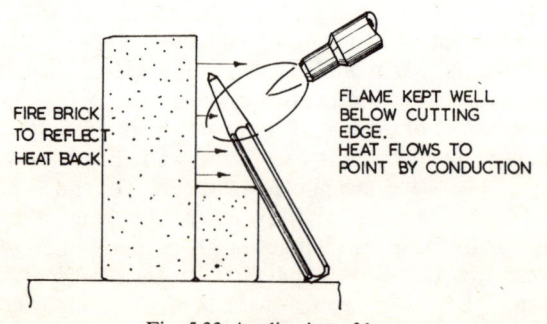

Fig. 5.33 Application of heat

Engineering Materials

The first is to ensure that the cutting edges are not overheated and this is very easily done if the flame is allowed to play directly on the cutting point as shown in Fig. 5·32B. Because of the small mass of metal at the points of both scriber and chisel, overheating is certain to take place, and when hardened the points will be too brittle to be of any practical use. The best plan is to use the technique illustrated in Fig. 5.33, where it may be seen that the flame is kept well below the cutting edge, and allow the point to be heated by conduction. When the cutting edge has reached a cherry red colour, quench it vertically in cold clean water, immersing about one third of the chisel, and moving it around in the water to ensure even cooling. On removal from the water it will be found that the cutting edge is extremely hard; a simple test using a smooth hand file demonstrates that the file is quite unable to file the hardened edge. In this condition the chisel is extremely hard and also extremely brittle, and is very likely to fracture at the cutting edge if used. A further heat treatment is now required to make the cutting edge less brittle, tougher and safer to use. This process is called tempering.

5.5.4 Tempering high carbon steel

The object of tempering is to make the chisel edge better able to stand up to working conditions which involve striking the chisel head with a hammer, the force of the blow being transmitted to the cutting edge. We need this cutting edge to be less hard and less brittle, or in other words we need to toughen it, so that it will not fracture when subjected to the shocks of the hammer blows.

A popular method of tempering hand tools is to make use of the microscopic oxide films that form on the surface of the polished steel when it is slowly heated. The colour of the oxide films changes accord-

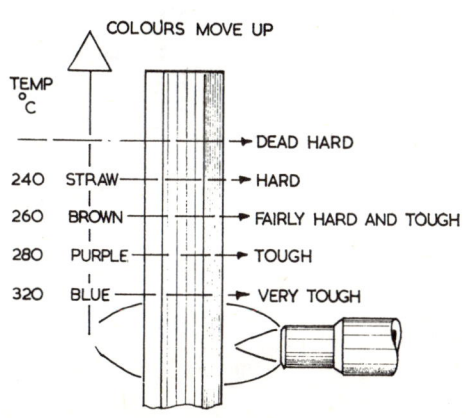

Fig. 5.34 Use of oxide films

ing to the temperature, allowing us to use the colour at the cutting edge as an indication of its temperature. Figure 5.34 shows the basic principles of tempering high carbon steel using the oxide film technique. Here we see a polished hexagonal high carbon steel bar heated at the lower end. Note the band of colours which moves upwards as heat travels up the bar by conduction. The diagram gives the colours, and the respective temperatures, together with the working conditions best suited to the colour obtained when the cutting edge is quenched. The following three factors are involved when tempering small hand tools:

(i) temperature,
(ii) colour of oxide film,
(iii) conditions the cutting edge will stand.

Table 5.6 sets out the connection between the above factors.

Table 5.6 Tempering high carbon steel

Temp. °C	Colour of oxide film	Conditions	Typical tools
230	Light straw	Some pressure No blows	Scrapers Scribers
240	Dark straw	Heavy pressure Light blows	Turning tools Drills
250	Light brown	Medium blows and torques	Punches Taps
260	Dark brown	Heavy blows	Chisels Heavy punches
280	Purple	Very heavy blows	Hammer heads Axes
320	Blue	Shock-loading	Springs

Although it is possible to harden and temper a cold chisel at one heating using a blacksmith's forge, it is a better plan to adopt two separate operations, namely hardening and tempering. Best results are obtained using a blowpipe as this permits more control over the heating process. It is essential when tempering to ensure that the flame is kept well below the cutting edge, because if applied directly to the cutting edge rapid heating of the outside surface takes place with the formation of the coloured oxide films. The inner structure, however, will be relatively unheated, and therefore not tempered, and very likely to fracture when put to use.

The proper technique is simply illustrated in Fig. 5.35. Heat is con-

Engineering Materials

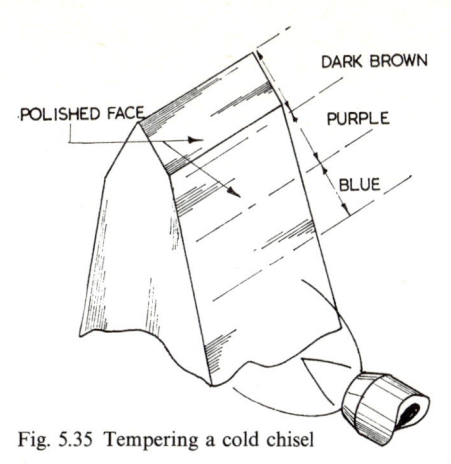

Fig. 5.35 Tempering a cold chisel

ducted to the cutting point, so that the colour of the oxide film is a true indication of the temperature of the chisel point.

Hotplate tempering
It is sometimes necessary to temper a component throughout its whole section; an example would be a drill bush made from high carbon steel. Such a bush is shown in Fig. 5.36A, and the hotplate technique is very suitable for this kind of component. It may be seen that the bush is placed on a thick steel plate which is heated with a blowpipe as shown in the diagram (Fig. 5.36B). The hardened bush absorbs heat slowly by conduction from the hotplate, and as soon as the correct colour appears on the surface of the bush, it is quenched in water or oil. As the drill bush is subject only to friction and pressure of the revolving drill, a surface colour for the oxide film could be light straw.

An alternative method is to wire the component and immerse it in a sand box (Fig. 5.36C). An even colouring effect is obtained, provided

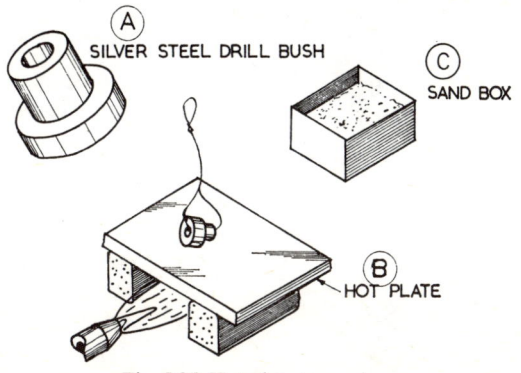

Fig. 5.36 Hot-plate tempering

the component is removed and checked from time to time, and then quenched when it reaches the desired colour.

Finally, it is necessary to test a hardened component before use, for example a cold chisel, and the technique is shown in Fig. 5.37.

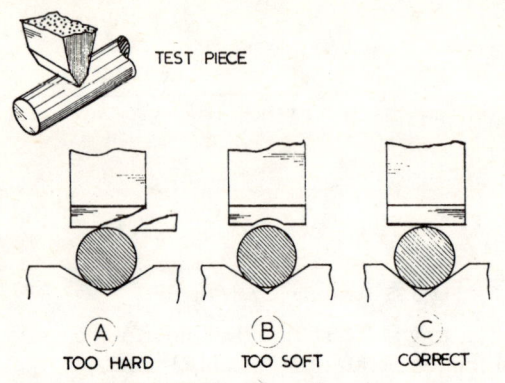

Fig. 5.37 Testing the cutting edge

Protective goggles must be worn during the test. If the chisel has been properly hardened and tempered no damage will be caused to the cutting edge. This means that the chisel will be quite safe to use, and it is unlikely that injuries will be inflicted from pieces flying off the cutting edge.

5.6 The use of non-ferrous metals and alloys

An important use of non-ferrous metals and alloys in engineering is the manufacture of plain or friction bearings. A bearing may be considered as a device which both locates and guides a revolving shaft as shown in Fig. 5.38A. Where there is metal to metal contact, that is to

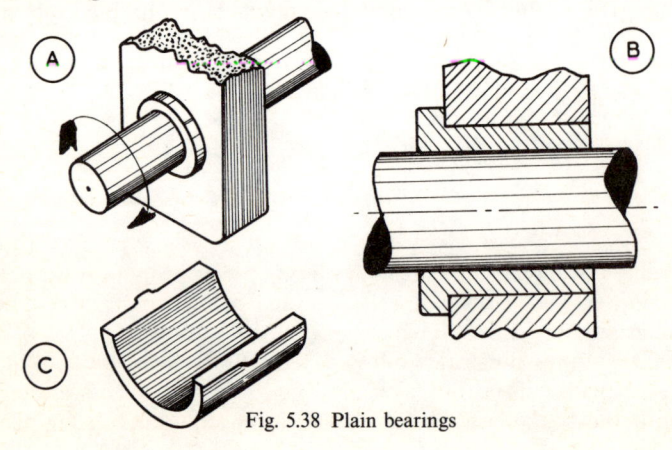

Fig. 5.38 Plain bearings

Engineering Materials

say the shaft is rubbing directly on the bush, the bearing is known as a friction bearing; the principle is simply shown in Fig. 5.38B. The use of revolving shafts to transmit torque or motion is extensive in all types of mechanical assemblies such as motor cars and machine tools, and a range of bearing metals are available according to the speed of rotation and load carried by the revolving shaft.

5.6.1 Bearing metals

Phosphor bronze
This is an alloy of copper, tin and phosphorus, and a typical composition would be as follows:

copper	89·0%
tin	10·0%
phosphorus	1·0%

Supplied in the form of sand-cast rods or hollow tubes, this is a low-tin bronze suitable for small bearings carrying medium loads at medium speeds. For heavy duty work, where high pressures are likely to be encountered, such as a bearing for a heavy turntable, a high-tin phosphor bronze is used, having the following composition:

copper	81·0%
tin	18·0%
phosphorus	1·0%

Both the above bronzes are capable of severe work-hardening and it is important that the shaft be in true alignment with the bearing when assembled, as the phosphor bronze is incapable of any 'give'. All phosphor bronzes however, make excellent bearing metals; they will not corrode to the shaft during periods of non-use, and are best used for medium to heavy loads at relatively low speeds. An improved bearing bronze, more suitable for higher speeds is known as *leaded* bronze, having the following composition:

copper	75·0%
tin	5·0%
lead	20·0%

This type of bronze is widely used for a special kind of bearing known as a *shell* bearing, much used in motor car engines, and providing an example of how engineers continually seek to effect economies in materials used in the engineering industry. The principle of shell bearings is simply illustrated in Fig. 5.38C, and involves the coating of a mild steel shell with a thin layer of leaded bronze, thus combining the strength and cheapness of mild steel with the superior bearing properties of leaded bronze. These shell bearings are relatively cheap to produce, and because of their high lead content are suitable for light loads at medium to fast speeds.

White metals
A disadvantage of phosphor bronze as a bearing material is that there is the danger of the bearing and shaft *seizing* as a result of temperature rise through lack of lubrication. Very serious damage to machinery can result from a seized shaft, and for conditions of high speeds and light to medium loads it is standard practice to use white metal as a bearing material. All white metals have a low melting point and in the event of oil failure causing a rise in temperature, the white metal melts and runs out of the bearing. A loud knocking noise gives ample warning of the bearing failure, and allows the machinery to be stopped before any serious damage occurs.

The composition of a typical lead-base white metal is given below.

Lead	80·0%
Antimony	15·0%
Tin	5·0%

For more exacting conditions, involving higher speeds and greater loads, a tin-base white metal may be used, with the following composition:

Tin	60%
Lead	27%
Antimony	10%
Copper	3%

Owing to the high tin content, this white metal is more expensive than the lead-base metal.

Porous bearings
These bearings are widely used in assemblies where it is difficult to ensure that a reasonable supply of oil is applied to the bearing from time to time. A typical example is the starter motor of a motor car, which is situated in a most inaccessible position, making it an awkward matter to apply even a few drops of oil. The remedy in cases like these, is to use porous phosphor bronze graphitised bearings, which are mass produced using a process known as powder metallurgy. This consists of compresssng a mixture of tin, copper, phosphorus and graphite using mechanical presses and high pressures. The shape of the mould is the shape of the bush, which is then heated or sintered and quenched in oil. Figure 5.39 illustrates the process of manufacturing porous phosphor bronze bushes on a production basis. Very high output rates are possible when the process is fully automated.

Engineering Materials 287

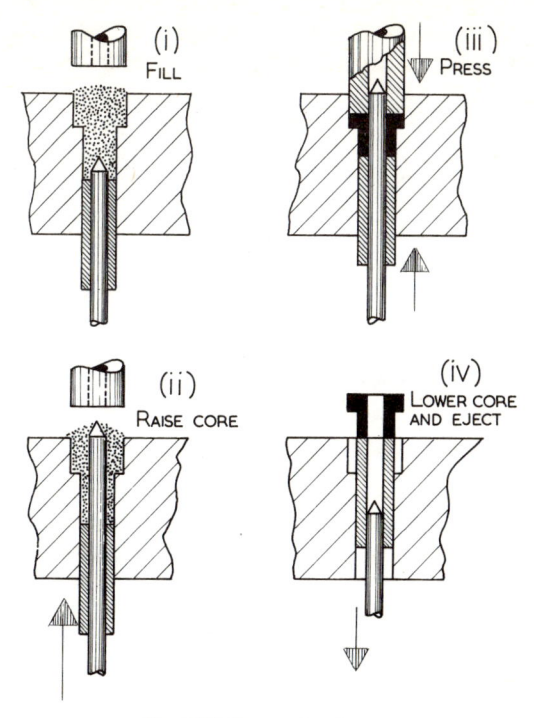

Fig. 5.39 Porous bearings

A typical composition for a porous phosphor bronze bearing would be as follows:

Copper	89%
Tin	10%
Graphite	1%

The addition of the graphite gives the bearing a built-in lubricating property, while further impregnation with oil increases the anti-friction properties. Many years of trouble-free life can be expected from this type of bearing.

Nylon bearings
A nylon bearing is the best substitute when the use of oil is not permissible or not possible. For example, most food products are produced using highly automated or mechanised equipment with many moving parts, including revolving shafts. The use of oil or grease for lubricants would most certainly result in contamination of the food, and they are therefore not permitted. Nylon however, is very suitable as a bearing material because it has self-lubricating properties, and a stainless steel

shaft will rotate for an indefinite period with no need for lubrication of any kind, although water may be used as a lubricant under certain conditions.

Cast iron and brass bearings
Both grey cast iron and brass can be used as cheap bearing materials for use with steel shafts, and provided they are well lubricated they will give reasonable service. They lack the superior bearing properties possessed by the phosphor bronzes and the tin-based white metals, and this is due to the differences between the structures of the metals. Figure 5.40A shows a greatly magnified section of a plain brass bear-

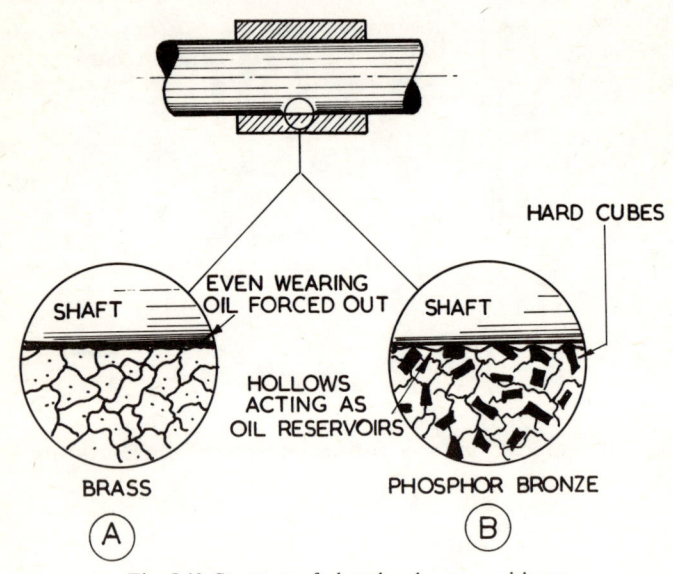

Fig. 5.40 Structure of phosphor bronze and brass

ing carrying a revolving shaft, whilst at B we see a similar view of a section of a phosphor bronze bearing. Note in B the presence of minute cubes of copper–tin phosphide, an intermetallic compound which is considerably harder than the parent metal. These cubes, interspaced throughout the structure of the phosphor bronze, tend to produce an uneven wearing surface, with the softer parent metal wearing more than the harder cubes, causing minute depressions or hollows. These hollows tend to retain oil, and so act as useful oil reservoirs from which the revolving shaft tends to pick up a certain amount of lubricant. On the other hand, a plain brass bearing possesses no hard cubes of an intermetallic compound, with the result that the shaft tends to squeeze or force the oil out from the bearing surface as shown in Fig. 5.40A.

5.6.2 The use of aluminium alloys

Engineers have little use for pure aluminium because of its low tensile strength; about 90 N/mm^2 in the normal state, increasing to 135 N/mm^2 when work hardened by cold-rolling. It is however, a very malleable metal, and easily cold-rolled into very thin sheet or foil, finding extensive use as a packaging material for cigarettes and chocolate. Aluminium is also a good conductor of both heat and electricity, and because of its low relative density it is, mass for mass, a better conductor than copper. It is cheaper than copper, and is much used in the conducting cables used for high voltage electrical power transmission. Because of the distance between the pylons used as supports, the aluminium cable would be unable to support its own weight, and it therefore is given a high tensile steel core to provide the additional strength needed. Figure 5.41 shows a comparison between

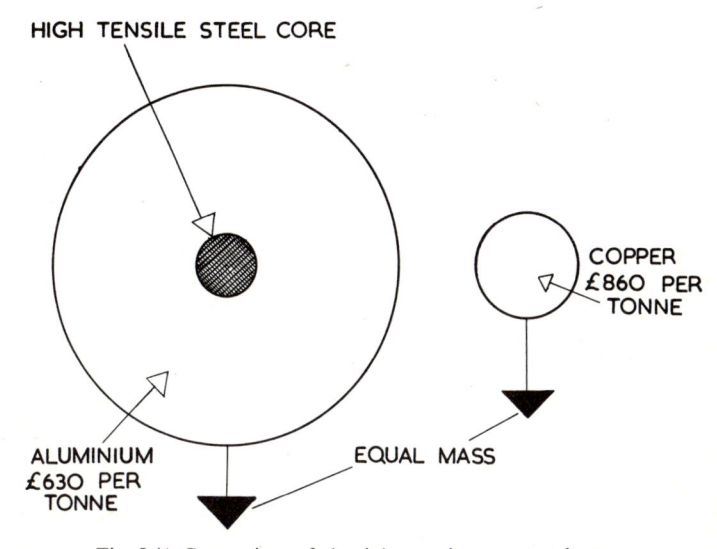

Fig. 5.41 Comparison of aluminium and copper conductors

cables of equal mass and shows the savings achieved by using a high tensile steel cored aluminium cable.

Casting aluminium alloys
A typical aluminium alloy suitable for the production of castings would have the following composition:

        Aluminium   88·0%
        Silicon   12·0%

This casting alloy has an UTS of about 150 N/mm^2, and is widely used for castings for motor car engines and gear boxes, produced either as sand castings, gravity die castings or pressure die castings.

Structural aluminium alloys
A structural aluminium alloy may be considered as one which is to be subjected to stresses and loads whilst in use. A typical high strength aluminium alloy, more commonly known as *duralumin*, has the following composition:

Copper	4·1%
Magnesium	0·8%
Manganese	0·7%
Silicon	0·5%
Aluminium	93·9%

This alloy has the ability to *age-harden*, that is to say its tensile strength and hardness will increase over a period of four days from the time the alloy is heated to 480°C.

5.6.3 The use of magnesium alloys

Magnesium is a very light metal, having a relative density of 1·7. In its pure state it is fairly weak, with an UTS of about 170 N/mm^2; one third the strength of mild steel. However, when alloyed with aluminium, manganese and zinc, a very light and reasonably strong material is produced, which is very suitable for components which are subject to some stressing in use, yet need to be as light as possible, for example wheels for aircraft.

A typical magnesium alloy suitable for casting purposes, would have the following composition:

Aluminium	8·0%
Zinc	0·7%
Manganese	0·3%
Magnesium	91·0%

After age or precipitation hardening the ultimate tensile strength reaches a value of about 200 N/mm^2.

5.6.4 The use of zinc-based alloys

We see from Fig. 5.7 that zinc is one of the low-melting point metals, which means that it can be cast without much difficulty. Modern casting techniques make use of the pressure die-casting principle, by which molten metal is forced into the dies under the influence of external pressure; the resultant casting possesses a dense strong structure. Quickly cooled by the high thermal conductivity of the dies, which are sometimes water cooled, the metal has a fine close grain which improves its mechanical properties.

It is an essential condition of all zinc-based alloys that only high purity zinc is used. The presence of minute quantities of tin, cadmium or lead in a zinc-base alloy leads to severe brittleness in a pressure die-casting; this condition is further hastened if the casting is exposed to damp conditions. The tendency at the present time is to restrict the number of zinc-based alloys to the trade, with strict control kept over the zinc purity and alloying elements in the casting metals that are available.

In general, only *two* zinc-based pressure die-casting alloys are now in use, each carrying the trade-name *mazak* (a term derived from the initials of the following metals which make up the alloy):

> Magnesium
> Aluminium
> Zinc
> 'Kopper'

The compositions and main mechanical properties of the two main zinc-based alloys are given in Table 5.7.

Table 5.7 Composition and mechanical properties of Mazak 3 and 5

Alloy	Aluminium	Copper	Magnesium	UTS N/mm^2	Percentage elongation on 50 mm
Mazak 3	4·0	–	0·05	280	13
Mazak 5	4·0	1·0	0·05	340	8·5

It can be seen that Mazak 5 is both harder and stronger than Mazak 3, and is therefore used for those components that are likely to be subject to some degree of stressing under service conditions. Mazak 3 may be considered as a general purpose alloy with good dimensional stability; for example very suitable as a casting metal for motor car carburettors. Because of the use of pressure when forcing the metal into the dies, the finish and detail are excellent, making possible very intricate castings which may require no machining, and which can be chromium plated if a decorative effect is required on the finished product.

Forms of supply

Reference back to Fig. 5.24, shows that mild steel is available in a wide

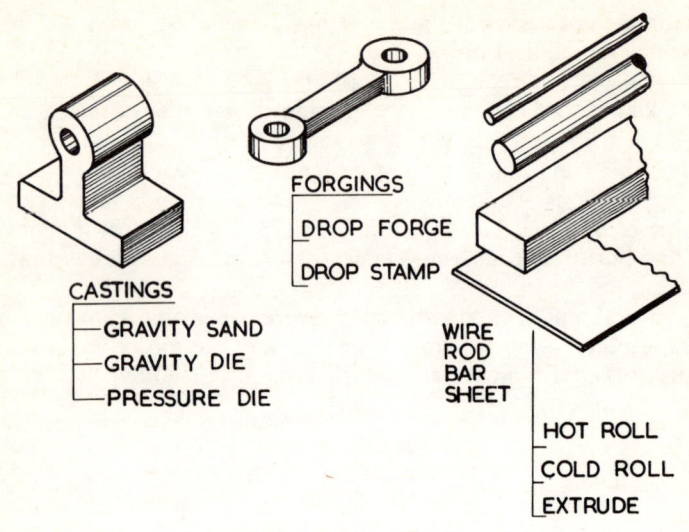

Fig. 5.42 Forms of supply

range of shapes and sections. In other words a varied form of supply exists, for in addition to the standard sections shown in the diagram, mild steel is available also as forgings and castings. It is a great advantage to have the particular metal or alloy from which components are to be manufactured delivered in the precise thickness of sheet or diameter of bar and tube, together with the correct sizes and proportions of channel and angle sections. Figure 5.42 shows the standard forms of supply for the main metals and alloys in engineering use.

5.7 Use of plastics

Plastics are being used increasingly in all types of manufactured products. They are relatively cheap and lend themselves easily to mass production processes such as moulding and extrusion, and have high output rates. Most plastics can be drilled, turned, welded and hot-worked, and products as diversified as children's toys and 20-metre ships' hulls can be manufactured from plastics. The type of plastic chosen for a particular product depends on many factors, which include working stresses and working temperatures, but in general the choice must be made from one of the three main types listed below:

(i) Thermoplastics,
(ii) thermosetting plastics,
(iii) cold-setting plastics.

5.7.1 Thermoplastics

Thermoplastics are those which soften when heated, allowing them to be shaped, formed or extruded. Because they will always soften when heated, they are not suitable for any application where temperature rises may be experienced, for example as containers for boiling liquids. Their best and obvious use is for those applications where no significant temperature rise is likely, for example cold water fittings, cosmetic tubes and containers for foodstuffs.

Low density polyethylene (LDPE)
This is a translucent plastic with excellent electrical insulating properties and good resistance to most chemicals, but it will soften at about 85°C. It is a fairly cheap material, and finds extensive use as a protective covering material, and is available in rolls of varying widths and thicknesses. It is used also for the manufacture of kitchen bowls and waste bags.

High density polyethylene (HDPE)
This is a stronger and more durable type of polyethylene, and is often referred to as *rigid* polythene. Best used for components that require a degree of strength and support, typical examples include crates for bottles, carboys and dust bins. The resistance to temperature rise is superior to that of LDPE; boiling water has no effect on high density polyethylene, and it requires a temperature of about 120°C before it begins to soften.

Polypropylene
Although similar in many respects to polyethylene, it has exceptional resistance to flexing and fatigue, and so is used increasingly as a material for cheap ropes, wire and fibres. It will not soften below a temperature of around 140°C and is quite capable of withstanding boiling water and steam.

Polyvinyl-chloride (PVC)
Polyvinyl-chloride, more commonly known as PVC is perhaps the most well known of the plastics family. Available either as a rigid material, or as a soft, pliable and flexible material, PVC finds popular use as a covering for electrical cables, or as handles for screw drivers. It has excellent electrical insulating properties, is resistant to most liquids and solvents and will not soften at temperatures below 90°C.

Polystyrene
This is an inexpensive thermoplastic, much used in the food industry for light cheap containers. When dropped on a hard surface poly-

styrene gives off a metallic ring, and when toughened is suitable for such components as portable radio transistor cabinets. It has excellent resistance to most liquids, solvents and acids, and has very good electrical insulating properties, but tends to soften at temperatures of around 90°C.

Expanded polystyrene is a very light or low density material, much used for ceiling tiles and as a packaging or protective material, and may also be used as a lightweight material for small dinghies or boats. It must be remembered that both polystyrene and expanded polystyrene are attacked or dissolved by petrol or benzine, and for this reason must *never* be used as petrol containers. Expanded polystyrene is a most efficient heat insulator, and is often used in granular or blanket form to provide additional insulation, as for example in the attics of houses in order to reduce the heat losses by conduction and convection.

Nylon
Another thermoplastic which can be cast, extruded and machined. Softening at a higher temperature than most of the other thermoplastics, at about 220°C, nylon finds considerable engineering use for bushes, gears and other mechanical devices. Needing no lubrication it is widely used in the mechanical handling equipment to be found in the food manufacturing industries, where the presence of even small quantities of lubricants such as oil and grease cannot be permitted because of the possible contamination risk to the food.

Polyacetal
This is the plastic best suited for small mechanical devices such as gear wheels, levers and other fittings that are to be found in such assemblies as typewriters, washing machines and calculating machines. Its smooth hard surface, with good resistance to wear and impact, makes it ideal for all moving parts, and as its softening temperature is in excess of 170°C it is unaffected by the normal operating temperatures of most industrial and domestic equipment.

Polytetrafluoroethylene (*PTFE*)
This is a more expensive type of low coefficient plastic with complete resistance to most solvents and acids. This is the material used for the coating given to frying pans which carry a *non-stick* warranty. It is also an excellent electrical insulator and does not soften at temperatures below 280°C. Engineering uses include gaskets, valve packings and bearings for assemblies liable to temperature rises of not more than about 250°C.

5.7.2 Thermosetting plastics

Compression moulding
A thermosetting plastic may be defined as one which having been set or cured, cannot be brought back to its original shape or structure. Such curing may be brought about by a combination of heat and pressure, or by the addition of an activator. When heat and pressure are used to cure the plastic, the process is more commonly known as plastic moulding, and Fig. 5.43A, shows the basic principle underlying

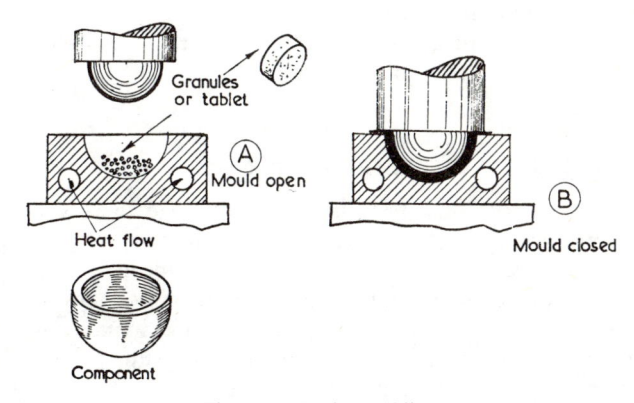

Fig. 5.43 Plastic moulding

the principle of plastic moulding. Note the highly polished alloy-steel moulds in the open position, with heat supplied to the bottom mould. A measured quantity of a thermosetting plastic, in the form of a tablet or granules, enters the bottom mould, and following closure of the moulds involving both heat and pressure, structural changes take place, causing the plastic to flow into the cavity between the top and bottom moulds. On cooling, which takes only a short time, a strong and firm moulding results, which cannot be re-softened in any way.

The most widely used thermosetting materials in general use are the *amino* Resins, of which there are two main types:

Urea formaldehyde resin,
Melamine formaldehyde resin.

The main applications for urea formaldehyde resins are in connection with the adhesive required for the bonding of plywood and chipboard, and moulding powders for the cheaper kind of domestic product which is not likely to be subject to undue stressing or loading when in use. Typical examples include bottle tops, and electrical fittings such as sockets and adaptors, all of which are easily broken or fractured if excessive force is used. For example, the ultimate tensile strength of a granular urea formaldehyde moulding powder is about 40 Newtons per square millimetre (40 N/mm^2); this is about twelve times weaker than mild steel, and makes this thermosetting plastic a relatively weak and brittle material. When compression moulded, a temperature of about 160°C is required, with a moulding pressure of around 30 N/mm^2.

The melamine formaldehyde powders are used for those components that require superior strength and resistance to rough usage, for example high quality electrical fittings that are to resist impact and conditions of humidity and high temperature. A typical melamine powder would have an UTS of about 75 N/mm^2, making it twice as strong as its urea counterpart.

5.8 Protection of metal surfaces

The ill effects of corrosion need no emphasising to the owners of bicycles or motor vehicles. We have seen, in earlier work that carbon steels are widely used for the manufacture of engineering components; their application to such components as paper clips, battleships, gun barrels and bridges is due to their superior mechanical properties. Unfortunately however, all the carbon steels are subject to corrosion when exposed to the atmosphere and this is due to the affinity or chemical attraction of oxygen to the surface of the steel. It needs to be remembered that the corrosion of steel components may be accelerated under certain environmental conditions. For example the exhaust system of a motor vehicle has a short life due to the corrosive nature of the exhaust gases, and the barrel of a shotgun is very rapidly corroded by the gases liberated by the explosion of the cartridge. In the same way the corrosion of carbon steel rails in a tunnel may be fourteen times as rapid as that of rails laid in open countryside.

The nature of corrosion
We may define corrosion as attack by gases or acids on the surface layers of metal components. This is strictly a chemical action, whereby the surface layers, in the case of carbon steel, decompose to produce the familiar scaling or rusting effect. Apart from the unsightly appearance of rust there exists the grave fact that the rusted steel loses size or volume and is therefore weakened. For example, a steel bridge, left in an unpainted state suffers rapid corrosion, and will eventually be unsafe and unable to stand up to the stresses and loads for which it was originally designed.

Clearly then, the prevention of corrosion is an important part of all engineering manufacturing except where plastic materials are used, when the problem becomes one of ultimate disposal, and the following

Engineering Materials

notes are intended to outline the basic principles adopted by engineers to protect the surface layers of metal components.

Principles of metal protection
Figure 5.44 illustrates in simple form the main techniques adopted. Note that the more conventional processes of painting, enamelling and

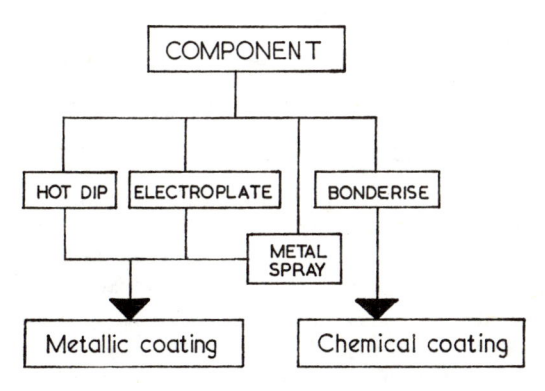

Fig. 5.44 Methods of metal protection

varnishing are not included. We see that the following protective processes are in use:

  (i) hot dip,
 (ii) electroplating,
(iii) metal spraying.

All these are methods of *adding* a protective coat to the surface of the steel, whilst the remaining processes of bonderising or anodising, *modify* the surface of the steel by a chemical interaction and are essentially rust-proofing techniques.

Surface preparation
The success of any protective coating to be applied to the surface of a metal component depends on the quality of the pre-treatment or preparation of the surface. The actual type of pre-treatment process depends on the sort of protective covering to be applied, and Fig. 5.45 gives the particular pre-treatments applicable to a finishing process. In all cases, absolute cleanliness of the surface is vital. In order to give a clear picture of the pre-treatment techniques, together with their associated protective coating processes, we will consider the following examples.

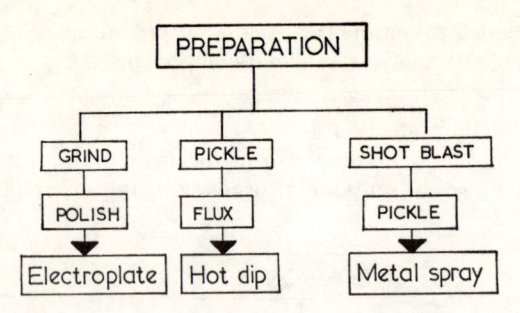

Fig. 5.45 Surface preparation

5.8.1 Hot-Dip Galvanising

Figure 5.46A shows a commonplace yet indispensable utensil. Subject to much misuse, and certain to spend most of its working life exposed to the atmosphere, the dustbin illustrated needs to be manufactured from a cheap, strong and workable metal. Mild steel is the metal used and, if the dustbin is to have a long and useful life, a protective coating is needed to prevent the oxygen in the atmosphere attacking the steel immediately it is exposed.

We see from the diagram that the dustbin is prefabricated, that is to say several operations such as rolling, riveting and seaming have been carried out, together with drilling or piercing in order that the lifting handles may be attached to the bin.

The protective coating given to the surface of the bin is pure zinc, and the process is referred to as *galvanising*. It is essential that total immersion takes place, to ensure that the seamed rivets and handles are given complete protection. The sequence of operations is illustrated in Fig. 5.46B, where we see the bin first immersed in a bath of

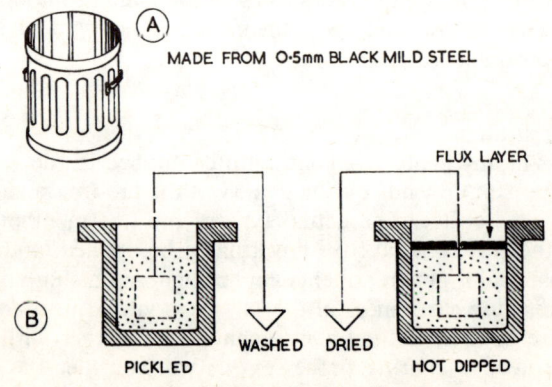

Fig. 5.46 Hot-dip galvanising

Engineering Materials

acid, the process known as 'pickling'. This pickling process, not only removes any dirt, scale or impurities, but also etches or roughens the surface, thus providing a grip or key for the molten zinc. The acid is generally a solution of sulphuric acid, and after immersion the dustbins are thoroughly washed in clean water to remove all traces of acid. Before their hot-dip into a bath of molten zinc, the bins are quickly dipped in a solution of dilute hydrochloric acid and dried, following which the bin is quickly lowered into a bath of molten zinc, as shown in the diagram.

Note the layer of flux floating on the liquid zinc; this flux assists the 'taking' of the zinc to the surface of the steel. On removal from the zinc bath, the bin is shaken to remove any surplus zinc. This hot-dipping process has the disadvantage that the presence of the flux tends to contaminate the molten zinc, and this has led to the introduction of a 'dry-fluxing' technique, whereby a suitable flux is dried on the surface of the steel prior to immersion in the bath of molten zinc. Reference back to Table 5.1 (p. 253), shows that zinc is a fairly cheap metal, and the galvanising technique is a popular method of surface protection for large or small components that are exposed to the atmosphere. Typical examples include mild steel chains, corrugated roof coverings, and all types of steel fixings used in the building and construction industries.

5.8.2 Tinning

The coating of sheet iron with a thin layer of tin has been carried out for about 300 years. The great advantage possessed by tin is that it is non-toxic, as well as having an excellent resistance to corrosion. This makes possible the use of tin-coated mild steel for containers in which foodstuffs such as sardines, salmon and soups, and a host of other products may be kept over a long period. The tinning technique differs little from the galvanising process just described, and may be summarised as follows:

   (i) pickling and rinsing,
   (ii) immersing in a flux bath,
   (iii) immersing in a molten tin bath,
   (iv) immersing in a palm oil bath,
   (v) passing through rollers.

Figure 5.47A illustrates in a simple manner the sequence of operations, but it needs to be remembered that for large scale production, such as that carried on at a steelworks, the tinning of the steel sheet is automated to a high degree. A much enlarged section of a tinned steel sheet is shown at Fig. 5.47B. Note that the thickness of the tin coating is very thin indeed, because tin is one of the most expensive metals used

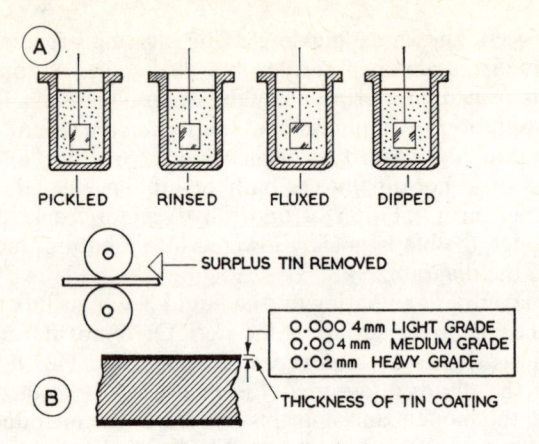

Fig. 5.47 Hot-dip tinning

by engineers. Three grades of tinning are available as shown in the diagram. The presence of the tin coating in no way affects the working properties of the tinned steel, which can be bent or rolled with no ill effects. When the tinned steel is to be used as containers for slightly acid products such as fruit or alcoholic drinks, it is necessary to apply a thin coat of non-toxic clear varnish to the inside of the container, and this is achieved usually by automatic means.

5.8.3 Electroplating

A very wide range of components are given a protective coating using the electroplating technique. Figure 5.48 shows some typical examples, and it may be seen that the type of coating depends on the nature

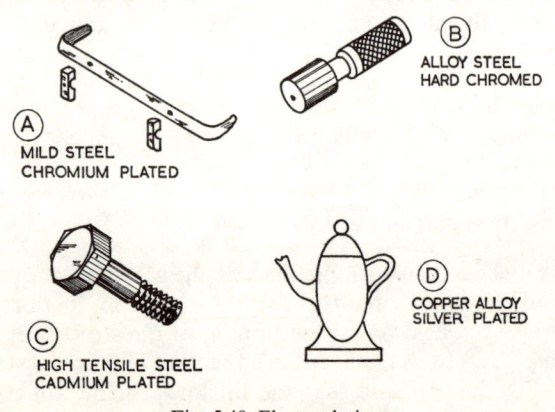

Fig. 5.48 Electroplating

or purpose of the component. At A we see the bumper for a motor vehicle; it is made from bright mild steel, and its appearance and resistance to corrosion are greatly increased by the application of a thin protective coat of chromium. At Fig. 5.48B we see a worn plug gauge which has been given a coating of hard chromium; this allows it to be re-ground to size as well as providing a hard wearing surface.

Fig. 5.48C shows a high tensile steel bolt used in aircraft construction. This bolt has been cadmium plated to give increased resistance to corrosion; the thin film of cadmium on the surface of the bolt providing excellent resistance to the corrosive attacks of salt water conditions. The cadmium possesses also, a self-healing property, that is to say if the bolt is accidently scratched or marked there is a tendency for the scratch to heal, and resist further corrosion in the scratch area.

Finally at Fig. 5.48D, we see an attractive coffee pot. Made from copper, or a copper alloy, the finished article is silver plated, giving an attractive and untarnishable protective coating to the coffee pot. We see now that the electroplating technique covers a wide range of applications, designed not only to improve the corrosion resisting properties but also to give an attractive appearance to the finished article; these applications are summarised in Fig. 5.49.

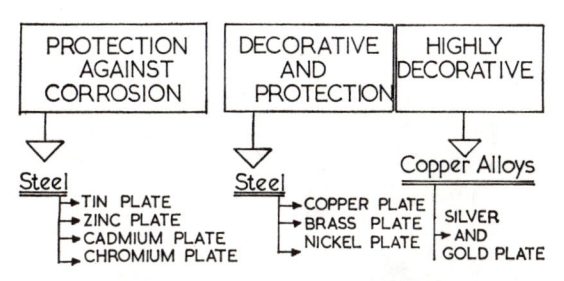

Fig. 5.49 Applications of electroplating

Most modern electroplating plants are automatic in operation and highly integrated, but all operate on the principle of electrolysis simply illustrated in Fig. 5.50. Essentially the process makes use of a suitable solution, together with an anode, a cathode and a source of d.c. current. In the example shown, copper atoms are taken from the anode and deposited on the cathode. The nature of the anode and the type of solution depends on the metal that is to be deposited; for example, if an article is to be gold plated, the anode needs to be pure gold.

5.8.4 Metal spraying

The spraying of metal is now a commonplace technique, offering many advantages when a protective metallic coating is required. There is

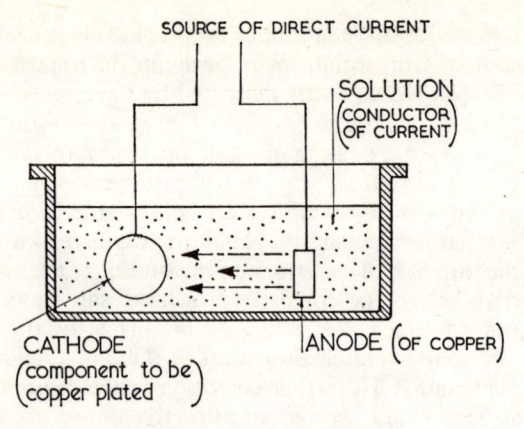

Fig. 5.50 Principle of electroplating

practically no limit to the metals that can be sprayed, and Fig. 5.51 illustrates some typical examples of the metal spraying process. At A we see a large carbon steel crankshaft which has been badly scored in use, receiving a coating of carbon steel deposited on the damaged area by a spray gun, whilst the shaft rotates in a centre lathe. It is standard

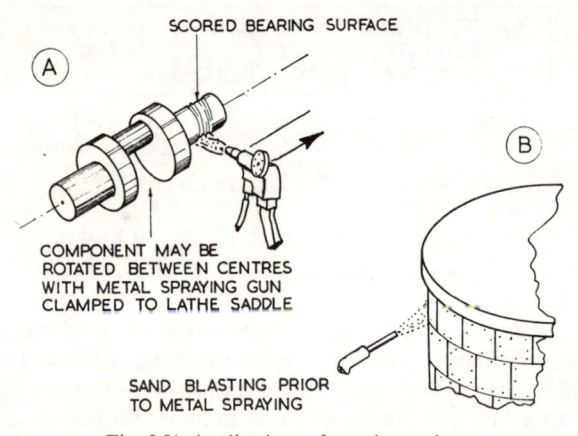

Fig. 5.51 Applications of metal spraying

procedure to ensure that the area to be sprayed is first rough-machined to provide a good key for the sprayed metal. On completion of the metal spraying, the crankshaft is machined to remove surplus metal and bring the re-surfaced area to the correct dimensional size. A further example of applying a protective metal coating by spraying is shown in Fig. 5.51B, which shows a large steel container receiving a protective coating of zinc or aluminium. As in all forms of protective coating it is essential that the surface to be coated is properly prepared, and in the case of zinc or aluminium spraying, the usual technique is to shot or sand blast the surface.

5.8.5 Principle of metal spraying

The basic principle of metal spraying consists in melting a metal wire using a gas–air flame, atomising the liquid metal with a blast of high pressure gas and directing the resultant spray of molten metal droplets on the surface to be coated. Figure 5.52A illustrates a typical metal-

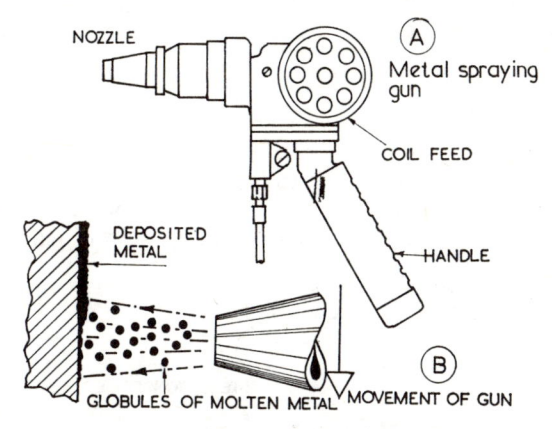

Fig. 5.52 Principle of metal spraying

spraying gun. Note the automatic coil feed which supplies wire to the nozzle of the gun, and the gas–air supply pipes. A simple picture of the coating action is seen in Fig. 5.52B, where the molten droplets adhere to the metal surface on impact. The distance between the gun nozzle and the surface to be coated is important; because the bond or adhesion between the sprayed and parent metal is a mechanical one, adhesion and retention are achieved through the combined effects of impact and a rough surface. It is essential that adequate safety precautions are taken; for example both lead and tin are *toxic* metals, and masks must be worn when spraying them. Dangers also exist when sand- or shot blasting, because the abrasive particles used are small and hard, and leave the spraying hose with considerable velocity. Once again it is essential that a protective mask is worn, not only to afford protection to eyes, but also to ensure that no injurious particles are inhaled by the operator.

It must not be thought that the brief descriptions given above completely cover the field of metal coating. The business of engineers is not only to manufacture reliable components at competitive prices, but also to ensure that the components resist the corrosive attack of the atmosphere and yet possess an attractive and durable surface. The extraction of metals from ores is a difficult and expensive matter, and by giving cheap metals a thin coating of a more expensive metal considerable economy is achieved. This procedure provides an important and valuable contribution both to the ever-increasing need for material conservation, and also the important matter of reducing the huge amount of money spent making good the ravages caused by the rusting of steel components. It is certain that plastic-coated steel sheet will come into increasing use in the future, and that there will be new developments in both chemical and electrical methods of metal deposition.

5.8.6 Rust proofing

This is a chemical process designed to increase the rust-resisting properties of steel components, and to facilitate further enamelling, painting or cellulosing. If we consider the finishing of motor car bodies, then reference to Fig. 5.53A shows that the application of paint or enamel

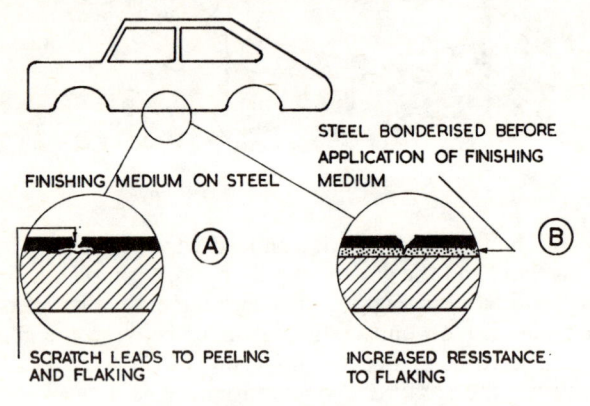

Fig. 5.53 The bonderising process

directly onto the surface of steel may lead to two undesirable conditions.

First, some activity of a chemical nature takes place between the steel and the finishing medium. This is known as *subcutaneous* corrosion, which causes premature peeling or flaking of the painted surface.

Second, as may be seen from the diagram, a scratch tends to admit the atmosphere, with subsequent subcutaneous corrosion between the paint film and the steel surface.

Rust proofing is the process adopted to minimise the above defects. Essentially the process consists of first cleaning the surface using nor-

Engineering Materials

mal methods and then immersing the component in a bath containing a ferric phosphate solution for about half an hour. On removal the surface is dried and takes on a grey colour, caused by a chemical interaction on the surface of the steel. The presence of this chemical film tends to prevent subcutaneous corrosion and the technique is sometimes referred to as phosphating, or chemical deposition.

Several types of rust-proofing processes are now in use, all based on similar principles, and include:

(i) anodising,
(ii) parkerising,
(iii) granodising.

Where high production rates are required, for example in the manufacture of motor car components, the rust-proofing processes are automated, and spray guns may be used in preference to immersion in the phosphate solution.

Chapter 5 Assimilation Exercises

1. Write down three ferrous metals used in engineering manufacture and sketch a component made from each.
2. Write down three non-ferrous metals used in engineering manufacture and sketch a component made from each.
3. Define the following terms:
    (i) Ultimate tensile strength,
    (ii) elastic limit,
    (iii) density,
    (iv) electrical conductivity.
4. State the difference between the following Hardness Tests:
    (i) Brinell,
    (ii) Rockwell.
5. Outline briefly the principles underlying the Izod Impact Test.
6. Sketch a component made from grey cast iron, giving three reasons why this metal has been used for its manufacture.
7. Explain why wrought iron is ideally suited for the manufacture of chains required for marine use.
8. Give three reasons for the popularity of mild steel as an engineering material.
9. Arrange the following metals in their cost per tonne order, with the most expensive at the head of the list: lead, copper, tin, aluminium, cast iron, mild steel.
10. Write down the four main types of carbon steels, sketching an engineering component made from each.

11. Explain the circumstances that require the use of an alloy steel instead of a carbon steel.
12. Write down the names of four heat treatment processes carried out on carbon steels.
13. Outline briefly how the following heat treatments are carried out: normalising, annealing, hardening.
14. Explain why it is necessary to temper a hardened high carbon steel component, and outline the technique involved when tempering a hand scriber.
15. Explain why a mild steel bush is unsuitable for a rotating mild steel shaft.
16. Explain why phosphor bronze is widely used as a bearing metal for shafts carrying heavy loads at medium speeds.
17. Outline the circumstances that would require the use of white metal as a bearing material.
18. Explain what is meant by a porous bearing.
19. State the circumstances where nylon could be used as a bearing material.
20. Make a neat sketch of an engineering component made from each of the following alloys, stating the main reason for the choice of each particular alloy:

    (i) zinc-base alloy,
    (ii) aluminium-base alloy,
    (iii) magnesium-base alloy.

21. Explain the essential difference between a thermoplastic and a thermosetting plastic.
22. State the plastics identified by the following abbreviations; LDPE, HDPE, PVC, and give a typical example of a component made from each type.
23. With the aid of a neat diagram illustrate the principle of Compression Moulding using thermosetting plastics.
24. Sketch two engineering components that are subject to corrosion, and describe the effects of the corrosion on the appearance and performance of the component.
25. A number of mild steel brackets are to be used for outside distribution lines supporting electrical cables. Describe a process that will give these brackets increased resistance to corrosion.
26. Explain what is meant by the term 'tinplate', and explain why it is widely used to make food containers.
27. With a neat diagram show the basic principle of metal spraying and describe a typical application of this technique.
28. Name two rust-proofing techniques, and explain the basic difference between these processes and electroplating.

6
WORKING IN PLASTICS

Objectives—The principles and applications of:
1. Joining plastics materials
2. Bending and casting plastics materials
3. Machining plastics materials
4. The use of encapsulating techniques

The manipulation of plastics, that is to say the processes used to bring plastics materials into the shapes and forms of the end products, is an ever-increasing engineering activity. The processes adopted differ little in principle from those used for the more common ferrous and non-ferrous metals, and are shown in Fig. 6.1. It may be noted that machin-

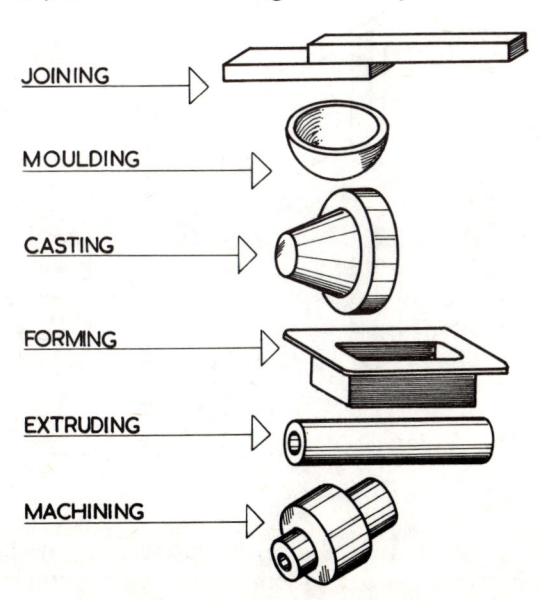

Fig. 6.1 Manipulation of plastics

ing is at the bottom of the list, and this is as it should be, for machining is a costly process requiring expensive machine tools together with skilled setters or operators. One of the advantages to be gained by the use of plastics materials is the ease with which they can be moulded or formed into the finished article; the amount of heat and pressure required is considerably less than that required to form or shape mild steel, copper or aluminium.

6.1 Joining plastics

The techniques adopted to join plastics materials are very similar to those used for metal-joining, and may be summarised as follows:

(i) adhesive bonding,
(ii) solvent welding,
(iii) heat welding.

6.1.1 Adhesive bonding

This is still a popular method of joining plastics, and a typical example of adhesive bonding is illustrated in Fig. 6.2A, where we see a pictorial

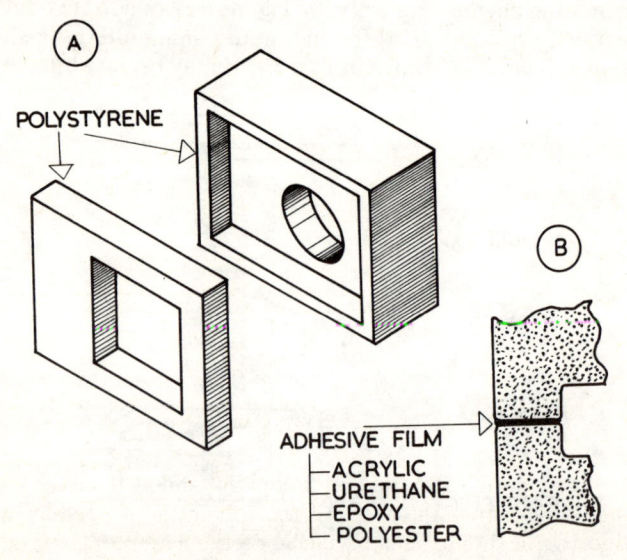

Fig. 6.2 Adhesive bonding of plastics

view of two halves of a clock case. When the clock mechanism has been inserted, it is necessary to join the two parts permanently, and this is achieved by the application of a thin film of adhesive on the joint line. Depending on the size of the job, the adhesive may be applied by hand, spray or dipped, but it is essential to ensure that a thin film only be applied. A large number of adhesives are available, and in all cases strong and reliable joints result when the correct adhesive is used. Provided a reasonable degree of automation is used, the adhesive bonding of plastics components can be carried out on a mass production basis at very reasonable cost.

6.1.2 Solvent welding

The assembly of model kits provides a typical example of the use of solvents as a means of joining plastic components. It is important to appreciate that a solvent has the ability to dissolve the plastic material it is intended to join, with the result that a weld rather than a bonded joint is obtained. Figure 6.3A shows a typical example of the use of a

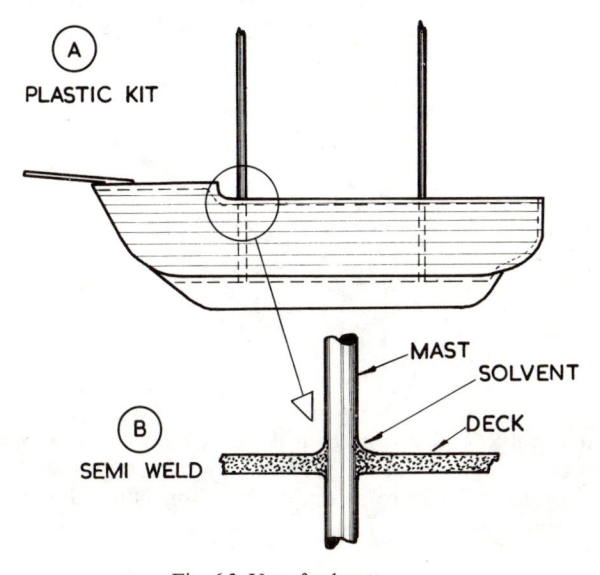

Fig. 6.3 Use of solvents

solvent when joining the parts of a model kit, and at B we see a section of the joint. Note that the effect of the solvent is to produce a semi-weld, due to the homogeneous effect of the parent plastic and the solvent, and great care needs to be taken not to allow the solvent to make contact with other partsof the model in case permanent disfiguration results. Most solvents are quick drying but excessive force should not be applied to a solvent joint until they are completely hardened.

6.1.3 Welding

All true welded joints in plastics require the application of heat in order to melt the plastics at the joint interface. Several different techniques are employed, although it may be stated that all welding techniques apply only to the welding or joining of relatively thin sections such as polythene sheet or film. The main welding processes are as follows:

(i) Heated tool welding,
(ii) high-frequency welding,

Heated tool welding
This technique makes use of electrically heated tools or blades which conduct heat to the joint interface as shown in Fig. 6.4. On removal,

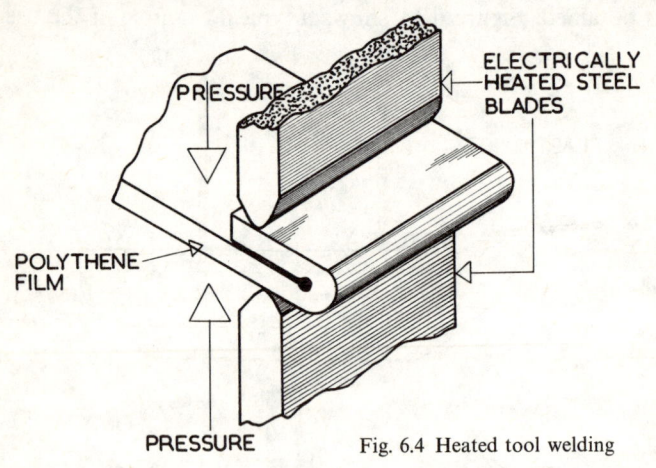

Fig. 6.4 Heated tool welding

the joint is subject to pressure until the plastics have cooled. In addition to the equipment being fixed and used for automated assembly, small portable devices are available enabling hand closure of small plastic containers.

High frequency welding
This technique is applied to the high-speed welding of relatively thin plastic sheets such as PVC. Figure 6.5 illustrates in a simple manner, the basic principle underlying high frequency welding, and the similarity to the spot welding of mild steel may be seen by reference back to Fig. 4.22. The resistance by the plastics, to the passage of the high frequency current causes a temperature rise at the joint interface as shown in the diagram, and at the correct moment pressure is applied to make the weld permanent on cooling. This is a popular method of sealing food containers and is sometimes referred to as *hot sealing*.

Working in Plastics

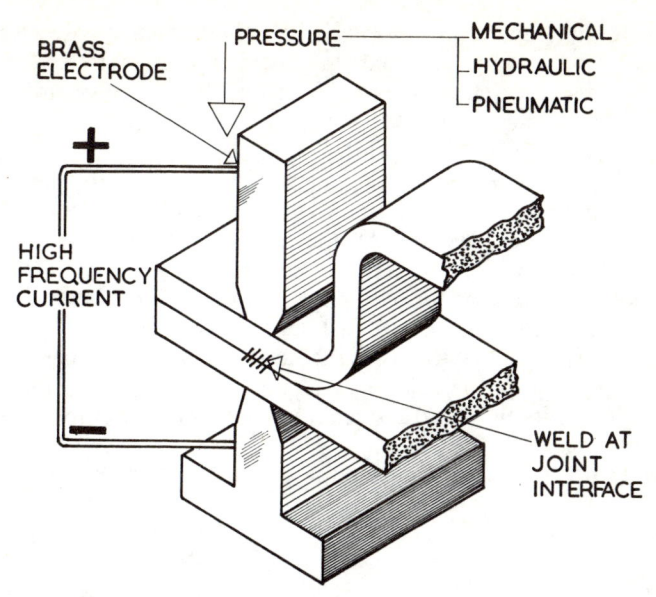

Fig. 6.5 High frequency welding

6.2 Bending and casting

All thermoplastics can be bent with relative ease provided they are first slowly heated to the appropriate temperature. Perspex, for example, is best immersed in boiling water for about twenty minutes, and then quickly bent to the required shape, great care being taken not to damage or mark the surface of the perspex whilst it is in the soft state.

Table 6.1 Softening temperatures for thermoplastics

Thermoplastic	*Temperature °C*
Low density polyethylene	90
High density polyethylene	100
Cellulose acetate	75
Polystyrene	80
Flexible PVC	115
Rigid PVC	90
Nylon	110

The softening temperatures of some of the thermoplastics are given in Table 6.1, and it may be noted that there is some variation in the softening temperatures, although all are very low when compared to the low melting point metals.

Under no circumstances should thermoplastics be heated using a naked flame, since there is a risk of damage to the surface, the possibility of toxic vapours and gases forming, and the dangers of combustion should the heat be excessive. All the plastics given in Table 6.1 are readily softened by a period of immersion in boiling water, the time of immersion being dependent on the mass of the thermoplastics.

It is not often that plastics are cast into the required shape, although provided the plastic material is available as a liquid which has reasonable free-flowing properties there is no reason why it cannot be poured into a suitable mould and removed on solidification, to produce a plastic casting.

Figure 6.6A shows a typical example of the need to produce a plastic

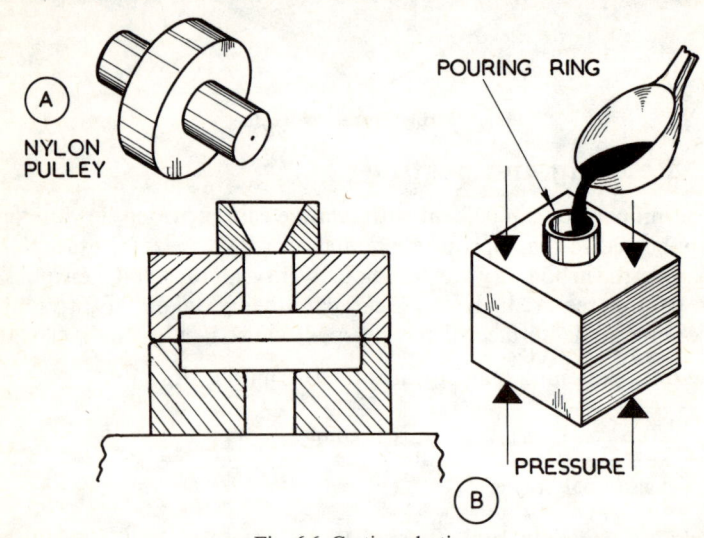

Fig. 6.6 Casting plastics

casting, the plastic in this case being nylon with a melting point of about 230°C. At B we see a simple mould used for the casting process; a suitable hard wood or plaster would be an acceptable material for the manufacture of the mould.

6.3 Machining plastics

Having cast the pulley shown in Fig. 6.6, it is certain that it will require machining using conventional machine tools such as centre lathes, drilling and milling machines. Whilst the machining of plastics presents no particular difficulty, it is not often that machining is carried out to produce a finished component. This is due to the relative ease with which accurate and well-finished products are produced in plastics from such forming processes as:

    (i) Compression moulding,
   (ii) transfer moulding,
  (iii) injection moulding,
  (iv) blow moulding,
   (v) vacuum moulding,
  (vi) extruding.

In all the above processes, the finished product is a high quality article, with regard to both dimensional stability and surface finish. Nevertheless it is still necessary on occasions to machine plastics, for example a large nylon brush which is to be a press fit, and the following notes are intended to outline the basic precautions that need to be taken.

6.3.1 Drilling

Although special twist drills with highly polished flutes are available for drilling plastics, good results are possible with standard high-speed twist drills provided the following precautions are taken. In general a fairly high spindle speed is required, especially for small diameter drills, and the feed must be on the light side, with continual uplift of the drill to remove swarf. It is essential that a piece of hard wood be placed beneath the plastics being drilled to prevent material breaking away from the hole edge as the drill breaks through, feed being eased off at this point.

Figure 6.7 shows the re-grinding of the drill point required to meet special circumstances. At A we see a small flat ground on the cutting edges of the drill; this tends to give a negative rake effect which is preferred for plastics drilling, and is similar to the technique adopted when modifying a twist drill point for drilling grey cast iron.

Figure 6.7B shows a drill point re-ground for drilling soft plastics such as polyethylene, this reduction of the point angle improves the finish, especially where the drill breaks through. When thin plastics need to be drilled, it is advisable that the full diameter of the drill has entered before the point breaks through, and this may require a large point angle as shown in Fig. 6.7C. This modification applies only to twist drills of reasonably large diameter.

Perhaps the essential point to keep in mind when drilling plastics, is to ensure that no undue temperature rise takes place during the drill-

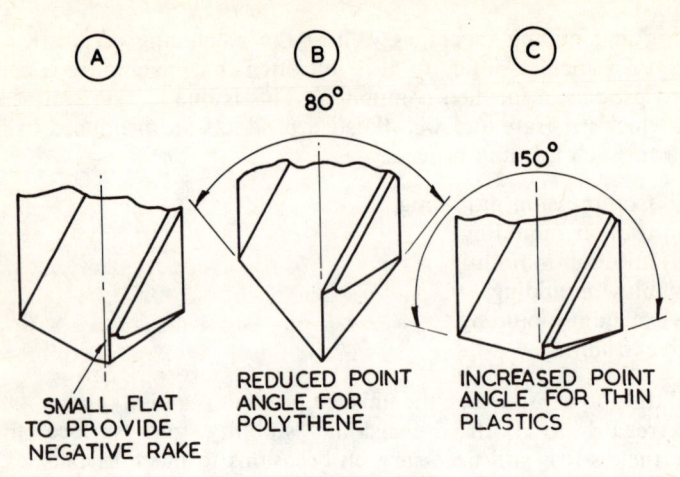

Fig. 6.7 Drilling plastics

ing operation. Reference back to Table 6.1, makes clear the fact that most plastics will soften at relatively low temperatures.

6.3.2 Turning plastics

As in drilling, fairly high cutting speeds need to be employed, and a reasonable figure would be about five times that used for turning mild steel, say about 180 metres per minute. Thus, when turning a 25-mm-

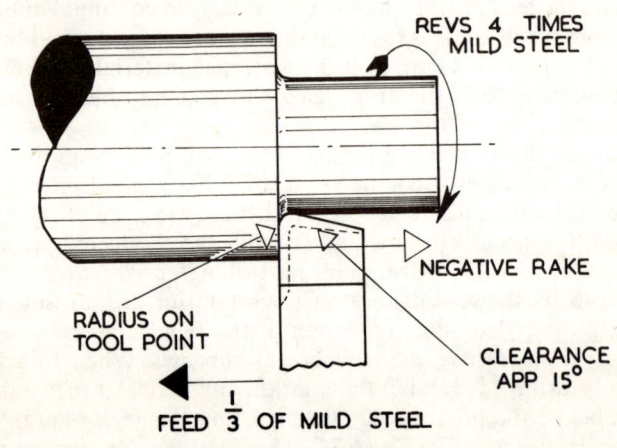

Fig. 6.8 Turning plastics

diameter nylon bar on a centre lathe, the spindle speed would be calculated as follows:

Working in Plastics

$$\text{rev/min} = \frac{1000 \times \text{Cutting speed}}{\pi \times \text{Diameter}}$$

$$= \frac{1000 \times 180}{\pi 25}$$

$$= 2292 \text{ rev/min}$$

Once again negative rake is preferred to positive rake with a good radius on the tool point as shown in Fig. 6.8. Clearance angles need to be slightly greater than those used when machining metals, and a value of about 15° is acceptable. It is important that all cutting tools are maintained in a sharp condition, and that feed rates are less than those normally used when turning mild steel; the object being to ensure that no undue temperature rise takes place. The Table 6.2 below gives a general guide when turning and drilling plastics:

Table 6.2 Rake angle and cutting speeds when turning and drilling plastics

Type of plastics	Rake angle	Cutting speed
Nylon, Polypropylene	0°– 5°	152–300
Moulded thermosetting	0° –10°	76–243
Rigid PVC	0°– 5°	90–300
Polystyrene	0°– 5°	90–300
	Degrees	Metres/minute

6.4 Encapsulating

We may define encapsulating as the technique of protecting electrical assemblies by means of a case, coat or film of plastic material. Perhaps the simplest example of encapsulation is the electric light bulb shown in Fig. 6.9A. The passage of an electric current is resisted by the tungsten filament which reaches a high temperature and becomes white hot. With a melting point of 3380°C there is little possibility of the tungsten filament melting, but it is vital that the white-hot filament is prevented from coming into contact with the oxygen present in the atmosphere. Exposure to the atmosphere would result in rapid deterioration of the filament giving it a very short life. The purpose of the glass bulb is therefore to protect the filament by ensuring that it not only keeps out the atmosphere, but also retains the inert gas which is

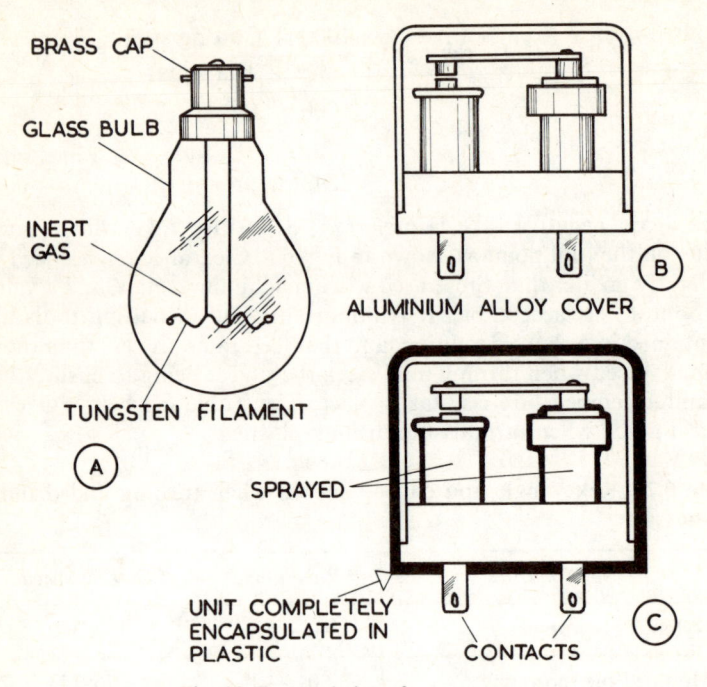

Fig. 6.9 Encapsulation of components

introduced into the bulb during manufacture. Clearly the encapsulating material for a light bulb needs to be a rigid transparent material, and glass is ideal for this purpose, whilst a thermoplastic would be totally unsuitable because of the softening effect of the heat given off by the white-hot tungsten filament. There are however, examples where plastics can be used as encapsulating materials, in particular for electrical assemblies which require a measure of protection from any of the following:

  (i) heat,
 (ii) shock,
(iii) moisture
(iv) vibration,
 (v) corrosion,
(vi) tampering.

Figure 6.9B shows a sensitive microswitch designed to operate when subject to very small variations in the electrical supply. With no protection of the different parts of the assembly, constant exposure to the atmosphere is certain to involve oxidisation, condensation and electrolysis, with the result that accurate functioning is quickly affected, and the unit soon rendered useless. Encapsulation of the unit will

Working in Plastics 317

prevent any deterioration due to moisture and oxidisation, and will also prevent any tampering with the adjustment.

Epoxy resins are widely used as encapsulating materials, and can be applied by dipping or spraying, or cast into a protective sheath into which the component is placed, prior to sealing by dipping or spraying. Figure 6.9C shows the total encapsulation of the microswitch, except for the contacts required to make the necessary electrical connections. Note that both the coil and the metal parts are sprayed or dipped before assembly, and this tends to inhibit any electrolytic action between the different metals present. It is worth while remembering that both spraying and dipping with a plastics material are forms of encapsulation, designed to ensure that the assembly functions correctly over a long period of time.

Chapter 6 Assimilation Exercises

1. Explain the essential technique adopted when joining two plastics by adhesive bonding.
2. Sketch a typical example of the use of a solvent when joining plastics. In which way does solvent welding differ from the use of adhesives?
3. Outline the principles underlying the application of the following welding techniques:
    (i) heated tool welding,
    (ii) high frequency welding.
4. Outline the main precautions to be taken when drilling 12-mm-diameter holes in 4-mm-thick perspex sheet.
5. A centre lathe which has been set-up to turn a mild steel bar of 20 mm diameter is to be used for turning nylon bars of 50-mm diameter. Outline the changes needed with regard to spindle speeds and feed rates.
6. A perspex sheet of 5-mm thickness is to be bent at an angle of 90°. Describe the procedure adopted to both heat and bend this sheet.
7. Make a neat sketch of a plastic component that would be produced as a casting, specifying the type of plastics used and the material used for the mould.
8. Explain the main purpose of encapsulating an engineering component or assembly.

INDEX

Numbers in **bold** type refer to illustrations

Active flux 227
Alloy steels 250, 275, **5.27**
Aluminium 251, **5.41**
Aluminium alloys 289
Angular dimensions 45
Annealing 277, **5.29**
Arc welding 233, **4.21**
Automatic lathe 164, **3.56**

Balanced dial 67, **2.48**
Beaded edge 108, **2.94**
Bearing surfaces 120, **3.10**
Bench shears 107, **2.93**
Bending allowance 103, **2.91**
Bent tool 203, **3.107**
Bonderising 304, **5.53**
Boring tools 140, **3.42**
Brass 252
Brazing 230, **4.18**
Brinell Hardness Test 263, **5.16**
Bronze 252
Butt welding 234, **4.22**
Button die 191, **3.88**

Caliper principle 47, **2.27**, **2.28**
Capstan lathe 164, **3.56**
Carbon steels 250, 271, **5.2–5.26**
Cemented joints 237, **4.26**
Cementite 274
Centre punches 22–3, **2.7**
Chasing dial 157, **3.49**
Chasing tool 157, **3.50**
Clapper box 176, **3.70**

Clearance angle 188, **3.85**
Collet chucks 146, **3.38–3.39**
Combination gauging 74, **2.54**
Compound gear train 154, **3.47**
Compound slide 126, **3.16**
Compound table drilling machine 91, **2.72**
Coolants 206
Copper 251
Core drills 97, **2.81**
Counterbores 100, **2.85**
Countersinks 100, **2.84**
Crank tool 208, **3.107**
Cross slide 127, **3.17**
Cross-cut chisel 30, **2.15**
Cutting fluids 205–7
Cutting friction 206, **3.111**
Cutting speed 137, **3.29**

Datum faces 40–3, **2.22**
Density 255, **5.8**
Dial test indicator 66, **2.48**
Diamond point chisel 31, **2.16**
Digital comparator 75, **2.55**
Dovetail slide 123, **3.13**
Draw filing 27–8, **3.13**
Drill holding 78–9, **2.57–2.58**
Drill vice 79, **2.59**
Drilling plastics 102, **2.87**
Ductility 261, **5.14**
Duralumin 252

Electrical comparator 75, **2.55**

Electrical conductivity 257, **5.9**
Electroplating 300, **5.48**, **5.50**
Emergency stop button 12
Encapsulating 315, **6.9**
End standards 68, **2.50**, **2.51**
Engineer's dividers 21, **2.5**
Engineer's plain protractor 50, **2.30**
Engineer's steel rule 46, **2.26**
Epoxy resins 238, **4.27**
Eye protection 12

Faceplate 138, **3.26**
Feet protection 13, **1.7**
Files 109, **2.10–2.11**
Flat chisel 29, **2.14**
Flush joint 109, **2.95**
Forming 114
Four-jaw chuck 132, **3.24**
Functional dimensions 45

Galvanising 298, **5.46**
Generating 114, **3.4–3.5**
Grey cast iron 249, 267, **5.2–5.21**
Grinding wheels 6, **1.2**
Grooved joint 109, **2.95**
Guideways 120, **3.10**

Hacksaw blades 24–6, **2.9**
Hand scriber 20, **2.4**
Hand shears 105, **2.92**
Hand soldering 229, **4.18**
Hardening 280, **5.32**
Head protection 13, **1.7**
Head slide 168, **3.58**
Health and Safety Act 5
Heated tool welding 310, **6.4**
High frequency welding 310, **6.5**

Identifiable hazards 3–4, **1.1**
Indexing dial 124, **3.14**
ISO Course Series Threads 151
Isolating switches 10–11, **1.6**
Izod Impact Test 266, **5.20**

Jobber twist drills 95

Knife-edge lathe tools 199, **3.100**

Lap joint 109, **2.95**
Lathe bed 120, **3.9**
Lathe headstock 128, **3.20**
Lathe quadrant 130, **3.21**
Lathe toolpost 125, **3.15**
Lead 251
Lead screw 124, **3.14**
Lever-type comparator 68, **3.20**
Linear dimensions 45, **2.25**
Locking devices 216–19, **4.7**
Long Series twist drills 96
Loose dies 34, **2.19**

Machine vice 184, **3.80**
Magnesium alloys 290
Master square 40, **2.29**
Measurement by comparison 64, **2.47**
Mechanical adhesion 236, **4.24**
Mechanical properties 267
Melting points 255, **5.7**
Metal spraying 301, **5.51**
Micrometer principle 60, **2.4–2.42**
Morse tapers 87
Morse taper shank drills 96
Mushroom head chisel 31, **2.16**

Non-identifiable hazards 3–4, **1.1**
Non-linear functions 45
Normalising 278, **5.31**
Nylon bearings 287

Oblique cutting 197, **3.97**
Odd-leg calipers 22, **2.6**
Off-set 190, **3.87**
Offsetting tailstock 159, **3.51**
Orthogonal cutting 197, **3.95–3.97**
Oxy-acetylene welding 231, **4.20**

Parting-off tool 201, **3.102**
Passive flux 227
Pearlite 271

Index

Pedestal grinding machine 5
Percentage elongation 261, **5.15**
Phosphor bronze 285
Pig iron 248, **5.2**
Pillar drilling machines 86, **2.66**
Plastics materials 292–4
Plumber's solder 228, **4.17**
Plunger-type comparator 66, **2.48**
Porous bearings 286, **5.39**
Potter's wheel 116, **3.5**
Powered hand tools 35–7, **2.20**
Protector slips 71, **2.50**

Quick-return motion 172, **3.64**

Radial cutting 194, **3.92**
Radial drilling machines 89, **2.68**
Rake angles 188, 194, **3.85**
Reamers 98, **2.82–2.83**
Resistance welding 233, **4.22**
Riveting 220, **4.9–4.10**
Rockwell Hardness Test 265, **5.18**
Round-nose chisel 31, **2.16**
Round-nose lathe tool 201, **3.103**

Safety at the drilling machine 8, **1.3**
Safety at the lathe 7, **1.3**
Safety at the milling machine 9, **1.5**
Safety at Work Booklets 5
Scrapers 28, **2.13**
Screw cutting 32, **2.17**
Scribing blocks 19, **2.3**
Sclerescope Hardness Test 265
Sensitive drilling machine 81–4, **2.61**
Sheet metalwork 2.5, **2.88**
Silver soldering 229, **4.18**
Simple gear train 154
Skin protection 14, **1.7**
Slip gauge accessories 73, **2.52**
Slip gauges 60
Soft soldering 222, **4.12–4.13**
Soldering iron 224, **4.15**

Specific adhesion 237, **4.25**
Spot welding 234, **4.22**
Spotfacing 101, **2.86**
Stocks and dies 33, **2.18**
Stop and start buttons 11
Stub drills 96
Surface plates 18, **2.2**
Sweating 226, **4.16**
Swivel plate 178, **3.72**

Tailstock 128, **3.18–3.19**
Tangential cutting 195, **3.93**
Taper turning 158, **3.51–3.53**
Tapping drills 33
Tempering 281, **5.34–5.36**
Tensile strength 250
Thermal conductivity 257, **5.10**
Thermo-setting plastics 295, **5.43**
Threaded fasteners 212–16
Three-jaw chuck 131, **3.22**
Tin 251
Tinman's solder 229
Tinning 225, 229, **5.47**
Tolerance pointers 67, **2.48**
Toolmaker's flat 19
Try square 40, **2.29**
Twist drills 93, **2.75–2.77**
Turning between centres 144, **3.35–3.36**

Vee blocks 23, 43, **2.8**, **2.24**
Vernier calipers 51, **2.31**
Vernier depth gauge 55, **2.35**
Vernier micrometer 63, **2.45**
Vernier principle 51
Vernier protractor 57, **2.37**
Vicker's Hardness Test 264, **5.17**

Welding by hand 231, **4.19**
White cast iron 249, **5.2**
White metals 286
Wired edge 108, **2.94**
Wrought iron 240, **2.71–5.23**

Zinc 251
Zinc-base alloys 291